90

ALSO BY WALTER ISAACSON

Steve Jobs

American Sketches

Einstein: His Life and Universe

A Benjamin Franklin Reader

Benjamin Franklin: An American Life

Kissinger: A Biography

The Wise Men: Six Friends and the World They Made (with Evan Thomas)

Pro and Con

HOW A GROUP OF

HACKERS, GENIUSES,

AND GEEKS CREATED

THE DIGITAL

REVOLUTION

THE INNOVATORS

WALTER ISAACSON

SIMON & SCHUSTER

NEW YORK LONDON TORONTO SYDNEY NEW DELHI

Simon & Schuster
1230 Avenue of the Americas
New York, NY 10020

First Simon & Schuster hardcover edition October 2014

For information about special discounts for bulk purchases, please contact Simon & Schuster Special Sales at 1-866-506-1949 or business@simonandschuster.com.

The Simon & Schuster Speakers Bureau can bring authors to your live event. For more information or to book an event contact the Simon & Schuster Speakers Bureau at 1-866-248-3049 or visit our website at www.simonspeakers.com.

Photo research and editing by Laura Wyss, Wyssphoto, Inc., with the assistance of Elizabeth Seramur, Amy Hikida, and Emily Vinson, and by Jonathan Cox.

Interior design by Ruth Lee-Mui
Jacket design by Pete Garceau
Jacket art (from top to bottom):
Ada Lovelace by Margaret Carpenter/Hultpn Archive/Getty Images;
Steve Jobs by Justin Sullivan /Stringer/Getty Images News;
Bill Gates by Theo Wargo/WireImage for OMEGA /Getty Images;
Alan Turing by Private Collection/Prismatic Pictures/Bridgeman Images

Manufactured in the United States of America

1 3 5 7 9 10 8 6 4 2

Library of Congress Cataloging-in-Publication Data
Isaacson, Walter.
The innovators : how a group of inventors, hackers, geniuses, and geeks created the digital revolution / Walter Isaacson. — First Simon & Schuster hardcover edition.
pages cm
1. Computer scientists—Biography. 2. Computer science—History. 3. Internet—History. I. Title.
QA76.2.A2187 2014
004.092'2—dc23
[B] 2014021391
ISBN 978-1-4767-0869-0
ISBN 978-1-4767-0871-3 (ebook)

Photo credits appear on page 525.

CONTENTS

THE INNOVATORS

1843

Ada, Countess of Lovelace, publishes "Notes" on Babbage's Analytical Engine.

1847

George Boole creates a system using algebra for logical reasoning.

1890

The census is tabulated with Herman Hollerith's punch-card machines.

1931

Vannevar Bush devises the Differential Analyzer, an analog electromechanical computer.

1935

Tommy Flowers pioneers use of vacuum tubes as on-off switches in circuits.

1937

Alan Turing publishes "On Computable Numbers," describing a universal computer.

Claude Shannon describes how circuits of switches can perform tasks of Boolean algebra.

Bell Labs' George Stibitz proposes a calculator using an electric circuit.

Howard Aiken proposes construction of large digital computer and discovers parts of Babbage's Difference Engine at Harvard.

John Vincent Atanasoff puts together concepts for an electronic computer during a long December night's drive.

1938

William Hewlett and David Packard form company in Palo Alto garage.

1939

Atanasoff finishes model of electronic computer with mechanical storage drums.

Turing arrives at Bletchley Park to work on breaking German codes.

1941

Konrad Zuse completes Z3, a fully functional electromechanical programmable digital computer.

John Mauchly visits Atanasoff in Iowa, sees computer demonstrated.

1800

1942

Atanasoff completes partly working computer with three hundred vacuum tubes, leaves for Navy.

1943

Colossus, a vacuum-tube computer to break German codes, is completed at Bletchley Park.

1944

Harvard Mark I goes into operation.

John von Neumann goes to Penn to work on ENIAC.

1945

Von Neumann writes "First Draft of a Report on the EDVAC" describing a stored-program computer.

Six women programmers of ENIAC are sent to Aberdeen for training.

Vannevar Bush publishes "As We May Think," describing personal computer.

Bush publishes "Science, the Endless Frontier," proposing government funding of academic and industrial research.

ENIAC is fully operational.

1947

Transistor invented at Bell Labs.

1950

Turing publishes article describing a test for artificial intelligence.

1952

Grace Hopper develops first computer compiler.

Von Neumann completes modern computer at the Institute for Advanced Study.

UNIVAC predicts Eisenhower election victory.

1952

1954

Turing commits suicide.

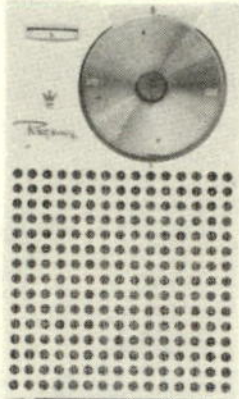

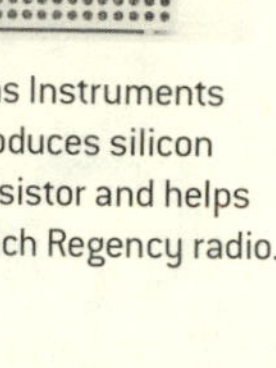

Texas Instruments introduces silicon transistor and helps launch Regency radio.

1956

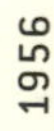

Shockley Semiconductor founded.

First artificial intelligence conference.

1957

Robert Noyce, Gordon Moore, and others form Fairchild Semiconductor.

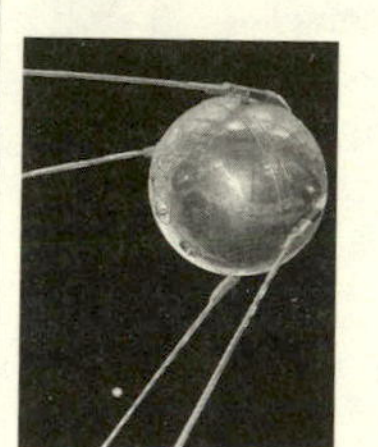

Russia launches *Sputnik*.

1958

Advanced Research Projects Agency (ARPA) announced.

Jack Kilby demonstrates integrated circuit, or microchip.

1959

Noyce and Fairchild colleagues independently invent microchip.

1960

J. C. R. Licklider publishes "Man-Computer Symbiosis."

Paul Baran at RAND devises packet switching.

1961

President Kennedy proposes sending man to the moon.

1962

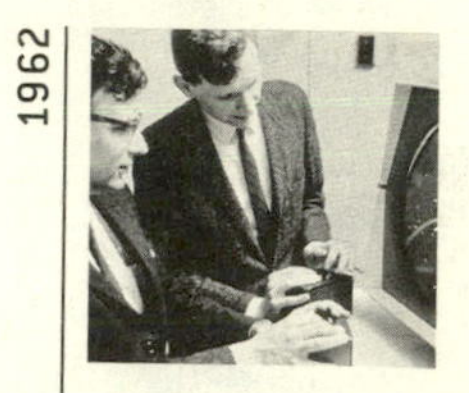

MIT hackers create *Spacewar* game.

Licklider becomes founding director of ARPA's Information Processing Techniques Office.

Doug Engelbart publishes "Augmenting Human Intellect."

1963

Licklider proposes an "Intergalactic Computer Network."

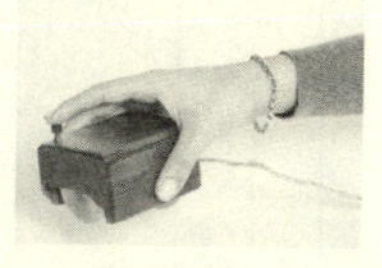

Engelbart and Bill English invent the mouse.

1954

1964

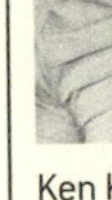

Ken Kesey and the Merry Pranksters take bus trip across America.

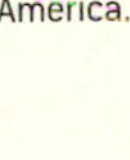

1965

Ted Nelson publishes first article about "hypertext."

Moore's Law predicts microchips will double in power each year or so.

1966

Stewart Brand hosts Trips Festival with Ken Kesey.

Bob Taylor convinces ARPA chief Charles Herzfeld to fund ARPANET.

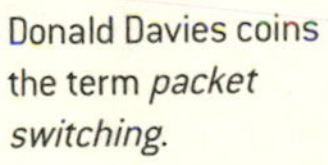

Donald Davies coins the term *packet switching*.

1967

ARPANET design discussions in Ann Arbor and Gatlinburg.

1968

Larry Roberts sends out request for bids to build the ARPANET's IMPs.

Noyce and Moore form Intel, hire Andy Grove.

Brand publishes first *Whole Earth Catalog*.

Engelbart stages the Mother of All Demos with Brand's help.

1969

First nodes of ARPANET installed.

1971

Don Hoefler begins column for *Electronic News* called "Silicon Valley USA."

Demise party for *Whole Earth Catalog*.

Intel 4004 microprocessor unveiled.

Ray Tomlinson invents email.

1972

Nolan Bushnell creates Pong at Atari with Al Alcorn.

1972

1973

Alan Kay helps to create the Alto at Xerox PARC.

Ethernet developed by Bob Metcalfe at Xerox PARC.

Community Memory shared terminal set up at Leopold's Records, Berkeley.

Vint Cerf and Bob Kahn complete TCP/IP protocols for the Internet.

1974

Intel 8080 comes out.

1975

HOW TO "READ" FM TUNER SPECIFICATIONS

Popular Electronics

PROJECT BREAKTHROUGH!

World's First Minicomputer Kit to Rival Commercial Models...

"ALTAIR 8800" SAVE OVER $1000

ALSO IN THIS ISSUE:

- An Under-$90 Scientific Calculator Project
- CCD's—TV Camera Tube Successor?
- Thyristor-Controlled Photoflashers

TEST REPORTS:

Technics 200 Speaker System

Altair personal computer from MITS appears.

Paul Allen and Bill Gates write BASIC for Altair, form Microsoft.

First meeting of Homebrew Computer Club.

Steve Jobs and Steve Wozniak launch the Apple I.

1977

The Apple II is released.

1978

First Internet Bulletin Board System.

1979

Usenet newsgroups invented.

Jobs visits Xerox PARC.

1980

IBM commissions Microsoft to develop an operating system for PC.

1981

Hayes modem marketed to home users.

1983

Microsoft announces Windows.

Richard Stallman begins developing GNU, a free operating system.

1973

1984

Apple introduces Macintosh.

1985

Stewart Brand and Larry Brilliant launch The WELL.

CVC launches Q-Link, which becomes AOL.

1991

Linus Torvalds releases first version of Linux kernel.

Tim Berners-Lee announces World Wide Web.

1993

Marc Andreessen announces Mosaic browser.

Steve Case's AOL offers direct access to the Internet.

1994

Justin Hall launches Web log and directory.

HotWired and Time Inc.'s Pathfinder become first major magazine publishers on Web.

1995

Ward Cunningham's Wiki Wiki Web goes online.

1997

IBM's Deep Blue beats Garry Kasparov in chess.

1998

Larry Page and Sergey Brin launch Google.

1999

Ev Williams launches Blogger.

2001

Jimmy Wales, with Larry Sanger, launches Wikipedia.

2011

IBM's computer Watson wins *Jeopardy!*

2011

INTRODUCTION

HOW THIS BOOK CAME TO BE

The computer and the Internet are among the most important inventions of our era, but few people know who created them. They were not conjured up in a garret or garage by solo inventors suitable to be singled out on magazine covers or put into a pantheon with Edison, Bell, and Morse. Instead, most of the innovations of the digital age were done collaboratively. There were a lot of fascinating people involved, some ingenious and a few even geniuses. This is the story of these pioneers, hackers, inventors, and entrepreneurs—who they were, how their minds worked, and what made them so creative. It's also a narrative of how they collaborated and why their ability to work as teams made them even *more* creative.

The tale of their teamwork is important because we don't often focus on how central that skill is to innovation. There are thousands of books celebrating people we biographers portray, or mythologize, as lone inventors. I've produced a few myself. Search the phrase "the man who invented" on Amazon and you get 1,860 book results. But we have far fewer tales of collaborative creativity, which is actually more important in understanding how today's technology revolution was fashioned. It can also be more interesting.

We talk so much about innovation these days that it has become a buzzword, drained of clear meaning. So in this book I set out to report on how innovation actually happens in the real world. How did the most imaginative innovators of our time turn disruptive ideas into realities? I focus on a dozen or so of the most significant breakthroughs of the digital age and the people who made them. What ingredients produced their creative leaps? What skills proved most useful? How did they lead and collaborate? Why did some succeed and others fail?

I also explore the social and cultural forces that provide the atmosphere for innovation. For the birth of the digital age, this included a research ecosystem that was nurtured by government spending and managed by a military-industrial-academic collaboration. Intersecting with that was a loose alliance of community organizers, communal-minded hippies, do-it-yourself hobbyists, and homebrew hackers, most of whom were suspicious of centralized authority.

Histories can be written with a different emphasis on any of these factors. An example is the invention of the Harvard/IBM Mark I, the first big electromechanical computer. One of its programmers, Grace Hopper, wrote a history that focused on its primary creator, Howard Aiken. IBM countered with a history that featured its teams of faceless engineers who contributed the incremental innovations, from counters to card feeders, that went into the machine.

Likewise, what emphasis should be put on great individuals versus on cultural currents has long been a matter of dispute; in the mid-nineteenth century, Thomas Carlyle declared that "the history of the world is but the biography of great men," and Herbert Spencer responded with a theory that emphasized the role of societal forces. Academics and participants often view this balance differently. "As a professor, I tended to think of history as run by impersonal forces," Henry Kissinger told reporters during one of his Middle East shuttle missions in the 1970s. "But when you see it in practice, you see the difference personalities make."[1] When it comes to digital-age innovation, as with Middle East peacemaking, a variety of personal and cultural forces all come into play, and in this book I sought to weave them together.

The Internet was originally built to facilitate collaboration. By contrast, personal computers, especially those meant to be used at home, were devised as tools for individual creativity. For more than a decade, beginning in the early 1970s, the development of networks and that of home computers proceeded separately from one another. They finally began coming together in the late 1980s with the advent of modems, online services, and the Web. Just as combining the steam engine with ingenious machinery drove the Industrial Revolution, the combination of the computer and distributed networks led to a digital revolution that allowed anyone to create, disseminate, and access any information anywhere.

Historians of science are sometimes wary about calling periods of great change *revolutions,* because they prefer to view progress as evolutionary. "There was no such thing as the Scientific Revolution, and this is a book about it," is the wry opening sentence of the Harvard professor Steven Shapin's book on that period. One method that Shapin used to escape his half-joking contradiction is to note how the key players of the period "vigorously expressed the view" that they were part of a revolution. "Our sense of radical change afoot comes substantially from them."[2]

Likewise, most of us today share a sense that the digital advances of the past half century are transforming, perhaps even revolutionizing the way we live. I can recall the excitement that each new breakthrough engendered. My father and uncles were electrical engineers, and like many of the characters in this book I grew up with a basement workshop that had circuit boards to be soldered, radios to be opened, tubes to be tested, and boxes of transistors and resistors to be sorted and deployed. As an electronics geek who loved Heathkits and ham radios (WA5JTP), I can remember when vacuum tubes gave way to transistors. At college I learned programming using punch cards and recall when the agony of batch processing was replaced by the ecstasy of hands-on interaction. In the 1980s I thrilled to the static and screech that modems made when they opened for you the weirdly magical realm of online services and bulletin boards, and in the early 1990s I helped to run a digital division at *Time* and Time Warner that launched new Web and broadband Internet services. As Wordsworth said of the enthusiasts who were present at the beginning of the French Revolution, "Bliss was it in that dawn to be alive."

I began work on this book more than a decade ago. It grew out of my fascination with the digital-age advances I had witnessed and also from my biography of Benjamin Franklin, who was an innovator, inventor, publisher, postal service pioneer, and all-around information networker and entrepreneur. I wanted to step away from doing biographies, which tend to emphasize the role of singular individuals, and once again do a book like *The Wise Men*, which I had coauthored with a colleague about the creative teamwork of six friends who shaped America's cold war policies. My initial plan was to focus on the teams that invented the Internet. But when I interviewed Bill Gates, he convinced me that the simultaneous emergence of the Internet and the personal computer made for a richer tale. I put this book on hold early in 2009, when I began working on a biography of Steve Jobs. But his story reinforced my interest in how the development of the Internet and computers intertwined, so as soon as I finished that book, I went back to work on this tale of digital-age innovators.

The protocols of the Internet were devised by peer collaboration, and the resulting system seemed to have embedded in its genetic code a propensity to facilitate such collaboration. The power to create and transmit information was fully distributed to each of the nodes, and any attempt to impose controls or a hierarchy could be routed around. Without falling into the teleological fallacy of ascribing intentions or a personality to technology, it's fair to say that a system of open networks connected to individually controlled computers tended, as the printing press did, to wrest control over the distribution of information from gatekeepers, central authorities, and institutions that employed scriveners and scribes. It became easier for ordinary folks to create and share content.

The collaboration that created the digital age was not just among peers but also between generations. Ideas were handed off from one cohort of innovators to the next. Another theme that emerged from my research was that users repeatedly commandeered digital innovations to create communications and social networking tools. I also became interested in how the quest for artificial intelligence—machines that think on their own—has consistently proved less fruitful than creating ways to forge a partnership or symbiosis between people and machines. In other words, the collaborative creativity that marked the digital age included collaboration between humans and machines.

Finally, I was struck by how the truest creativity of the digital age came from those who were able to connect the arts and sciences. They believed that beauty mattered. "I always thought of myself as a humanities person as a kid, but I liked electronics," Jobs told me when I embarked on his biography. "Then I read something that one of my heroes, Edwin Land of Polaroid, said about the importance of people who could stand at the intersection of humanities and sciences, and I decided that's what I wanted to do." The people who were comfortable at this humanities-technology intersection helped to create the human-machine symbiosis that is at the core of this story.

Like many aspects of the digital age, this idea that innovation resides where art and science connect is not new. Leonardo da Vinci was the exemplar of the creativity that flourishes when the humanities and sciences interact. When Einstein was stymied while working out General Relativity, he would pull out his violin and play Mozart until he could reconnect to what he called the harmony of the spheres.

When it comes to computers, there is one other historical figure, not as well known, who embodied the combination of the arts and sciences. Like her famous father, she understood the romance of poetry. Unlike him, she also saw the romance of math and machinery. And that is where our story begins.

Lord Byron (1788–1824), Ada's father, in Albanian dress, painted by Thomas Phillips in 1835.

Ada, Countess of Lovelace (1815–52), painted by Margaret Sarah Carpenter in 1836.

Charles Babbage (1791–1871), photograph taken circa 1837.

CHAPTER ONE

ADA, COUNTESS OF LOVELACE

POETICAL SCIENCE

In May 1833, when she was seventeen, Ada Byron was among the young women presented at the British royal court. Family members had worried about how she would acquit herself, given her high-strung and independent nature, but she ended up behaving, her mother reported, "tolerably well." Among those Ada met that evening were the Duke of Wellington, whose straightforward manner she admired, and the seventy-nine-year-old French ambassador Talleyrand, who struck her as "an old monkey."[1]

The only legitimate child of the poet Lord Byron, Ada had inherited her father's romantic spirit, a trait that her mother tried to temper by having her tutored in mathematics. The combination produced in Ada a love for what she took to calling "poetical science," which linked her rebellious imagination to her enchantment with numbers. For many, including her father, the rarefied sensibilities of the Romantic era clashed with the techno-excitement of the Industrial Revolution. But Ada was comfortable at the intersection of both eras.

So it was not surprising that her debut at court, despite the glamour of the occasion, made less impression on her than her attendance a few weeks later at another majestic event of the London season, at which she met Charles Babbage, a forty-one-year-old widowed science and math eminence who had established himself as a luminary on London's social circuit. "Ada was more pleased with a party she was at on Wednesday than with any of the assemblages in the grand monde," her mother reported to a friend. "She met there a few scientific people—amongst them Babbage, with whom she was delighted."[2]

Babbage's galvanizing weekly salons, which included up to three hundred guests, brought together lords in swallow-tail coats and ladies in brocade gowns with writers, industrialists, poets, actors, statesmen, explorers, botanists, and other "scientists," a word that Babbage's friends had recently coined.[3] By bringing scientific scholars into this exalted realm, said one noted geologist, Babbage "successfully asserted the rank in society due to science."[4]

The evenings featured dancing, readings, games, and lectures accompanied by an assortment of seafood, meat, fowl, exotic drinks, and iced desserts. The ladies staged *tableaux vivants*, in which they dressed in costume to re-create famous paintings. Astronomers set up telescopes, researchers displayed their electrical and magnetic contrivances, and Babbage allowed guests to play with his mechanical dolls. The centerpiece of the evenings—and one of Babbage's many motives for hosting them—was his demonstration of a model portion of his Difference Engine, a mammoth mechanical calculating contraption that he was building in a fireproof structure adjacent to his home. Babbage would display the model with great drama, cranking its arm as it calculated a sequence of numbers and, just as the audience began to get bored, showed how the pattern could suddenly change based on instructions that had been coded into the machine.[5] Those who were especially intrigued would be invited through the yard to the former stables, where the complete machine was being constructed.

Babbage's Difference Engine, which could solve polynomial equations, impressed people in different ways. The Duke of Wellington commented that it could be useful in analyzing the variables a general might face before going into battle.[6] Ada's mother, Lady Byron, marveled that it was a "*thinking* machine." As for Ada, who would later famously note that machines could never truly *think*, a friend who went with them to the demonstration reported, "Miss Byron, young as she was, understood its working, and saw the *great beauty* of the invention."[7]

Ada's love of both poetry and math primed her to see beauty in a computing machine. She was an exemplar of the era of Romantic science, which was characterized by a lyrical enthusiasm for invention and discovery. It was a period that brought "imaginative intensity and excitement to scientific work," Richard Holmes wrote in *The Age of Wonder*. "It was driven by a common ideal of intense, even reckless, personal commitment to discovery."[8]

In short, it was a time not unlike our own. The advances of the Industrial Revolution, including the steam engine, mechanical loom, and telegraph, transformed the nineteenth century in much the same way that the advances of the Digital Revolution—the computer, microchip, and Internet—have transformed our own. At the heart of both eras were innovators who combined imagination and passion with wondrous technology, a mix that produced Ada's poetical science and what the twentieth-century poet Richard Brautigan would call "machines of loving grace."

LORD BYRON

Ada inherited her poetic and insubordinate temperament from her father, but he was not the source of her love for machinery. He was, in fact, a Luddite. In his maiden speech in the House of Lords, given in February 1812 when he was twenty-four, Byron defended the followers of Ned Ludd, who were rampaging against mechanical weaving machines. With sarcastic scorn Byron mocked the mill owners of Nottingham, who were pushing a bill that would make destroying automated looms a crime punishable by death. "These machines were to them an advantage, inasmuch as they superseded the necessity of employing a number of workmen, who were left in consequence to starve," Byron declared. "The rejected workmen, in the blindness of their ignorance, instead of rejoicing at these improvements in arts so beneficial to mankind, conceived themselves to be sacrificed to improvements in mechanism."

Two weeks later, Byron published the first two cantos of his epic poem *Childe Harold's Pilgrimage*, a romanticized account of his wanderings through Portugal, Malta, and Greece, and, as he later remarked, "awoke one morning and found myself famous." Beautiful, seductive, troubled, brooding, and sexually adventurous, he was living the life of a Byronic hero while creating the archetype in his poetry. He became the toast of literary London and was feted at three parties each day, most memorably a lavish morning dance hosted by Lady Caroline Lamb.

Lady Caroline, though married to a politically powerful aristocrat who was later prime minister, fell madly in love with Byron. He thought she was "too thin," yet she had an unconventional sexual ambiguity (she liked to dress as a page boy) that he found enticing. They had a turbulent affair, and after it ended she stalked him obsessively. She famously declared him to be "mad, bad, and dangerous to know," which he was. So was she.

At Lady Caroline's party, Lord Byron had also noticed a reserved young woman who was, he recalled, "more simply dressed." Annabella Milbanke, nineteen, was from a wealthy and multi-titled family. The night before the party, she had read *Childe Harold* and had mixed feelings. "He is rather too much of a mannerist," she wrote. "He excels most in the delineation of deep feeling." Upon seeing him across the room at the party, her feelings were conflicted, dangerously so. "I did not seek an introduction to him, for all the women were absurdly courting him, and trying to deserve the lash of his Satire," she wrote her mother. "I am not desirous of a place in his lays. I made no offering at the shrine of Childe Harold, though I shall not refuse the acquaintance if it comes my way."[9]

That acquaintance, as it turned out, did come her way. After he was introduced to her formally, Byron decided that she might make a suitable wife. It was, for him, a rare display of reason over romanticism. Rather than arousing his passions, she seemed to be the sort of woman who might tame those passions and protect him from his excesses—as well as help pay off his burdensome debts. He proposed to her halfheartedly by letter. She sensibly declined. He wandered off to far less appropriate liaisons, including one with his half sister,

Augusta Leigh. But after a year, Annabella rekindled the courtship. Byron, falling more deeply in debt while grasping for a way to curb his enthusiasms, saw the rationale if not the romance in the possible relationship. "Nothing but marriage and a *speedy one* can save me," he admitted to Annabella's aunt. "If your niece is obtainable, I should prefer her; if not, the very first woman who does not look as if she would spit in my face."[10] There were times when Lord Byron was not a romantic. He and Annabella were married in January 1815.

Byron initiated the marriage in his Byronic fashion. "Had Lady Byron on the sofa before dinner," he wrote about his wedding day.[11] Their relationship was still active when they visited his half sister two months later, because around then Annabella got pregnant. However, during the visit she began to suspect that her husband's friendship with Augusta went beyond the fraternal, especially after he lay on a sofa and asked them both to take turns kissing him.[12] The marriage started to unravel.

Annabella had been tutored in mathematics, which amused Lord Byron, and during their courtship he had joked about his own disdain for the exactitude of numbers. "I know that two and two make four—and should be glad to prove it too if I could," he wrote, "though I must say if by any sort of process I could convert two and two into five it would give me much greater pleasure." Early on, he affectionately dubbed her the "Princess of Parallelograms." But when the marriage began to sour, he refined that mathematical image: "We are two parallel lines prolonged to infinity side by side but never to meet." Later, in the first canto of his epic poem *Don Juan*, he would mock her: "Her favourite science was the mathematical. . . . She was a walking calculation."

The marriage was not saved by the birth of their daughter on December 10, 1815. She was named Augusta Ada Byron, her first name that of Byron's too-beloved half sister. When Lady Byron became convinced of her husband's perfidy, she thereafter called her daughter by her middle name. Five weeks later she packed her belongings into a carriage and fled to her parents' country home with the infant Ada.

Ada never saw her father again. Lord Byron left the country that April after Lady Byron, in letters so calculating that she earned his sobriquet of "Mathematical Medea," threatened to expose his alleged incestuous and homosexual affairs as a way to secure a separation agreement that gave her custody of their child.[13]

The opening of canto 3 of *Childe Harold*, written a few weeks later, invokes Ada as his muse:

> *Is thy face like thy mother's, my fair child!*
> *Ada! sole daughter of my house and of my heart?*
> *When last I saw thy young blue eyes they smiled,*
> *And then we parted.*

Byron wrote these lines in a villa by Lake Geneva, where he was staying with the poet Percy Bysshe Shelley and Shelley's future wife, Mary. It rained relentlessly. Trapped inside for days, Byron suggested they write horror stories. He produced a fragment of a tale about a vampire, one of the first literary efforts on that subject, but Mary's story was the one that became a classic: *Frankenstein, or The Modern Prometheus*. Playing on the ancient Greek myth of the hero who crafted a living man out of clay and snatched fire from the gods for human use, *Frankenstein* was the story of a scientist who galvanized a man-made assemblage into a thinking human. It was a cautionary tale about technology and science. It also raised the question that would become associated with Ada: Can man-made machines ever truly think?

The third canto of *Childe Harold* ends with Byron's prediction that Annabella would try to keep Ada from knowing about her father, and so it happened. There was a portrait of Lord Byron at their house, but Lady Byron kept it securely veiled, and Ada never saw it until she was twenty.[14]

Lord Byron, by contrast, kept a picture of Ada on his desk wherever he wandered, and his letters often requested news or portraits of her. When she was seven, he wrote to Augusta, "I wish you would obtain from Lady B some accounts of Ada's disposition. . . . Is the girl imaginative? . . . Is she passionate? I hope that the Gods have made her anything save *poetical*—it is enough to have one such fool in the family." Lady Byron reported that Ada had an imagination that was "chiefly exercised in connection with her mechanical ingenuity."[15]

Around that time, Byron, who had been wandering through Italy, writing and having an assortment of affairs, grew bored and decided to enlist in the Greek struggle for independence from the Ottoman Empire. He sailed for Missolonghi, where he took command of part of the rebel army and prepared to attack a Turkish fortress. But before he could engage in battle, he caught a violent cold that was made worse by his doctor's decision to treat him by bloodletting. On April 19, 1824, he died. According to his valet, among his final words were "Oh, my poor dear child!—my dear Ada! My God, could I have seen her! Give her my blessing."[16]

ADA

Lady Byron wanted to make sure that Ada did not turn out like her father, and part of her strategy was to have the girl rigorously study math, as if it were an antidote to poetic imagination. When Ada, at age five, showed a preference for geography, Lady Byron ordered that the subject be replaced by additional arithmetic lessons, and her governess soon proudly reported, "She adds up sums of five or six rows of figures with accuracy." Despite these efforts, Ada developed some of her father's propensities. She had an affair as a young teenager with one of her tutors, and when they were caught and the tutor banished, she tried to run away from home to be with him. In addition, she had mood swings that took her from feelings of grandiosity to despair, and she suffered various maladies both physical and psychological.

Ada accepted her mother's conviction that an immersion in math could help tame her Byronic tendencies. After her dangerous liaison with her tutor, and inspired by Babbage's Difference Engine, she decided on her own, at eighteen, to begin a new series of lessons. "I must cease to think of living for pleasure or self-gratification," she wrote her new tutor. "I find that nothing but very close and intense application to subjects of a scientific nature now seems to keep my imagination from running wild. . . . It appears to me that the first thing is to go through a course of Mathematics." He agreed with the prescription: "You are right in supposing that your chief resource and safeguard at the present is in a course of severe intellectual study. For this purpose there is no subject to be compared to Mathematics."[17] He prescribed Euclidean geometry, followed by a dose of trigonometry and algebra. That should cure anyone, they both thought, from having too many artistic or romantic passions.

Her interest in technology was stoked when her mother took her on a trip through the British industrial midlands to see the new factories and machinery. Ada was particularly impressed with an automated weaving loom that used punch cards to direct the creation of the desired fabric patterns, and she drew a sketch of how it worked. Her father's famous speech in the House of Lords had defended the Luddites who had smashed such looms because of their fear of what technology might inflict on humanity. But Ada waxed poetical about them and saw the connection with what would someday be called computers. "This Machinery reminds me of Babbage and his gem of all mechanism," she wrote.[18]

Ada's interest in applied science was further stimulated when she met one of Britain's few noted female mathematicians and scientists, Mary Somerville. Somerville had just finished writing one of her great works, *On the Connexion of the Physical Sciences*, in which she tied together developments in astronomy, optics, electricity, chemistry, physics, botany, and geology.* Emblematic of the time, it provided a unified sense of the extraordinary endeavors of discovery that were under way. She proclaimed in her opening sentence, "The progress of modern science, especially within the last five years, has been remarkable for a tendency to simplify the laws of nature and to unite detached branches by general principles."

Somerville became a friend, teacher, inspiration, and mentor to Ada. She met with Ada regularly, sent her math books, devised problems for her to solve, and patiently explained the correct answers. She was also a good friend of Babbage's, and during the fall of 1834 she

---

*It was in a review of this book that one of Babbage's friends, William Whewell, coined the term *scientist* to suggest the connection among these disciplines.

and Ada would often visit his Saturday-evening salons. Somerville's son, Woronzow Greig, aided Ada's efforts to settle down by suggesting to one of his former classmates at Cambridge that she would make a suitable—or at least interesting—wife.

William King was socially prominent, financially secure, quietly intelligent, and as taciturn as Ada was excitable. Like her, he was a student of science, but his focus was more practical and less poetic: his primary interests were crop rotation theories and advances in livestock breeding techniques. He proposed marriage within a few weeks of meeting Ada, and she accepted. Her mother, with motives that only a psychiatrist could fathom, decided it was imperative to tell William about Ada's attempted elopement with her tutor. Despite this news, William was willing to proceed with the wedding, which was held in July 1835. "Gracious God, who has so mercifully given you an opportunity of turning aside from the dangerous paths, has given you a friend and guardian," Lady Byron wrote her daughter, adding that she should use this opportunity to "bid adieu" to all of her "peculiarities, caprices, and self-seeking."[19]

The marriage was a match made in rational calculus. For Ada, it offered the chance to adopt a more steady and grounded life. More important, it allowed her to escape dependence on her domineering mother. For William, it meant having a fascinating, eccentric wife from a wealthy and famous family.

Lady Byron's first cousin Viscount Melbourne (who had the misfortune of having been married to Lady Caroline Lamb, by then deceased) was the prime minister, and he arranged that, in Queen Victoria's coronation list of honors, William would become the Earl of Lovelace. His wife thus became Ada, Countess of Lovelace. She is therefore properly referred to as Ada or Lady Lovelace, though she is now commonly known as Ada Lovelace.

That Christmas of 1835, Ada received from her mother the family's life-size portrait of her father. Painted by Thomas Phillips, it showed Lord Byron in romantic profile, gazing at the horizon, dressed in traditional Albanian costume featuring a red velvet jacket, ceremonial sword, and headdress. For years it had hung over Ada's grandparents' mantelpiece, but it had been veiled by a green cloth from the day her parents had separated. Now she was trusted not only to see it but to possess it, along with his inkstand and pen.

Her mother did something even more surprising when the Lovelaces' first child, a son, was born a few months later. Despite her disdain for her late husband's memory, she agreed that Ada should name the boy Byron, which she did. The following year Ada had a daughter, whom she dutifully named Annabella, after her mother. Ada then came down with yet another mysterious malady, which kept her bedridden for months. She recovered well enough to have a third child, a son named Ralph, but her health remained fragile. She had digestive and respiratory problems that were compounded by being treated with laudanum, morphine, and other forms of opium, which led to mood swings and occasional delusions.

Ada was further unsettled by the eruption of a personal drama that was bizarre even by the standards of the Byron family. It involved Medora Leigh, the daughter of Byron's half sister and occasional lover. According to widely accepted rumors, Medora was Byron's daughter. She seemed determined to show that darkness ran in the family. She had an affair with a sister's husband, then ran off with him to France and had two illegitimate children. In a fit of self-righteousness, Lady Byron went to France to rescue Medora, then revealed to Ada the story of her father's incest.

This "most strange and dreadful history" did not seem to surprise Ada. "I am not in the least astonished," she wrote her mother. "You merely confirm what I have for years and years felt scarcely a doubt about."[20] Rather than being outraged, she seemed oddly energized by the news. She declared that she could relate to her father's defiance of authority. Referring to his "misused genius," she wrote to her mother, "If he has transmitted to me any portion of that genius, I would use it to bring out great truths and principles. I think he has bequeathed this task to me. I have this feeling strongly, and there is a pleasure attending it."[21]

Once again Ada took up the study of math in order to settle herself, and she tried to convince Babbage to become her tutor. "I have a peculiar way of learning, and I think it must be a peculiar man to teach me successfully," she wrote him. Whether due to her opiates or her breeding or both, she developed a somewhat outsize opinion of her own talents and began to describe herself as a genius. In her letter to Babbage, she wrote, "Do not reckon me conceited, . . . but I believe I have the power of going just as far as I like in such pursuits, and where there is so decided a taste, I should almost say a passion, as I have for them, I question if there is not always some portion of natural genius even."[22]

Babbage deflected Ada's request, which was probably wise. It preserved their friendship for an even more important collaboration, and she was able to secure a first-rate math tutor instead: Augustus De Morgan, a patient gentleman who was a pioneer in the field of symbolic logic. He had propounded a concept that Ada would one day employ with great significance, which was that an algebraic equation could apply to things other than numbers. The relations among symbols (for example, that $a + b = b + a$) could be part of a logic that applied to things that were not numerical.

Ada was never the great mathematician that her canonizers claim, but she was an eager pupil, able to grasp most of the basic concepts of calculus, and with her artistic sensibility she liked to visualize the changing curves and trajectories that the equations were describing. De Morgan encouraged her to focus on the rules for working through equations, but she was more eager to discuss the underlying concepts. Likewise with geometry, she often asked for visual ways to picture problems, such as how the intersections of circles in a sphere divide it into various shapes.

Ada's ability to appreciate the beauty of mathematics is a gift that eludes many people, including some who think of themselves as intellectual. She realized that math was a lovely language, one that describes the harmonies of the universe and can be poetic at times. Despite her mother's efforts, she remained her father's daughter, with a poetic sensibility that allowed her to view an equation as a brushstroke that painted an aspect of nature's physical splendor, just as she could visualize the "wine-dark sea" or a woman who "walks in beauty, like the night." But math's appeal went even deeper; it was spiritual. Math "constitutes the language through which alone we can adequately express the great facts of the natural world," she said, and it allows us to portray the "changes of mutual relationship" that unfold in creation. It is "the instrument through which the weak mind of man can most effectually read his Creator's works."

This ability to apply imagination to science characterized the Industrial Revolution as well as the computer revolution, for which Ada was to become a patron saint. She was able, as she told Babbage, to understand the connection between poetry and analysis in ways that transcended her father's talents. "I do not believe that my father was (or ever could have been) such a Poet as I shall be an Analyst; for with me the two go together indissolubly," she wrote.[23]

Her reengagement with math, she told her mother, spurred her creativity and led to an "immense development of *imagination*, so much so that I feel no doubt if I continue my studies I shall in due time be a *Poet*."[24] The whole concept of imagination, especially as it was applied to technology, intrigued her. "What is imagination?" she asked in an 1841 essay. "It is the Combining faculty. It brings together things, facts, ideas, conceptions in new, original, endless, ever-varying combinations. . . . It is that which penetrates into the unseen worlds around us, the worlds of Science."[25]

By then Ada believed she possessed special, even supernatural abilities, what she called "an intuitive perception of hidden things." Her exalted view of her talents led her to pursue aspirations that were unusual for an aristocratic woman and mother in the early Victorian age. "I believe myself to possess a most singular combination of qualities exactly fitted to make me pre-eminently a discoverer of the hidden realities of nature," she explained in a letter to her mother in 1841. "I can throw rays from every quarter of the universe into one vast focus."[26]

It was while in this frame of mind that she decided to engage again with Charles Babbage, whose salons she had first attended eight years earlier.

CHARLES BABBAGE AND HIS ENGINES

From an early age, Charles Babbage had been interested in machines that could perform human tasks. When he was a child, his mother took him to many of the exhibition halls and museums of wonder that were springing up in London in the early 1800s. At one in Hanover Square, a proprietor aptly named Merlin invited him up to the attic workshop where there was a variety of mechanical dolls, known as "automata." One was a silver female dancer, about a foot tall, whose arms moved with grace and who held in her hand a bird that could wag its tail, flap its wings, and open its beak. The Silver Lady's ability to display feelings and personality captured the boy's fancy. "Her eyes were full of imagination," he recalled. Years later he discovered the Silver Lady at a bankruptcy auction and bought it. It served as an amusement at his evening salons where he celebrated the wonders of technology.

At Cambridge Babbage became friends with a group, including John Herschel and George Peacock, who were disappointed by the way math was taught there. They formed a club, called the Analytical Society, which campaigned to get the university to abandon the calculus notation devised by its alumnus Newton, which relied on dots, and replace it with the one devised by Leibniz, which used *dx* and *dy* to represent infinitesimal increments and was thus known as "d" notation. Babbage titled their manifesto "The Principles of pure D-ism in opposition to the Dot-age of the University."[27] He was prickly, but he had a good sense of humor.

One day Babbage was in the Analytical Society's room working on a table of logarithms that was littered with discrepancies. Herschel asked him what he was thinking. "I wish to God these calculations had been executed by steam," Babbage answered. To this idea of a mechanical method for tabulating logarithms Herschel replied, "It is quite possible."[28] In 1821 Babbage turned his attention to building such a machine.

Over the years, many had fiddled with making calculating contraptions. In the 1640s, Blaise Pascal, the French mathematician and philosopher, created a mechanical calculator to reduce the drudgery of his father's work as a tax supervisor. It had spoked metal wheels with the digits 0 through 9 on their circumference. To add or subtract numbers, the operator used a stylus to dial a number, as if using a rotary phone, then dialed in the next number; an armature carried or borrowed a 1 when necessary. It became the first calculator to be patented and sold commercially.

Thirty years later, Gottfried Leibniz, the German mathematician and philosopher, tried to improve upon Pascal's contraption with a "stepped reckoner" that had the capacity to multiply and divide. It had a hand-cranked cylinder with a set of teeth that meshed with counting wheels. But Leibniz ran into a problem that would be a recurring theme of the digital age. Unlike Pascal, an adroit engineer who could combine scientific theories with mechanical genius, Leibniz had little engineering skill and did not surround himself with those who did. So, like many great theorists who lacked practical collaborators, he was unable to produce reliably working versions of his device. Nevertheless, his core concept, known as the Leibniz wheel, would influence calculator design through the time of Babbage.

Babbage knew of the devices of Pascal and Leibniz, but he was trying to do something more complex. He wanted to construct a mechanical method for tabulating logarithms, sines, cosines, and tangents.* To do so, he adapted an idea that the French mathematician Gaspard de Prony came up with in the 1790s. In order to create logarithm and trigonometry tables, de Prony broke down the operations into very simple steps that involved only addition and subtraction. Then he provided easy instructions so that scores of human laborers, who knew little math, could perform these simple tasks and pass along their answers to the next set of laborers. In other words, he created an assembly line, the great industrial-age innovation that was memorably analyzed by Adam Smith in his description of the division of labor in a pin-making factory. After a trip to Paris in which he heard of de Prony's method, Babbage wrote, "I conceived all of a sudden the idea of applying the same method to the immense work with which I had been burdened, and to manufacture logarithms as one manufactures pins."[29]

Even complex mathematical tasks, Babbage realized, could be broken into steps that came down to calculating "finite differences"

---

* Specifically, he wanted to use the method of divided differences to closely approximate logarithmic and trigonometric functions.

Replica of the Difference Engine.

Replica of the Analytical Engine.

The Jacquard loom.

Silk portrait of Joseph-Marie Jacquard (1752–1834) woven by a Jacquard loom.

through simple adding and subtracting. For example, in order to make a table of squares—1^2, 2^2, 3^2, 4^2, and so on—you could list the initial numbers in such a sequence: 1, 4, 9, 16. . . . This would be column A. Beside it, in column B, you could figure out the differences between each of these numbers, in this case 3, 5, 7, 9. . . . Column C would list the difference between each of column B's numbers, which is 2, 2, 2, 2. . . . Once the process was thus simplified, it could be reversed and the tasks parceled out to untutored laborers. One would be in charge of adding 2 to the last number in column B, and then would hand that result to another person, who would add that result to the last number in column A, thus generating the next number in the sequence of squares.

Babbage devised a way to mechanize this process, and he named it the Difference Engine. It could tabulate any polynomial function and provide a digital method for approximating the solution to differential equations.

How did it work? The Difference Engine used vertical shafts with disks that could be turned to any numeral. These were attached to cogs that could be cranked in order to add that numeral to (or subtract it from) a disk on an adjacent shaft. The contraption could even "store" the interim results on another shaft. The main complexity was how to "carry" or "borrow" when necessary, as we do with pencils when we calculate 36 + 19 or 42 - 17. Drawing on Pascal's devices, Babbage came up with a few ingenious contrivances that allowed the cogs and shafts to handle the calculation.

The machine was, in concept, a true marvel. Babbage even figured out a way to get it to create a table of prime numbers up to 10 million. The British government was impressed, at least initially. In 1823 it gave him seed money of £1,700 and would eventually sink more than £17,000, twice the cost of a warship, into the device during the decade Babbage spent trying to build it. But the project ran into two problems. First, Babbage and his hired engineer did not quite have the skills to get the device working. Second, he began dreaming up something better.

Babbage's new idea, which he conceived in 1834, was a general-purpose computer that could carry out a variety of different operations based on programming instructions given to it. It could perform one task, then be made to switch and perform another. It could even tell itself to switch tasks—or alter its "pattern of action," as Babbage explained—based on its own interim calculations. Babbage named this proposed machine the Analytical Engine. He was one hundred years ahead of his time.

The Analytical Engine was the product of what Ada Lovelace, in her essay on imagination, had called "the Combining Faculty." Babbage had combined innovations that had cropped up in other fields, a trick of many great inventors. He had originally used a metal drum that was studded with spikes to control how the shafts would turn. But then he studied, as Ada had, the automated loom invented in 1801 by a Frenchman named Joseph-Marie Jacquard, which transformed the silk-weaving industry. Looms create a pattern by using hooks to lift selected warp threads, and then a rod pushes a woof thread underneath. Jacquard invented a method of using cards with holes punched in them to control this process. The holes determined which hooks and rods would be activated for each pass of the weave, thus automating the creation of intricate patterns. Each time the shuttle was thrown to create a new pass of the thread, a new punch card would come into play.

On June 30, 1836, Babbage made an entry into what he called his "Scribbling Books" that would represent a milestone in the prehistory of computers: "Suggested Jacquard's loom as a substitute for the drums."[30] Using punch cards rather than steel drums meant that an unlimited number of instructions could be input. In addition, the sequence of tasks could be modified, thus making it easier to devise a general-purpose machine that was versatile and reprogrammable.

Babbage bought a portrait of Jacquard and began to display it at his salons. It showed the inventor sitting in an armchair, a loom in the background, holding a pair of calipers over rectangular punch cards. Babbage amused his guests by asking them to guess what it was. Most thought it a superb engraving. He would then reveal that it was actually a finely woven silk tapestry, with twenty-four thousand rows of threads, each controlled by a different punch card. When Prince Albert, the husband of Queen Victoria, came to one of Babbage's salons, he asked Babbage why he found the tapestry so interesting. Babbage replied, "It will greatly assist in explaining the nature of my calculating machine, the Analytical Engine."[31]

Few people, however, saw the beauty of Babbage's proposed new machine, and the British government had no inclination to fund it. Try as he might, Babbage could generate little notice in either the popular press or scientific journals.

But he did find one believer. Ada Lovelace fully appreciated the concept of a general-purpose machine. More important, she envisioned an attribute that might make it truly amazing: it could potentially process not only numbers but any symbolic notations, including musical and artistic ones. She saw the poetry in such an idea, and she set out to encourage others to see it as well.

She barraged Babbage with letters, some of which verged on cheeky, even though he was twenty-four years her senior. In one, she described the solitaire game using twenty-six marbles, where the goal is to execute jumps so that only one marble remains. She had mastered it but was trying to derive a "mathematical formula . . . on which the solution depends, and which can be put into symbolic language." Then she asked, "Am I too imaginative for you? I think not."[32]

Her goal was to work with Babbage as his publicist and partner in trying to get support to build the Analytical Engine. "I am very anxious to talk to you," she wrote in early 1841. "I will give you a hint on what. It strikes me that at some future time . . . my head may be made by you subservient to some of your purposes and plans. If so, if ever I could be worthy or capable of being used by you, my head will be yours."[33]

A year later, a tailor-made opportunity presented itself.

LADY LOVELACE'S NOTES

In his quest to find support for his Analytical Engine, Babbage had accepted an invitation to address the Congress of Italian Scientists in Turin. Taking notes was a young military engineer, Captain Luigi Menabrea, who would later serve as prime minister of Italy. With Babbage's help, Menabrea published a detailed description of the machine, in French, in October 1842.

One of Ada's friends suggested that she produce a translation of Menabrea's piece for *Scientific Memoirs*, a periodical devoted to scientific papers. This was her opportunity to serve Babbage and show her talents. When she finished, she informed Babbage, who was pleased but also somewhat surprised. "I asked why she had not herself written an original paper on a subject with which she was so intimately acquainted," Babbage said.[34] She replied that the thought had not occurred to her. Back then, women generally did not publish scientific papers.

Babbage suggested that she add some notes to Menabrea's memoir, a project that she embraced with enthusiasm. She began working on a section she called "Notes by the Translator" that ended up totaling 19,136 words, more than twice the length of Menabrea's original article. Signed "A.A.L.," for Augusta Ada Lovelace, her "Notes" became more famous than the article and were destined to make her an iconic figure in the history of computing.[35]

As she worked on the notes at her country estate in Surrey in the summer of 1843, she and Babbage exchanged scores of letters, and in the fall they had numerous meetings after she moved back to her London home. A minor academic specialty and gender-charged debate has grown up around the issue of how much of the thinking was hers rather than his. In his memoirs, Babbage gives her much of the credit: "We discussed together the various illustrations that might be introduced: I suggested several but the selection was entirely her own. So also was the algebraic working out of the different problems, except, indeed, that relating to the numbers of Bernoulli, which I had offered to do to save Lady Lovelace the trouble. This she sent back to me for an amendment, having detected a grave mistake which I had made in the process."[36]

In her "Notes," Ada explored four concepts that would have historical resonance a century later when the computer was finally born. The first was that of a general-purpose machine, one that could not only perform a preset task but could be programmed and reprogrammed to do a limitless and changeable array of tasks. In other words, she envisioned the modern computer. This concept was at the core of her "Note A," which emphasized the distinction between Babbage's original Difference Engine and his proposed new Analytical Engine. "The particular function whose integral the *Difference* Engine was constructed to tabulate is $\Delta^7 u_x = 0$," she began, explaining that its purpose was the computation of nautical tables. "The *Analytical* Engine, on the contrary, is not merely adapted for tabulating the results of one particular function and of no other, but for developing and tabulating any function whatever."

This was done, she wrote, by "the introduction into it of the principle which Jacquard devised for regulating, by means of punched cards, the most complicated patterns in the fabrication of brocaded stuffs." Even more than Babbage, Ada realized the significance of this. It meant that the machine could be like the type of computer we now take for granted: one that does not merely do a specific arithmetic task but can be a general-purpose machine. She explained:

> The bounds of arithmetic were outstepped the moment the idea of applying cards had occurred. The Analytical Engine does not occupy common ground with mere "calculating machines." It holds a position wholly its own. In enabling a mechanism to combine together *general* symbols, in successions of unlimited variety and extent, a uniting link is established between the operations of matter and the abstract mental processes.[37]

Those sentences are somewhat clotted, but they are worth reading carefully. They describe the essence of modern computers. And Ada enlivened the concept with poetic flourishes. "The Analytical Engine weaves algebraical patterns just as the Jacquard loom weaves flowers and leaves," she wrote. When Babbage read "Note A," he was thrilled and made no changes. "Pray do not alter it," he said.[38]

Ada's second noteworthy concept sprang from this description of a general-purpose machine. Its operations, she realized, did not need to be limited to math and numbers. Drawing on De Morgan's extension of algebra into a formal logic, she noted that a machine such as the Analytical Engine could store, manipulate, process, and act upon anything that could be expressed in symbols: words and logic and music and anything else we might use symbols to convey.

To explain this idea, she carefully defined what a computer operation was: "It may be desirable to explain that by the word 'operation,' we mean any process which alters the mutual relation of two or more things, be this relation of what kind it may." A computer operation, she noted, could alter the relationship not just between numbers but between any symbols that were logically related. "It might act upon other things besides number, were objects found whose mutual fundamental relations could be expressed by those of the abstract science of operations." The Analytical Engine could, in theory, even perform operations on musical notations: "Supposing, for instance, that the fundamental relations of pitched sounds in the science of harmony and of musical composition were susceptible of such expression and adaptations, the engine might compose elaborate and scientific pieces of music of any degree of complexity." It was the ultimate Ada-like "poetical science" concept: an elaborate and scientific piece of music composed by a machine! Her father would have shuddered.

This insight would become the core concept of the digital age: any piece of content, data, or information—music, text, pictures, numbers, symbols, sounds, video—could be expressed in digital form and manipulated by machines. Even Babbage failed to see this fully; he focused on numbers. But Ada realized that the digits on the cogs could represent things other than mathematical quantities. Thus did she make the conceptual leap from machines that were mere calculators to ones that we now call computers. Doron Swade, a computer historian who specializes in studying Babbage's engines, has declared this one of Ada's historic legacies. "If we are looking and sifting history for that transition, then that transition was made explicitly by Ada in that 1843 paper," he said.[39]

Ada's third contribution, in her final "Note G," was to figure out in step-by-step detail the workings of what we now call a computer program or algorithm. The example she used was a program to compute Bernoulli numbers,* an exceedingly complex infinite series that in various guises plays a role in number theory.

---

*Named after the seventeenth-century Swiss mathematician Jacob Bernoulli, who studied the sums of powers of consecutive integers, they play an intriguing role in number theory, mathematical analysis, and differential topology.

To show how the Analytical Engine could generate Bernoulli numbers, Ada described a sequence of operations and then made a chart showing how each would be coded into the machine. Along the way, she helped to devise the concepts of subroutines (a sequence of instructions that performs a specific task, such as computing a cosine or calculating compound interest, and can be dropped into larger programs as needed) and a recursive loop (a sequence of instructions that repeats itself).* These were made possible by the punch-card mechanism. Seventy-five cards were needed to generate each number, she explained, and then the process became iterative as that number was fed back into the process to generate the next one. "It will be obvious that the very same seventy-five variable cards may be repeated for the computation of every succeeding number," she wrote. She envisioned a library of commonly used subroutines, something that her intellectual heirs, including women such as Grace Hopper at Harvard and Kay McNulty and Jean Jennings at the University of Pennsylvania, would create a century later. In addition, because Babbage's engine made it possible to jump back and forth within the sequence of instruction cards based on the interim results it had calculated, it laid the foundation for what we now call conditional branching, changing to a different path of instructions if certain conditions are met.

Babbage helped Ada with the Bernoulli calculations, but the letters show her deeply immersed in the details. "I am doggedly attacking and sifting to the very bottom all the ways of deducing the Bernoulli numbers," she wrote in July, just weeks before her translation and notes were due at the printers. "I am in much dismay at having gotten so amazing a quagmire and botheration with these Numbers that I cannot possibly get the thing done today. . . . I am in a charming state of confusion."[40]

When it got worked out, she added a contribution that was primarily her own: a table and diagram showing exactly how the algorithm would be fed into the computer, step by step, including two recursive loops. It was a numbered list of coding instructions

---

*Ada's example involved tabulating polynomials using difference techniques as a subfunction, which required a nested loop structure with a varying range for the inner loop.

that included destination registers, operations, and commentary—something that would be familiar to any C++ coder today. "I have worked incessantly and most successfully all day," she wrote Babbage. "You will admire the Table and Diagram extremely. They have been made out with extreme care." From all of the letters it is clear that she did the table herself; the only help came from her husband, who did not understand the math but was willing to methodically trace in ink what she had done in pencil. "Lord L is at this moment kindly inking it all over for me," she wrote Babbage. "I had to do it in pencil."[41]

It was mainly on the basis of this diagram, which accompanied the complex processes for generating Bernoulli numbers, that Ada has been accorded by her fans the accolade of "the world's first computer programmer." That is a bit hard to defend. Babbage had already devised, at least in theory, more than twenty explanations of processes that the machine might eventually perform. But none of these was published, and there was no clear description of the way to sequence the operations. Therefore, it is fair to say that the algorithm and detailed programming description for the generation of Bernoulli numbers was the first computer program ever to be published. And the initials at the end were those of Ada Lovelace.

There was one other significant concept that she introduced in her "Notes," which harked back to the Frankenstein story produced by Mary Shelley after that weekend with Lord Byron. It raised what is still the most fascinating metaphysical topic involving computers, that of artificial intelligence: Can machines think?

Ada believed not. A machine such as Babbage's could perform operations as instructed, she asserted, but it could not come up with ideas or intentions of its own. "The Analytical Engine has no pretensions whatever to *originate* anything," she wrote in her "Notes." "It can do whatever we know how to order it to perform. It can follow analysis; but it has no power of anticipating any analytical relations or truths." A century later this assertion would be dubbed "Lady Lovelace's Objection" by the computer pioneer Alan Turing (see chapter 3).

Ada wanted her work to be regarded as a serious scientific paper and not merely a public advocacy piece, so at the outset of her "Notes" she stated that she would "offer no opinion" on the government's reluctance to continue funding Babbage's endeavors. This did not please Babbage, who proceeded to write a screed attacking the government. He wanted Ada to include it in her "Notes," without his name on it, as if it were her opinion. She refused. She did not want her work compromised.

Without informing her, Babbage sent his proposed appendage directly to *Scientific Memoirs*. The editors decided that it should appear separately and suggested that he "manfully" sign his name. Babbage was charming when he wished, but he could also be cranky, stubborn, and defiant, like most innovators. The proposed solution infuriated him, and he wrote Ada asking that she withdraw her work. Now it was her turn to become irate. Using a form of address typically used by male friends, "My Dear Babbage," she wrote that "withdrawing the translation and Notes" would "be dishonorable and unjustifiable." She concluded the letter, "Be assured that I am your best friend; but that I never can or will support you in acting on principles which I conceive to be not only wrong in themselves, but suicidal."[42]

Babbage backed down and agreed to have his piece published separately in another periodical. That day Ada complained to her mother:

> I have been harassed and pressed in a most perplexing manner by the conduct of Mr. Babbage. . . . I am sorry to come to the conclusion that he is one of the most impracticable, selfish, and intemperate persons one can have to do with. . . . I declared at once to Babbage that no power should induce me to lend myself to any of his quarrels or to become in any way his organ. . . . He was furious. I imperturbable and unmoved.[43]

Ada's response to the dispute was a bizarre sixteen-page letter to Babbage, poured forth in a frenzy, that vividly displayed her moodiness, exultations, delusions, and passions. She cajoled and berated him, praised and denigrated him. At one point she contrasted their motives. "My own uncompromising principle is to endeavour to love truth and God before fame and glory," she claimed. "Yours is to love truth and God; but to love fame, glory, honours yet more." She proclaimed that she saw her own inevitable fame as being of an exalted nature: "I wish to add my might toward expounding and interpreting the Almighty and his laws. . . . I should feel it no small glory if I were able to be one of his most noted prophets."[44]

Having laid that groundwork, she offered him a deal: they should forge a business and political partnership. She would apply her connections and persuasive pen to his endeavor to build his Analytical Engine if—and only if—he would let her have control over his business decisions. "I give you the first choice and offer of my services and my intellect," she wrote. "Do not lightly reject them." The letter read in parts like a venture capital term sheet or a prenuptial agreement, complete with the possibility of arbitrators. "You will undertake to abide wholly by the judgment of myself (or of any persons whom you may now please to name as referees, whenever we may differ) on all practical matters," she declared. In return, she promised, she would "lay before you in the course of a year or two explicit and honorable propositions for executing your engine."[45]

The letter would seem surprising were it not like so many others that she wrote. It was an example of how her grandiose ambitions sometimes got the best of her. Nevertheless, she deserves respect as a person who, rising above the expectations of her background and gender and defying plagues of family demons, dedicated herself diligently to complex mathematical feats that most of us never would or could attempt. (Bernoulli numbers alone would defeat many of us.) Her impressive mathematical labors and imaginative insights came in the midst of the drama of Medora Leigh and bouts of illness that would cause her to become dependent on opiates that amplified her mood swings. She explained at the end of her letter to Babbage, "My dear friend, if you knew what sad and direful experiences I have had, in ways of which you cannot be aware, you would feel that *some* weight is due to my feelings." Then, after a quick detour to raise a small point about using the calculus of finite differences to compute Bernoulli numbers, she apologized that "this letter is sadly blotted" and plaintively asked, "I wonder if you will choose to retain the lady-fairy in your service or not."[46]

Ada was convinced that Babbage would accept her offer to become entrepreneurial partners. "He has so strong an idea of the advantage of having my pen as his servant that he will probably yield; though I demand very strong concessions," she wrote her mother. "If he does consent to what I propose, I shall probably be enabled to keep him out of much hot water and to bring his engine to consummation."[47] Babbage, however, thought it wiser to decline. He went to see Ada and "refused all the conditions."[48] Although they never again collaborated on science, their relationship survived. "Babbage and I are I think more friends than ever," she wrote her mother the next week.[49] And Babbage agreed the next month to pay a visit to her country home, sending her a fond letter referring to her as "the Enchantress of Numbers" and "my dear and much admired Interpreter."

That month, September 1843, her translation and "Notes" finally appeared in *Scientific Memoirs*. For a while she was able to bask in acclaim from friends and to hope that, like her mentor Mary Somerville, she would be taken seriously in scientific and literary circles. Publication made her finally feel like "a completely professional person," she wrote to a lawyer. "I really have become as much tied to a profession as *you* are."[50]

It was not to be. Babbage got no more funding for his machines; they were never built, and he died in poverty. As for Lady Lovelace, she never published another scientific paper. Instead her life spiraled downward, and she became addicted to gambling and opiates. She had an affair with a gambling partner who then blackmailed her, forcing her to pawn her family jewels. During the final year of her life, she fought an exceedingly painful battle with uterine cancer accompanied by constant hemorrhaging. When she died in 1852, at age thirty-six, she was buried, in accordance with one of her last requests, in a country grave next to the poet father she never knew, who had died at the same age.

The Industrial Revolution was based on two grand concepts that were profound in their simplicity. Innovators came up with ways to simplify endeavors by breaking them into easy, small tasks that could be accomplished on assembly lines. Then, beginning in the textile industry, inventors found ways to mechanize steps so that they could be performed by machines, many of them powered by steam engines. Babbage, building on ideas from Pascal and Leibniz, tried to apply these two processes to the production of computations, creating a mechanical precursor to the modern computer. His most significant conceptual leap was that such machines did not have to be set to do only one process, but instead could be programmed and reprogrammed through the use of punch cards. Ada saw the beauty and significance of that enchanting notion, and she also described an even more exciting idea that derived from it: such machines could process not only numbers but anything that could be notated in symbols.

Over the years, Ada Lovelace has been celebrated as a feminist icon and a computer pioneer. For example, the U.S. Defense Department named its high-level object-oriented programming language Ada. However, she has also been ridiculed as delusional, flighty, and only a minor contributor to the "Notes" that bear her initials. As she herself wrote in those "Notes," referring to the Analytical Engine but in words that also describe her fluctuating reputation, "In considering any new subject, there is frequently a tendency, first, to overrate what we find to be already interesting or remarkable; and, secondly, by a sort of natural reaction, to undervalue the true state of the case."

The reality is that Ada's contribution was both profound and inspirational. More than Babbage or any other person of her era, she was able to glimpse a future in which machines would become partners of the human imagination, together weaving tapestries as beautiful as those from Jacquard's loom. Her appreciation for poetical science led her to celebrate a proposed calculating machine that was dismissed by the scientific establishment of her day, and she perceived how the processing power of such a device could be used on any form of information. Thus did Ada, Countess of Lovelace, help sow the seeds for a digital age that would blossom a hundred years later.

Vannevar Bush (1890–1974), with his Differential Analyzer at MIT.

Alan Turing (1912–54),
at the Sherborne School in 1928.

Claude Shannon (1916–2001) in 1951.

CHAPTER TWO

THE COMPUTER

Sometimes innovation is a matter of timing. A big idea comes along at just the moment when the technology exists to implement it. For example, the idea of sending a man to the moon was proposed right when the progress of microchips made it possible to put computer guidance systems into the nose cone of a rocket. There are other cases, however, when the timing is out of kilter. Charles Babbage published his paper about a sophisticated computer in 1837, but it took a hundred years to achieve the scores of technological advances needed to build one.

Some of those advances seem almost trivial, but progress comes not only in great leaps but also from hundreds of small steps. Take for example punch cards, like those Babbage saw on Jacquard's looms and proposed incorporating into his Analytical Engine. Perfecting the use of punch cards for computers came about because Herman Hollerith, an employee of the U.S. Census Bureau, was appalled that it took close to eight years to manually tabulate the 1880 census. He resolved to automate the 1890 count.

Drawing on the way that railway conductors punched holes in various places on a ticket in order to indicate the traits of each passenger (gender, approximate height, age, hair color), Hollerith devised punch cards with twelve rows and twenty-four columns that recorded the salient facts about each person in the census. The cards were then slipped between a grid of mercury cups and a set of spring-loaded pins, which created an electric circuit wherever there was a hole. The machine could tabulate not only the raw totals but also combinations of traits, such as the number of married males or foreign-born females. Using Hollerith's tabulators, the 1890 census was completed in one year rather than eight. It was the first major use of electrical circuits to process information, and the company that Hollerith founded became in 1924, after a series of mergers and acquisitions, the International Business Machines Corporation, or IBM.

One way to look at innovation is as the accumulation of hundreds of small advances, such as counters and punch-card readers. At places like IBM, which specialize in daily improvements made by teams of engineers, this is the preferred way to understand how innovation really happens. Some of the most important technologies of our era, such as the fracking techniques developed over the past six decades for extracting natural gas, came about because of countless small innovations as well as a few breakthrough leaps.

In the case of computers, there were many such incremental advances made by faceless engineers at places like IBM. But that was not enough. Although the machines that IBM produced in the early twentieth century could compile data, they were not what we would call computers. They weren't even particularly adroit calculators. They were lame. In addition to those hundreds of minor advances, the birth of the computer age required some larger imaginative leaps from creative visionaries.

DIGITAL BEATS ANALOG

The machines devised by Hollerith and Babbage were *digital*, meaning they calculated using digits: discrete and distinct integers such as 0, 1, 2, 3. In their machines, the integers were added and subtracted using cogs and wheels that clicked one digit at a time, like counters. Another approach to computing was to build devices that could mimic or model a physical phenomenon and then make measurements on the analogous model to calculate the relevant results. These were known as *analog* computers because they worked by analogy. Analog computers do not rely on discrete integers to make their calculations; instead, they use continuous functions. In analog computers, a variable quantity such as electrical voltage, the position of a rope on a pulley, hydraulic pressure, or a measurement of distance is employed as an analog for the corresponding quantities of the problem to be solved. A slide rule is analog; an abacus is digital. Clocks with sweeping hands are analog, and those with displayed numerals are digital.

Around the time that Hollerith was building his digital tabulator, Lord Kelvin and his brother James Thomson, two of England's most distinguished scientists, were creating an analog machine. It was designed to handle the tedious task of solving differential equations, which would help in the creation of tide charts and of tables showing the firing angles that would generate different trajectories of artillery shells. Beginning in the 1870s, the brothers devised a system that was based on a planimeter, an instrument that can measure the area of a two-dimensional shape, such as the space under a curved line on a piece of paper. The user would trace the outline of the curve with the device, which would calculate the area by using a small sphere that was slowly pushed across the surface of a large rotating disk. By calculating the area under the curve, it could thus solve equations by integration—in other words, it could perform a basic task of calculus. Kelvin and his brother were able to use this method to create a "harmonic synthesizer" that could churn out an annual tide chart in four hours. But they were never able to conquer the mechanical difficulties of linking together many of these devices in order to solve equations with a lot of variables.

That challenge of linking together multiple integrators was not mastered until 1931, when an MIT engineering professor, Vannevar (rhymes with beaver) Bush—remember his name, for he is a key character in this book—was able to build the world's first analog electrical-mechanical computer. He dubbed his machine a Differential Analyzer. It consisted of six wheel-and-disk integrators, not all that different from Lord Kelvin's, that were connected by an array of gears, pulleys, and shafts rotated by electric motors. It helped that

Bush was at MIT; there were a lot of people around who could assemble and calibrate complex contraptions. The final machine, which was the size of a small bedroom, could solve equations with as many as eighteen independent variables. Over the next decade, versions of Bush's Differential Analyzer were replicated at the U.S. Army's Aberdeen Proving Ground in Maryland, the Moore School of Electrical Engineering at the University of Pennsylvania, and Manchester and Cambridge universities in England. They proved particularly useful in churning out artillery firing tables—and in training and inspiring the next generation of computer pioneers.

Bush's machine, however, was not destined to be a major advance in computing history because it was an analog device. In fact, it turned out to be the last gasp for analog computing, at least for many decades.

New approaches, technologies, and theories began to emerge in 1937, exactly a hundred years after Babbage first published his paper on the Analytical Engine. It would become an annus mirabilis of the computer age, and the result would be the triumph of four properties, somewhat interrelated, that would define modern computing:

DIGITAL. A fundamental trait of the computer revolution was that it was based on digital, not analog, computers. This occurred for many reasons, as we shall soon see, including simultaneous advances in logic theory, circuits, and electronic on-off switches that made a digital rather than an analog approach more fruitful. It would not be until the 2010s that computer scientists, seeking to mimic the human brain, would seriously begin working on ways to revive analog computing.

BINARY. Not only would modern computers be digital, but the digital system they would adopt would be binary, or base-2, meaning that it employs just 0s and 1s rather than all ten digits of our everyday decimal system. Like many mathematical concepts, binary theory was pioneered by Leibniz in the late seventeenth century. During the

1940s, it became increasingly clear that the binary system worked better than other digital forms, including the decimal system, for performing logical operations using circuits composed of on-off switches.

ELECTRONIC. In the mid-1930s, the British engineer Tommy Flowers pioneered the use of vacuum tubes as on-off switches in electronic circuits. Until then, circuits had relied on mechanical and electromechanical switches, such as the clacking electromagnetic relays that were used by phone companies. Vacuum tubes had mainly been employed to amplify signals rather than as on-off switches. By using electronic components such as vacuum tubes, and later transistors and microchips, computers could operate thousands of times faster than machines that had moving electromechanical switches.

GENERAL PURPOSE. Finally, the machines would eventually have the ability to be programmed and reprogrammed—and even reprogram themselves—for a variety of purposes. They would be able to solve not just one form of mathematical calculation, such as differential equations, but could handle a multiplicity of tasks and symbol manipulations, involving words and music and pictures as well as numbers, thus fulfilling the potential that Lady Lovelace had celebrated when describing Babbage's Analytical Engine.

Innovation occurs when ripe seeds fall on fertile ground. Instead of having a single cause, the great advances of 1937 came from a combination of capabilities, ideas, and needs that coincided in multiple places. As often happens in the annals of invention, especially information technology invention, the time was right and the atmosphere was charged. The development of vacuum tubes for the radio industry paved the way for the creation of electronic digital circuits. That was accompanied by theoretical advances in logic that made circuits more useful. And the march was quickened by the drums of war. As nations began arming for the looming conflict, it became clear that computational power was as important as firepower. Advances fed on one another, occurring almost simultaneously and spontaneously, at Harvard and MIT and Princeton and Bell Labs and an apartment in Berlin and even, most improbably but interestingly, in a basement in Ames, Iowa.

Underpinning all of these advances were some beautiful—Ada might call them poetic—leaps of mathematics. One of these leaps led to the formal concept of a "universal computer," a general-purpose machine that could be programmed to perform any logical task and simulate the behavior of any other logical machine. It was conjured up as a thought experiment by a brilliant English mathematician with a life story that was both inspiring and tragic.

ALAN TURING

Alan Turing had the cold upbringing of a child born on the fraying fringe of the British gentry.[1] His family had been graced since 1638 with a baronetcy, which had meandered down the lineage to one of his nephews. But for the younger sons on the family tree, which Turing and his father and grandfather were, there was no land and little wealth. Most went into fields such as the clergy, like Alan's grandfather, and the colonial civil service, like his father, who served as a minor administrator in remote regions of India. Alan was conceived in Chhatrapur, India, and born on June 23, 1912, in London, while his parents were on home leave. When he was only one, his parents went back to India for a few years, and handed him and his older brother off to a retired army colonel and his wife to be raised in a seaside town on the south coast of England. "I am no child psychologist," his brother, John, later noted, "but I am assured that it is a bad thing for an infant in arms to be uprooted and put into a strange environment."[2]

When his mother returned, Alan lived with her for a few years and then, at age thirteen, was sent to boarding school. He rode there on his bicycle, taking two days to cover more than sixty miles, alone. There was a lonely intensity to him, reflected in his love of long-distance running and biking. He also had a trait, so common among innovators, that was charmingly described by his biographer Andrew

Hodges: "Alan was slow to learn that indistinct line that separated initiative from disobedience."[3]

In a poignant memoir, his mother described the son whom she doted upon:

> Alan was broad, strongly built and tall, with a square, determined jaw and unruly brown hair. His deep-set, clear blue eyes were his most remarkable feature. The short, slightly retroussé nose and humorous lines of his mouth gave him a youthful—sometimes a childlike—appearance. So much so that in his late thirties he was still at times mistaken for an undergraduate. In dress and habits he tended to be slovenly. His hair was usually too long, with an overhanging lock which he would toss back with a jerk of his head. . . . He could be abstracted and dreamy, absorbed in his own thoughts which on occasion made him seem unsociable. . . . There were times when his shyness led him into extreme gaucherie. . . . Indeed he surmised that the seclusion of a mediaeval monastery would have suited him very well.[4]

At the boarding school, Sherborne, he realized that he was homosexual. He became infatuated with a fair-haired, slender schoolmate, Christopher Morcom, with whom he studied math and discussed philosophy. But in the winter before he was to graduate, Morcom suddenly died of tuberculosis. Turing would later write Morcom's mother, "I simply worshipped the ground he trod on—a thing which I did not make much attempt to disguise, I am sorry to say."[5] In a letter to his own mother, Turing seemed to take refuge in his faith: "I feel that I shall meet Morcom again somewhere and that there will be work for us to do together there as I believed there was for us to do here. Now that I am left to do it alone, I must not let him down. If I succeed I shall be more fit to join his company than I am now." But the tragedy ended up eroding Turing's religious faith. It also turned him even more inward, and he never again found it easy to forge intimate relationships. His housemaster reported to his parents at Easter 1927, "Undeniably he's not a 'normal' boy; not the worse for that, but probably less happy."[6]

In his final year at Sherborne, Turing won a scholarship to attend King's College, Cambridge, where he went in 1931 to read mathematics. One of three books he bought with some prize money was *The Mathematical Foundations of Quantum Mechanics*, by John von Neumann, a fascinating Hungarian-born mathematician who, as a pioneer of computer design, would have a continuing influence on his life. Turing was particularly interested in the math at the core of quantum physics, which describes how events at the subatomic level are governed by statistical probabilities rather than laws that determine things with certainty. He believed (at least while he was young) that this uncertainty and indeterminacy at the subatomic level permitted humans to exercise free will—a trait that, if true, would seem to distinguish them from machines. In other words, because events at the subatomic level are not predetermined, that opens the way for our thoughts and actions not to be predetermined. As he explained in a letter to Morcom's mother:

> It used to be supposed in science that if everything was known about the Universe at any particular moment then we can predict what it will be through all the future. This idea was really due to the great success of astronomical prediction. More modern science however has come to the conclusion that when we are dealing with atoms and electrons we are quite unable to know the exact state of them; our instruments being made of atoms and electrons themselves. The conception then of being able to know the exact state of the universe then really must break down on the small scale. This means then that the theory which held that as eclipses etc. are predestined so were all our actions breaks down too. We have a will which is able to determine the action of the atoms probably in a small portion of the brain, or possibly all over it.[7]

For the rest of his life, Turing would wrestle with the issue of whether the human mind was fundamentally different from a deterministic machine, and he would gradually come to the conclusion that the distinction was less clear than he had thought.

He also had an instinct that, just as uncertainty pervaded the subatomic realm, there were also mathematical problems that could not be solved mechanically and were destined to be cloaked in indeterminacy. At the time, mathematicians were intensely focused on questions about the completeness and consistency of logical systems, partly due to the influence of David Hilbert, the Göttingen-based genius who, among many other achievements, had come up with the mathematical formulation of the theory of general relativity concurrently with Einstein.

At a 1928 conference, Hilbert posed three fundamental questions about any formal system of mathematics: (1) Was its set of rules complete, so that any statement could be proved (or disproved) using only the rules of the system? (2) Was it consistent, so that no statement could be proved true and also proved false? (3) Was there some procedure that could determine whether a particular statement was provable, rather than allowing the possibility that some statements (such as enduring math riddles like Fermat's last theorem,[*] Goldbach's conjecture,[†] or the Collatz conjecture[‡]) were destined to remain in undecidable limbo? Hilbert thought that the answer to the first two questions was yes, making the third one moot. He put it simply, "There is no such thing as an unsolvable problem."

Within three years, the Austrian-born logician Kurt Gödel, then twenty-five and living with his mother in Vienna, polished off the first two of these questions with unexpected answers: no and no. In his "incompleteness theorem," he showed that there existed statements that could be neither proved nor disproved. Among them, to oversimplify a bit, were those that were akin to self-referential statements such as "This statement is unprovable." If the statement is true, then it decrees that we can't prove it to be true; if it's false, that also leads to a logical contradiction. It is somewhat like the ancient Greek "liar's paradox," in which the truth of the statement "This statement

---

[*] For the equation $a^n + b^n = c^n$, in which a, b, and c are positive integers, there is no solution when n is greater than 2.

[†] Every even integer greater than 2 can be expressed as the sum of two primes.

[‡] A process in which a number is divided by 2 if it is even, and is tripled and the result added to 1 if odd, when repeated indefinitely, will always eventually lead to a result of 1.

is false" cannot be determined. (If the statement is true, then it's also false, and vice versa.)

By coming up with statements that could not be proved or disproved, Gödel showed that any formal system powerful enough to express the usual mathematics was incomplete. He was also able to produce a companion theorem that effectively answered no to Hilbert's second question.

That left the third of Hilbert's questions, that of decidability or, as Hilbert called it, the *Entscheidungsproblem* or "decision problem." Even though Gödel had come up with statements that could be neither proved nor disproved, perhaps that odd class of statements could somehow be identified and cordoned off, leaving the rest of the system complete and consistent. That would require that we find some method for *deciding* whether a statement was provable. When the great Cambridge math professor Max Newman taught Turing about Hilbert's questions, the way he expressed the *Entscheidungsproblem* was this: Is there a "mechanical process" that can be used to determine whether a particular logical statement is provable?

Turing liked the concept of a "mechanical process." One day in the summer of 1935, he was out for his usual solitary run along the Ely River, and after a couple of miles he stopped to lie down among the apple trees in Grantchester Meadows to ponder an idea. He would take the notion of a "mechanical process" literally, conjuring up a mechanical process—an imaginary machine—and applying it to the problem.[8]

The "Logical Computing Machine" that he envisioned (as a thought experiment, not as a real machine to be built) was quite simple at first glance, but it could handle, in theory, any mathematical computation. It consisted of an unlimited length of paper tape containing symbols within squares; in the simplest binary example, these symbols could be merely a 1 and a blank. The machine would be able to read the symbols on the tape and perform certain actions based on a "table of instructions" it had been given.[9]

The table of instructions would tell the machine what to do based on whatever configuration it happened to be in and what symbol, if any, it found in the square. For example, the table of instructions for a particular task might decree that if the machine was in configuration 1 and saw a 1 in the square, then it should move one square to the right and shift into configuration 2. Somewhat surprisingly, to us if not to Turing, such a machine, given the proper table of instructions, could complete any mathematical task, no matter how complex.

How might this imaginary machine answer Hilbert's third question, the decision problem? Turing approached the problem by refining the concept of "computable numbers." Any real number that was defined by a mathematical rule could be calculated by the Logical Computing Machine. Even an irrational number such as π could be calculated indefinitely using a finite table of instructions. So could the logarithm of 7, or the square root of 2, or the sequence of Bernoulli numbers that Ada Lovelace had helped produce an algorithm for, or any other number or series, no matter how challenging to compute, as long as its calculation was defined by a finite set of rules. All of these were, in Turing's parlance, "computable numbers."

Turing went on to show that *non*computable numbers also existed. This was related to what he called "the halting problem." There can be no method, he showed, to determine in advance whether any given instruction table combined with any given set of inputs will lead the machine to arrive at an answer or go into some loop and continue chugging away indefinitely, getting nowhere. The insolvability of the halting problem, he showed, meant that Hilbert's decision problem, the *Entscheidungsproblem*, was unsolvable. Despite what Hilbert seemed to hope, no mechanical procedure can determine the provability of every mathematical statement. Gödel's incompleteness theory, the indeterminacy of quantum mechanics, and Turing's answer to Hilbert's third challenge all dealt blows to a mechanical, deterministic, predictable universe.

Turing's paper was published in 1937 with the not so snappy title "On Computable Numbers, with an Application to the *Entscheidungsproblem*." His answer to Hilbert's third question was useful for the development of mathematical theory. But far more important was the by-product of Turing's proof: his concept of a Logical Computing Machine, which soon came to be known as a Turing machine. "It is possible to invent a single machine which can be used to compute any computable sequence," he declared.[10] Such a machine would be able to read the instructions of any other machine and carry out whatever task that machine could do. In essence, it embodied the dream of Charles Babbage and Ada Lovelace for a completely general-purpose universal machine.

A different and less beautiful solution to the *Entscheidungsproblem*, with the clunkier name "untyped lambda calculus," had been published earlier that year by Alonzo Church, a mathematician at Princeton. Turing's professor Max Newman decided that it would be useful for Turing to go there to study under Church. In his letter of recommendation, Newman described Turing's enormous potential. He also added a more personal appeal based on Turing's personality. "He has been working without any supervision or criticism from anyone," Newman wrote. "This makes it all the more important that he should come into contact as soon as possible with the leading workers on this line, so that he should not develop into a confirmed solitary."[11]

Turing did have a tendency toward being a loner. His homosexuality made him feel like an outsider at times; he lived alone and avoided deep personal commitments. At one point he proposed marriage to a female colleague, but then felt compelled to tell her that he was gay; she was unfazed and still willing to get married, but he believed it would be a sham and decided not to proceed. Yet he did not become "a confirmed solitary." He learned to work as part of a team, with collaborators, which was key to allowing his abstract theories to be reflected in real and tangible inventions.

In September 1936, while waiting for his paper to be published, the twenty-four-year-old doctoral candidate sailed to America in steerage class aboard the aging ocean liner RMS *Berengaria*, lugging with him a prized brass sextant. His office at Princeton was in the Mathematics Department building, which also then housed the Institute for Advanced Study, where Einstein, Gödel, and von Neumann held court. The cultivated and highly sociable von Neumann became particularly interested in Turing's work, despite their very different personalities.

The seismic shifts and simultaneous advances of 1937 were not directly caused by the publication of Turing's paper. In fact, it got little notice at first. Turing asked his mother to send out reprints of it to the mathematical philosopher Bertrand Russell and a half dozen other famous scholars, but the only major review was by Alonzo Church, who could afford to be flattering because he had been ahead of Turing in solving Hilbert's decision problem. Church was not only generous; he introduced the term *Turing machine* for what Turing had called a Logical Computing Machine. Thus at twenty-four, Turing's name became indelibly stamped on one of the most important concepts of the digital age.[12]

CLAUDE SHANNON AND GEORGE STIBITZ AT BELL LABS

There was another seminal theoretical breakthrough in 1937, similar to Turing's in that it was purely a thought experiment. This one was the work of an MIT graduate student named Claude Shannon, who that year turned in the most influential master's thesis of all time, a paper that *Scientific American* later dubbed "the Magna Carta of the Information Age."[13]

Shannon grew up in a small Michigan town where he built model planes and amateur radios, then went on to major in electrical engineering and math at the University of Michigan. In his senior year he answered a help-wanted listing tacked to a bulletin board, which offered a job at MIT working under Vannevar Bush helping to run the Differential Analyzer. Shannon got the job and was mesmerized by the machine—not so much the rods and pulleys and wheels that formed the analog components as the electromagnetic relay switches that were part of its control circuit. As electrical signals caused them to click open and clack closed, the switches created different circuit patterns.

During the summer of 1937, Shannon took a break from MIT and went to work at Bell Labs, a research facility run by AT&T. Located then in Manhattan on the Hudson River edge of Greenwich Village, it was a haven for turning ideas into inventions. Abstract theories intersected with practical problems there, and in the corridors and cafeterias eccentric theorists mingled with hands-on engineers, gnarly mechanics, and businesslike problem-solvers, encouraging the cross-fertilization of theory with engineering. This made Bell Labs an archetype of one of the most important underpinnings of digital-age innovation, what the Harvard science historian Peter Galison has called a "trading zone." When these disparate practitioners and theoreticians came together, they learned how to find a common parlance to trade ideas and exchange information.[14]

At Bell Labs, Shannon saw up close the wonderful power of the phone system's circuits, which used electrical switches to route calls and balance loads. In his mind, he began connecting the workings of these circuits to another subject he found fascinating, the system of logic formulated ninety years earlier by the British mathematician George Boole. Boole revolutionized logic by finding ways to express logical statements using symbols and equations. He gave true propositions the value 1 and false propositions a 0. A set of basic logical operations—such as *and*, *or*, *not*, *either/or*, and *if/then*—could then be performed using these propositions, just as if they were math equations.

Shannon figured out that electrical circuits could execute these logical operations using an arrangement of on-off switches. To perform an *and* function, for example, two switches could be put in sequence, so that both had to be on for electricity to flow. To perform an *or* function, the switches could be in parallel so that electricity would flow if either of them was on. Slightly more versatile switches called logic gates could streamline the process. In other words, you could design a circuit containing a lot of relays and logic gates that could perform, step by step, a sequence of logical tasks.

(A "relay" is simply a switch that can be opened and shut electrically, such as by using an electromagnet. The ones that clack open and closed are sometimes called *electromechanical* because they have moving parts. Vacuum tubes and transistors can also be used as switches in an electrical circuit; they are called *electronic* because they manipulate the flow of electrons but do not require the movement of any physical parts. A "logic gate" is a switch that can handle one or more inputs. For example, in the case of two inputs, an *and* logic gate switches on if *both* of the inputs are on, and an *or* logic gate switches on if *either* of the inputs is on. Shannon's insight was that these could be wired together in circuits that could execute the tasks of Boole's logical algebra.)

When Shannon returned to MIT in the fall, Bush was fascinated by his ideas and urged him to include them in his master's thesis. Entitled "A Symbolic Analysis of Relay and Switching Circuits," it showed how each of the many functions of Boolean algebra could be executed. "It is possible to perform complex mathematical operations by means of relay circuits," he summed up at the end.[15] This became the basic concept underlying all digital computers.

Shannon's ideas intrigued Turing because they neatly related to his own just-published concept of a universal machine that could use simple instructions, expressed in binary coding, to tackle problems not only of math but of logic. Also, since logic was related to the way human minds reason, a machine that performed logical tasks could, in theory, mimic the way humans think.

Working at Bell Labs at the same time was a mathematician named George Stibitz, whose job was to figure out ways to handle the increasingly complicated calculations needed by the telephone engineers. The only tools he had were mechanical desktop adding machines, so he set out to invent something better based on Shannon's insight that electronic circuits could perform mathematical and logical tasks. Late one evening in November, he went to the stockroom and took home some old electromagnetic relays and bulbs. At his kitchen table, he put the parts together with a tobacco tin and a few switches to form a simple logical circuit that could add binary numbers. A lit bulb represented a 1, and an unlit bulb represented a 0. His wife dubbed it the "K-Model," after the kitchen table. He took it into the office the next day and tried to convince his colleagues that, with enough relays, he could make a calculating machine.

One important mission of Bell Labs was to figure out ways to amplify a phone signal over long distances while filtering out static. The engineers had formulas that dealt with the amplitude and phase of the signal, and the solutions to their equations sometimes involved complex numbers (ones that include an imaginary unit that represents the square root of a negative number). Stibitz was asked by his supervisor if his proposed machine could handle complex numbers. When he said that it could, a team was assigned to help him build it. The Complex Number Calculator, as it was called, was completed in 1939. It had more than four hundred relays, each of which could open and shut twenty times per second. That made it both blindingly fast compared to mechanical calculators and painfully clunky compared to the all-electronic vacuum-tube circuits just being invented. Stibitz's computer was not programmable, but it showed the potential of a circuit of relays to do binary math, process information, and handle logical procedures.[16]

HOWARD AIKEN

Also in 1937 a Harvard doctoral student named Howard Aiken was struggling to do tedious calculations for his physics thesis using an adding machine. When he lobbied the university to build a more sophisticated computer to do the work, his department head mentioned that in the attic of Harvard's science center were some brass wheels from a century-old device that seemed to be similar to what he wanted. When Aiken explored the attic, he found one of six demonstration models of Charles Babbage's Difference Engine, which Babbage's son Henry had made and distributed. Aiken became fascinated by Babbage and moved the set of brass wheels into his office. "Sure enough, we had two of Babbage's wheels," he recalled. "Those were the wheels that I had later mounted and put in the body of the computer."[17]

That fall, just when Stibitz was cooking up his kitchen-table demonstration, Aiken wrote a twenty-two-page memo to his Harvard superiors and executives at IBM making the case that they should fund a modern version of Babbage's digital machine. "The desire to economize time and mental effort in arithmetical computations, and to eliminate human liability to error is probably as old as the science of arithmetic itself," his memo began.[18]

Aiken had grown up in Indiana under rough circumstances. When he was twelve, he used a fireplace poker to defend his mother against his drunk and abusive father, who then abandoned the family with no money. So young Howard dropped out of ninth grade to support the family by working as a telephone installer, then got a night job with the local power company so that he could attend a tech school during the day. He drove himself to be a success, but in the process he developed into a taskmaster with an explosive temper, someone who was described as resembling an approaching thunderstorm.[19]

Harvard had mixed feelings about building Aiken's proposed calculating machine or holding out the possibility that he might be granted tenure for a project that seemed to be more practical than academic. (In parts of the Harvard faculty club, calling someone practical rather than academic was considered an insult.) Supporting Aiken was President James Bryant Conant, who, as chairman of the National Defense Research Committee, was comfortable positioning Harvard as part of a triangle involving academia, industry, and the military. His Physics Department, however, was more purist. Its chairman wrote to Conant in December 1939, saying that the machine was "desirable if money can be found, but not necessarily more desirable than anything else," and a faculty committee said of Aiken, "It should be made quite clear to him that such activity did not increase his chances of promotion to a professorship." Eventually Conant prevailed and authorized Aiken to build his machine.[20]

In April 1941, as IBM was constructing the Mark I to Aiken's specifications at its lab in Endicott, New York, he left Harvard to serve in the U.S. Navy. For two years he was a teacher, with the rank of lieutenant commander, at the Naval Mine Warfare School in Virginia. One colleague described him as "armed to the teeth with room-length formulas and ivy-covered Harvard theories" and running "smack into a collection of Dixie dumbbells [none of whom] knew calculus from corn pone."[21] Much of his time was spent thinking about the Mark I, and he made occasional visits to Endicott wearing his full dress uniform.[22]

His tour of duty had one major payoff: at the beginning of 1944, as IBM was getting ready to ship the completed Mark I to Harvard, Aiken was able to convince the Navy to take over authority for the machine and assign him to be the officer in charge. That helped him circumnavigate the academic bureaucracy of Harvard, which was still balky about granting him tenure. The Harvard Computation Laboratory became, for the time being, a naval facility, and all of Aiken's staffers were Navy personnel who wore uniforms to work. He called them his "crew," they called him "commander," and the Mark I was referred to as "she," as if she were a ship.[23]

The Harvard Mark I borrowed a lot of Babbage's ideas. It was digital, although not binary; its wheels had ten positions. Along its fifty-foot shaft were seventy-two counters that could store numbers of up to twenty-three digits, and the finished five-ton product was eighty feet long and fifty feet wide. The shaft and other moving parts were turned electrically. But it was slow. Instead of electromagnetic relays, it used mechanical ones that were opened and shut by electric motors. That meant it took about six seconds to do a multiplication problem, compared to one second for Stibitz's machine. It did, however, have one impressive feature that would become a staple of modern computers: it was fully automatic. Programs and data were entered by paper tape, and it could run for days with no human intervention. That allowed Aiken to refer to it as "Babbage's dream come true."[24]

KONRAD ZUSE

Although they didn't know it, all of these pioneers were being beaten in 1937 by a German engineer working in his parents' apartment. Konrad Zuse was finishing the prototype for a calculator that was binary and could read instructions from a punched tape. However, at least in its first version, called the Z1, it was a mechanical, not an electrical or electronic, machine.

Like many pioneers in the digital age, Zuse grew up fascinated by both art and engineering. After graduating from a technical college, he got a job as a stress analyst for an aircraft company in Berlin, solving linear equations that incorporated all sorts of load and strength and elasticity factors. Even using mechanical calculators, it was almost impossible for a person to solve in less than a day more than six simultaneous linear equations with six unknowns. If there were twenty-five variables, it could take a year. So Zuse, like so many others, was driven by the desire to mechanize the tedious process of solving mathematical equations. He converted his parents' living room, in an apartment near Berlin's Tempelhof Airport, into a workshop.[25]

In Zuse's first version, binary digits were stored by using thin metal plates with slots and pins, which he and his friends made using a jigsaw. At first he used punched paper tape to input data and programs, but he soon switched to discarded 35 mm movie film, which not only was sturdier but happened to be cheaper. His Z1 was completed in 1938, and it was able to clank through a few problems, though not very reliably. All the components had been made by hand, and they tended to jam. He was handicapped by not being at a place like Bell Labs or part of a collaboration like Harvard had with IBM, which would have allowed him to team up with engineers who could have supplemented his talents.

The Z1 did, however, show that the logical concept Zuse had designed would work in theory. A college friend who was helping him, Helmut Schreyer, urged that they make a version using electronic vacuum tubes rather than mechanical switches. Had they done so right away, they would have gone down in history as the first inventors of a working modern computer: binary, electronic, and programmable. But Zuse, as well as the experts he consulted at the technical school, balked at the expense of building a device with close to two thousand vacuum tubes.[26]

So for the Z2 they decided instead to use electromechanical relay switches, acquired secondhand from the phone company, which were tougher and cheaper, although a lot slower. The result was a computer that used relays for the arithmetic unit. However, the memory unit was mechanical, using movable pins in a metal sheet.

In 1939 Zuse began work on a third model, the Z3, that used electromechanical relays both for the arithmetic unit and for the memory and control units. When it was completed in 1941, it became the first fully working all-purpose, programmable digital computer. Even though it did not have a way to directly handle conditional jumps and branching in the programs, it could theoretically perform as a universal Turing machine. Its major difference from later computers was that it used clunky electromagnetic relays rather than electronic components such as vacuum tubes or transistors.

Zuse's friend Schreyer went on to write a doctoral thesis, "The Tube Relay and the Techniques of Its Switching," that advocated using vacuum tubes for a powerful and fast computer. But when he and Zuse proposed it to the German Army in 1942, the commanders said they were confident that they would win the war before the two years it would take to build such a machine.[27] They were more interested in making weapons than computers. As a result, Zuse was pulled away from his computer work and sent back to engineering airplanes. In 1943 his computers and designs were destroyed in the Allied bombing of Berlin.

Zuse and Stibitz, working independently, had both come up with employing relay switches to make circuits that could handle binary computations. How did they develop this idea at the same time when war kept their two teams isolated? The answer is partly that advances in technology and theory made the moment ripe. Along with many other innovators, Zuse and Stibitz were familiar with the use of relays in phone circuits, and it made sense to tie that to binary operations of math and logic. Likewise, Shannon, who was also very familiar with phone circuits, made the related theoretical leap that electronic circuits would be able to perform the logical tasks of Boolean algebra. The idea that digital circuits would be the key to computing was quickly becoming clear to researchers almost everywhere, even in isolated places like central Iowa.

JOHN VINCENT ATANASOFF

Far from both Zuse and Stibitz, another inventor was also experimenting with digital circuits in 1937. Toiling in a basement in Iowa, he would make the next historic innovation: building a calculating device that, at least in part, used vacuum tubes. In some ways his machine was less advanced than the others. It wasn't programmable and multipurpose; instead of being totally electronic, he included some slow mechanical moving elements; and even though he built a model that was able to work in theory, he couldn't actually get the thing reliably operational. Nevertheless, John Vincent Atanasoff, known to his wife and friends as Vincent, deserves the distinction of being the pioneer who conceived the first partly electronic digital computer, and he did so after he was struck by inspiration during a long impetuous drive one night in December 1937.[28]

Atanasoff was born in 1903, the eldest of seven children of a Bulgarian immigrant and a woman descended from one of New England's oldest families. His father worked as an engineer in a New Jersey electric plant run by Thomas Edison, then moved the family to a town in rural Florida south of Tampa. At nine, Vincent helped his father wire their Florida house for electricity, and his father gave him a Dietzgen slide rule. "That slide rule was my meat," he recalled.[29] At an early age, he dove into the study of logarithms with an enthusiasm that seems a bit wacky even as he recounted it in earnest tones: "Can you imagine how a boy of nine, with baseball on his mind, could be transformed by this knowledge? Baseball was reduced to near zero as a stern study was made of logarithms." Over the summer, he calculated the logarithm of 5 to the base *e*, then, with his mother's help (she had once been a math teacher), he learned calculus while still in middle school. His father took him to the phosphate plant where he was an electrical engineer, showing him how the generators worked. Diffident, creative, and brilliant, young Vincent finished high school in two years, getting all A's in his double load of classes.

At the University of Florida he studied electrical engineering and displayed a practical inclination, spending time in the university's machine shop and foundry. He also remained fascinated by math and as a freshman studied a proof involving binary arithmetic. Creative and self-confident, he graduated with the highest grade point average of his time. He accepted a fellowship to pursue master's work in math and physics at Iowa State and, even though he later was admitted to Harvard, stuck with his decision to head up to the corn belt town of Ames.

Atanasoff went on to pursue a doctorate in physics at the University of Wisconsin, where he had the same experience as the other computer pioneers, beginning with Babbage. His work, which was on how helium can be polarized by an electric field, involved tedious calculations. As he struggled to solve the math using a desktop adding machine, he dreamed of ways to invent a calculator that could do more of the work. After returning to Iowa State in 1930 as an assistant professor, he decided that his degrees in electrical engineering, math, and physics had equipped him for the task.

There was a consequence to his decision not to stay at Wisconsin or to go to Harvard or a similar large research university. At Iowa State, where no one else was working on ways to build new calculators, Atanasoff was on his own. He could come up with fresh ideas, but he did not have around him people to serve as sounding boards or to help him overcome theoretical or engineering challenges. Unlike most innovators of the digital age, he was a lone inventor, drawing his inspiration during solo car trips and in discussions with one graduate student assistant. In the end, that would prove to be a drawback.

Atanasoff initially considered building an analog device; his love of slide rules led him to try to devise a supersize version using long strips of film. But he realized that the film would have to be hundreds of yards long in order to solve linear algebraic equations accurately enough to suit his needs. He also built a contraption that could shape a mound of paraffin so that it could calculate a partial differential equation. The limitations of these analog devices caused him to focus instead on creating a digital version.

The first problem he tackled was how to store numbers in a machine. He used the term *memory* to describe this feature: "At the time, I had only a cursory knowledge of the work of Babbage and so did not know he called the same concept 'store.' . . . I like his word, and perhaps if I had known, I would have adopted it; I like 'memory,' too, with its analogy to the brain."[30]

Atanasoff went through a list of possible memory devices: mechanical pins, electromagnetic relays, a small piece of magnetic material that could be polarized by an electric charge, vacuum tubes, and a small electrical condenser. The fastest would be vacuum tubes,

George Stibitz (1904–95) circa 1945.

Konrad Zuse (1910–95) with the Z4 computer in 1944.

John Atanasoff (1903–95)
at Iowa State, circa 1940.

Reconstruction of Atanasoff's computer.

but they were expensive. So he opted instead to use what he called condensers—what we now call capacitors—which are small and inexpensive components that can store, at least briefly, an electrical charge. It was an understandable decision, but it meant that the machine would be sluggish and clunky. Even if the adding and subtracting could be done at electronic speeds, the process of taking numbers in and out of the memory unit would slow things down to the speed of the rotating drum.

Once he had settled on the memory unit, Atanasoff turned his attention to how to construct the arithmetic and logic unit, which he called the "computing mechanism." He decided it should be fully electronic; that meant using vacuum tubes, even though they were expensive. The tubes would act as on-off switches to perform the function of logic gates in a circuit that could add, subtract, and perform any Boolean function.

That raised a theoretical math issue of the type he had loved since he was a boy: Should his digital system be decimal or binary or rely on some other numerical base? A true enthusiast for number systems, Atanasoff explored many options. "For a short time the base one-hundred was thought to have some promise," he wrote in an unpublished paper. "This same calculation showed that the base that theoretically gives the highest speed of calculation is *e*, the natural base."[31] But, balancing theory with practicality, he finally settled on base-2, the binary system. By late 1937, these and other ideas were jangling around in his head, a "hodgepodge" of concepts that wouldn't "jell."

Atanasoff loved cars; he liked to buy, if he could, a new one each year, and in December 1937, he had a new Ford with a powerful V8 engine. To relax his mind, he took it for a late-night spin for what would become a noteworthy moment in the history of computing:

> One night in the winter of 1937 my whole body was in torment from trying to solve the problems of the machine. I got in my car and drove at high speeds for a long while so I could control my emotions. It was my habit to do this for a few miles: I could gain control of

> myself by concentrating on driving. But that night I was excessively tormented, and I kept on going until I had crossed the Mississippi River into Illinois and was 189 miles from where I started.[32]

He turned off the highway and pulled into a roadhouse tavern. At least in Illinois, unlike in Iowa, he could buy a drink, and he ordered himself a bourbon and soda, then another. "I realized that I was no longer so nervous and my thoughts turned again to computing machines," he recalled. "I don't know why my mind worked then when it had not worked previously, but things seemed to be good and cool and quiet." The waitress was inattentive, so Atanasoff got to process his problem undisturbed.[33]

He sketched out his ideas on a paper napkin, then began to sort through some practical questions. The most important was how to replenish the charges in the condensers, which would otherwise drain after a minute or two. He came up with the idea of putting them on rotating cylinder drums, about the size of 46-ounce cans of V8 juice, so they would come into contact once a second with brushlike wires and have their charges refreshed. "During this evening in the tavern, I generated within my mind the possibility of the regenerative memory," he declared. "I called it 'jogging' at that time." With each turn of the rotating cylinder, the wires would jog the memory of the condensers and, when necessary, retrieve data from the condensers and store new data. He also came up with an architecture that would take numbers from two different cylinders of condensers, then use the vacuum-tube circuit to add or subtract them and put the result into memory. After a few hours of figuring everything out, he recalled, "I got in my car and drove home at a slower rate."[34]

By May 1939, Atanasoff was ready to begin construction of a prototype. He needed an assistant, preferably a graduate student with engineering experience. "I have your man," a friend on the faculty told him one day. Thus he struck up a partnership with another son of a self-taught electrical engineer, Clifford Berry.[35]

The machine was designed and hard-wired with a single purpose:

solving simultaneous linear equations. It could handle up to twenty-nine variables. With each step, Atanasoff's machine would process two equations and eliminate one of the variables, then print the resulting equations on 8 x 11 binary punch cards. This set of cards with the simpler equation would then be fed back into the machine for the process to begin anew, eliminating yet another variable. The process required a bit of time. The machine would (if they could get it to work properly) take almost a week to complete a set of twenty-nine equations. Still, humans doing the same process on desk calculators would require at least ten weeks.

Atanasoff demonstrated a prototype at the end of 1939 and, hoping to get funding to build a full-scale machine, typed up a thirty-five-page proposal, using carbon paper to make a few copies. "It is the main purpose of this paper to present a description and exposition of a computing machine which has been designed principally for the solution of large systems of linear algebraic equations," he began. As if to fend off criticism that this was a limited purpose for a big machine, Atanasoff specified a long list of problems that required solving such equations: "curve fitting . . . vibrational problems . . . electrical circuit analysis . . . elastic structures." He concluded with a detailed list of proposed expenditures, which added up to the grand sum of $5,330, which he ended up getting from a private foundation.[36] Then he sent one of the carbon copies of his proposal to a Chicago patent lawyer retained by Iowa State, who, in a dereliction of duty that would spawn decades of historical and legal controversy, never got around to filing for any patents.

By September 1942 Atanasoff's full-scale model was almost finished. It was the size of a desk and contained close to three hundred vacuum tubes. There was, however, a problem: the mechanism for using sparks to burn holes in the punch cards never worked properly, and there were no teams of machinists and engineers at Iowa State he could turn to for help.

At that point, work stopped. Atanasoff was drafted into the Navy and sent to its ordnance laboratory in Washington, DC, where he worked on acoustic mines and later attended the atomic bomb tests at Bikini Atoll. Shifting his focus from computers to ordnance engineering, he remained an inventor, earning thirty patents, including on a minesweeping device. But his Chicago lawyer never applied for patents on his computer.

Atanasoff's computer could have been an important milestone, but it was, both literally and figuratively, relegated to the dustbin of history. The almost-working machine was put into storage in the basement of the physics building at Iowa State, and a few years later no one seemed to remember what it did. When the space was needed for other uses in 1948, a graduate student dismantled it, not realizing what it was, and discarded most of the parts.[37] Many early histories of the computer age do not even mention Atanasoff.

Even if it had worked properly, his machine had limitations. The vacuum-tube circuit made lightning-fast calculations, but the mechanically rotated memory units slowed down the process enormously. So did the system for burning holes in the punch cards, even when it worked. In order to be truly fast, modern computers would have to be *all*-electronic, not just partly. Nor was Atanasoff's model programmable. It was geared to do just one thing: solve linear equations.

Atanasoff's enduring romantic appeal is that he was a lone tinkerer in a basement, with only his young sidekick Clifford Berry for a companion. But his tale is evidence that we shouldn't in fact romanticize such loners. Like Babbage, who also toiled in his own little workshop with just an assistant, Atanasoff never got his machine to be fully functional. Had he been at Bell Labs, amid swarms of technicians and engineers and repairmen, or at a big research university, a solution would likely have been found for fixing the card reader as well as the other balky parts of his contraption. Plus, when Atanasoff was called away to the Navy in 1942, there would have been team members left behind to put on the finishing touches, or at least to remember what was being built.

What saved Atanasoff from being a forgotten historical footnote is somewhat ironic, given the resentment he later felt about the event. It was a visit that he had in June 1941 from one of those people who, instead of toiling in isolation, loved visiting places and snatching up ideas and working with teams of people. John Mauchly's trip to Iowa would later be the subject of costly lawsuits, bitter accusations, and dueling historical narratives. But it is what saved Atanasoff from obscurity and moved the course of computer history forward.

JOHN MAUCHLY

In the early twentieth century, the United States developed, as Britain had earlier, a class of gentleman scientists who congregated at wood-paneled explorers' clubs and other rarefied institutes, where they enjoyed sharing ideas, listening to lectures, and collaborating on projects. John Mauchly was raised in that realm. His father, a physicist, was a research chief in the Department of Terrestrial Magnetism at the Washington-based Carnegie Institution, the nation's foremost foundation for promoting the advance and sharing of research. His specialty was recording electrical conditions in the atmosphere and relating them to the weather, a collegial endeavor that involved coordinating researchers from Greenland to Peru.[38]

Growing up in the Washington suburb of Chevy Chase, John was exposed to the area's growing scientific community. "Chevy Chase seemed to have practically all the scientists in Washington," he boasted. "The director of the Weights and Measures Division of the Bureau of Standards lived near us. So did the director of its Radio Division." The head of the Smithsonian was also a neighbor. John spent many weekends using a desktop adding machine to do calculations for his dad, and he developed a passion for data-driven meteorology. He also loved electrical circuits. With his young friends in his neighborhood, he laid intercom wires that connected their homes and built remote-control devices to launch fireworks for parties. "When I pressed a button, the fireworks would go off 50 feet away." At age fourteen he was earning money helping people in the neighborhood fix faulty wiring in their homes.[39]

While an undergraduate at Johns Hopkins University, Mauchly enrolled in a program for exceptional undergraduates to leap directly into a PhD program in physics. He did his thesis on light band spectroscopy because it combined beauty, experiments, and theory. "You had to know some theory to figure out what the band spectra was all about, but you couldn't do it unless you had the experimental photographs of that spectrum, and who's going to get it for you?" he said. "Nobody but you. So I got plenty of training in glass blowing, and drawing vacuums, finding the leaks etc."[40]

Mauchly had an engaging personality and a wonderful ability (and desire) to explain things, so it was natural that he would become a professor. Such posts were hard to come by in the Depression, but he managed to land one at Ursinus College, an hour's drive northwest from Philadelphia. "I was the only person teaching physics there," he said.[41]

An essential component of Mauchly's personality was that he liked to share ideas—usually with a broad grin and a sense of flair—which made him a wildly popular teacher. "He loved to talk and seemed to develop many of his ideas in the give-and-take of conversation," recalled a colleague. "John loved social occasions, liked to eat good food and drink good liquor. He liked women, attractive young people, the intelligent and the unusual."[42] It was dangerous to ask him a question, because he could discourse earnestly and passionately about almost anything, from theater to literature to physics.

In front of a class he played the showman. To explain momentum he would whirl around with his arms flung out and then pulled in, and to describe the concept of action and reaction he would stand on a homemade skateboard and lurch back and forth, a trick that one year resulted in his falling and breaking an arm. People used to drive miles to hear his end-of-term pre-Christmas lecture, which the college moved to its biggest auditorium to accommodate all the visitors. In it he explained how spectrography and other tools of physics could be used to determine what was inside a package without unwrapping it. According to his wife, "He measured it. He weighed it. He submerged it in water. He poked it with a long needle."[43]

Reflecting his boyhood fascination with meteorology, Mauchly's research focus in the early 1930s was on whether long-range weather patterns were related to solar flares, sunspots, and the rotation of the sun. The scientists at the Carnegie Institution and the U.S. Weather Bureau gave him twenty years of daily data from two hundred stations, and he set to work calculating correlations. He was able (this being the Depression) to buy used desk calculators cheaply from ailing banks and to hire a group of young people, through the New Deal's National Youth Administration, to do computations at fifty cents an hour.[44]

Like others whose work required tedious calculations, Mauchly yearned to invent a machine to do them. With his gregarious style, he set about finding out what others were doing and, in the tradition of great innovators, putting together a variety of ideas. In the IBM pavilion at the 1939 New York World's Fair, he saw an electric calculator that used punch cards, but he realized that relying on cards would be too slow, given the amount of data he had to crunch. He also saw an encryption machine that used vacuum tubes to code messages. Might the tubes be used for other logical circuits? He took his students on a field trip to Swarthmore College to see counting devices that used circuits made with vacuum tubes to measure bursts of cosmic-ray ionization.[45] He also took a night course in electronics and began to experiment with his own hand-wired vacuum-tube circuits to see what else they might do.

At a conference at Dartmouth College in September 1940, Mauchly saw a demonstration by George Stibitz of the Complex Number Calculator he had built at Bell Labs. What made the demonstration exciting was that Stibitz's computer was sitting at Bell's building in lower Manhattan, transmitting data over a Teletype line. It was the first computer to be used remotely. For three hours it solved problems submitted by the audience, taking about a minute for each. Among those at the demonstration was Norbert Wiener, a pioneer of information systems, who tried to stump Stibitz's machine by asking it to divide a number by zero. The machine didn't fall for the trap. Also present was John von Neumann, the Hungarian polymath who was soon to play a major role with Mauchly in the development of computers.[46]

When he decided to build a vacuum-tube computer of his own, Mauchly did what good innovators properly do: he drew upon all of the information he had picked up from his travels. Because Ursinus had no research budget, Mauchly paid for tubes out of his own pocket and tried to cadge them from manufacturers. He wrote the Supreme Instruments Corp. asking for components and declaring, "I am intending to construct an electrical calculating machine."[47] He discovered during a visit to RCA that neon tubes could also be used as switches; they were slower but cheaper than vacuum tubes, and he bought a supply at eight cents apiece. "Before November 1940," his wife later said, "Mauchly had successfully tested certain components of his proposed computer and convinced himself that it was possible to build a cheap, reliable digital device using only electronic elements." This occurred, she insisted, before he had even heard of Atanasoff.[48]

In late 1940 he confided in some friends that he hoped to pull together all of this information to make a digital electronic computer. "We are now considering construction of an electrical computing machine," he wrote that November to a meteorologist he had worked with. "The machine would perform its operations in about 1/200th second, using vacuum tube relays."[49] Even though he was collaborative and picking up information from many people, he began to exhibit a competitive urge to be the first to make a new type of computer. He wrote a former student in December, "For your own private information, I expect to have, in a year or so, when I can get the stuff and put it together, an electronic computing machine. . . . Keep this dark, since I haven't the equipment this year to carry it out and I would like to 'be the first.'"[50]

That month, December 1940, Mauchly happened to meet Atanasoff, setting off a series of events followed by years of disputes over Mauchly's propensity to gather information from different sources and his desire to "be the first." Atanasoff was attending a meeting at the University of Pennsylvania, and he dropped by a session at which Mauchly proclaimed his hope of building a machine to analyze weather data. Afterward Atanasoff came up to say that he had been building an electronic calculator at Iowa State. Mauchly jotted on his conference program a note that Atanasoff claimed to have devised a machine that could process and store data at a cost of only $2 per digit. (Atanasoff's machine could handle three thousand digits and cost about $6,000.) Mauchly was amazed. He estimated that the cost of a vacuum-tube computer would be almost $13 per digit. He said he would love to see how it was done, and Atanasoff invited him to come to Iowa.

Throughout the first half of 1941, Mauchly corresponded with Atanasoff and continued to marvel at the low cost he claimed for his machine. "Less than $2 per digit sounds next to impossible, and yet that is what I understood you to say," he wrote. "Your suggestion about visiting Iowa seemed rather fantastic when first made, but the idea grows on me." Atanasoff urged him to accept. "As an additional inducement I will explain the $2 per digit business," he promised.[51]

THE MAUCHLY-ATANASOFF VISIT

The fateful visit lasted four days in June 1941.[52] Mauchly drove from Washington and brought his six-year-old son, Jimmy, arriving late on Friday, June 13, much to the surprise of Atanasoff's wife, Lura, who had not yet prepared the guest room. "I had to fly around, go to the attic, get extra pillows, and everything," she later recalled.[53] She also made them supper, since the Mauchlys had arrived hungry. The Atanasoffs had three children of their own, but Mauchly seemed to assume that Lura would take care of Jimmy during the visit, so she did, grudgingly. She took a dislike to Mauchly. "I don't think he's honest," she told her husband at one point.[54]

Atanasoff was eager to show off his partly built machine, even as his wife worried that he was being too trusting. "You must be careful until this is patented," she warned. Nevertheless, Atanasoff took Mauchly, along with Lura and the children, to the physics building basement the next morning, proudly pulling off a sheet to reveal what he and Berry were cobbling together.

Mauchly was impressed by a few things. The use of condensers in the memory unit was ingenious and cost-effective, as was Atanasoff's method of replenishing their charge every second or so by putting them on rotating cylinders. Mauchly had thought about using condensers instead of more expensive vacuum tubes, and he appreciated how Atanasoff's method of "jogging their memory" made it workable. That was the secret behind how the machine could be constructed for $2 per digit. After reading Atanasoff's thirty-five-page memo detailing the machine, and taking notes, he asked if he could take a carbon copy home. That request Atanasoff denied, both because he had no extras to give away (photocopiers hadn't been invented) and because he was becoming worried that Mauchly was sucking in too much information.[55]

But for the most part, Mauchly was uninspired by what he saw in Ames—or at least that is what he insisted in retrospect. The foremost drawback was that Atanasoff's machine was not fully electronic but instead relied on the mechanical drums of condensers for memory. That made it inexpensive but also very slow. "I thought his machine was very ingenious, but since it was in part mechanical, involving rotating commutators for switching, it was not by any means what I had in mind," Mauchly remembered. "I no longer became interested in the details." Later, in his testimony at the trial over the validity of his patents, Mauchly called the semimechanical nature of Atanasoff's machine "a rather drastic disappointment" and dismissed it as "a mechanical gadget which uses some electronic tubes in operation."[56]

The second disappointment, Mauchly contended, was that Atanasoff's machine was designed for a single purpose and could not be programmed or modified to perform other tasks: "He had not done anything to plan for this machine to be anything but a single set purpose machine and to solve sets of linear equations."[57]

So Mauchly left Iowa not with a breakthrough concept for how to build a computer but rather with a handful of smaller insights to add to the basket of ideas he had been collecting, consciously and subconsciously, on his visits to conferences and colleges and fairs. "I came to Iowa with much the same attitude that I went to the World's Fair and other places," he testified. "Is there something here which would be useful to aid my computations or anyone else's?"[58]

Like most people, Mauchly gleaned insights from a variety of experiences, conversations, and observations—in his case at Swarthmore, Dartmouth, Bell Labs, RCA, the World's Fair, Iowa State, and elsewhere—then combined them into ideas he considered his own. "A new idea comes suddenly and in a rather intuitive way," Einstein once said, "but intuition is nothing but the outcome of earlier intellectual experience." When people take insights from multiple sources and put them together, it's natural for them to think that the resulting ideas are their own—as in truth they are. All ideas are born that way. So Mauchly considered his intuitions and thoughts about how to build a computer to be his own rather than a bag of ideas he had stolen from other people. And despite later legal findings, he was for the most part right, insofar as anyone can be right in thinking that his ideas are his own. That is the way the creative process—if not the patent process—works.

Unlike Atanasoff, Mauchly had the opportunity, and the inclination, to collaborate with a team filled with varied talents. As a result, instead of producing a machine that didn't quite work and was abandoned in a basement, he and his team would go down in history as the inventors of the first electronic general-purpose computer.

As he was preparing to leave Iowa, Mauchly got a piece of pleasant news. He had been accepted into an electronics course at the University of Pennsylvania, one of the many around the country being funded on an emergency basis by the War Department. It was a chance to learn more about using vacuum tubes in electronic circuits, which he was now convinced was the best way to make computers. It also showed the importance of the military in driving innovation in the digital age.

During this ten-week course in the summer of 1941, Mauchly got the chance to work with a version of the MIT Differential Analyzer, the analog computer designed by Vannevar Bush. The experience amped up his interest in building his own computer. It also made him realize that the resources to do so at a place like Penn were far greater than at Ursinus, so he was thrilled to accept an instructor's position at the university when it was offered at the end of the summer.

Mauchly conveyed the good news in a letter to Atanasoff, which also contained hints of a plan that unnerved the Iowa professor. "A

Howard Aiken (1900–1973) at Harvard in 1945.

John Mauchly (1907–80) circa 1945.

J. Presper Eckert (1919–95) circa 1945.

Eckert (*touching machine*), Mauchly (*by pillar*), Jean Jennings (*in back*), and Herman Goldstine (*by Jennings*) with ENIAC in 1946.

number of different ideas have come to me recently anent computing circuits—some of which are more or less hybrids, combining your methods with other things, and some of which are nothing like your machine," Mauchly wrote, truthfully. "The question in my mind is this: is there any objection, from your point of view, to my building some sort of computer which incorporates some of the features of your machine?"[59] It's hard to tell from the letter, or from the subsequent explanations, depositions, and testimony over the ensuing years, whether Mauchly's innocent tone was sincere or feigned.

Either way, the letter upset Atanasoff, who had still not succeeded in prodding his lawyer into filing any patent claims. He responded to Mauchly rather brusquely within a few days: "Our attorney has emphasized the need of being careful about the dissemination of information about our device until a patent application is filed. This should not require too long, and, of course, I have no qualms about having informed you about our device, but it does require that we refrain from making public any details for the time being."[60] Amazingly, this exchange still did not provoke Atanasoff or the lawyer to make a filing for patents.

Mauchly proceeded to forge ahead during that fall of 1941 with his own design for a computer, which he correctly believed drew ideas from a wide variety of sources and was very different from what Atanasoff had built. In his summer course, he met the right partner to join him in the endeavor: a graduate student with a perfectionist's passion for precision engineering, who knew so much about electronics that he served as Mauchly's lab instructor, even though he was twelve years younger (at twenty-two) and didn't yet have his PhD.

J. PRESPER ECKERT

John Adam Presper Eckert Jr., known formally as J. Presper Eckert and informally as Pres, was the only child of a millionaire real estate developer in Philadelphia.[61] One of his great-grandfathers, Thomas Mills, invented the machines that made salt water taffy in Atlantic City and, as important, created a business to manufacture and sell them. As a young boy, Eckert was driven by his family's chauffeur to the William Penn private school, founded in 1689. But his success came not from the privileges of birth but from his own talents. He won a citywide science fair at age twelve by building a guidance system for model boats using magnets and rheostats, and at fourteen he devised an innovative way to use household current to eliminate troublesome batteries for the intercom system in one of his father's buildings.[62]

In high school Eckert dazzled his classmates with his inventions, and he made money by building radios, amplifiers, and sound systems. Philadelphia, the city of Benjamin Franklin, was then a great electronics center, and Eckert spent time at the research lab of Philo Farnsworth, one of the inventors of television. Although he was accepted by MIT and wanted to go there, his parents did not wish him to leave. Pretending to have suffered financial setbacks because of the Depression, they pressured him to go to Penn and live at home. He did rebel, however, against their desire that he study business; instead he enrolled in the university's Moore School of Electrical Engineering because he found the subject more interesting.

Eckert's social triumph at Penn was creating what he called an "Osculometer" (from the Latin word for mouth), which purported to measure the passion and romantic electricity of a kiss. A couple would hold the handles of the device and then kiss, their lip contact completing an electric circuit. A row of bulbs would light up, the goal being to kiss passionately enough to light up all ten and set off a blast from a foghorn. Smart contestants knew that wet kisses and sweaty palms increased the circuit's conductivity.[63] Eckert also invented a device that used a light-modulating method to record sound on film, for which he successfully applied for a patent at age twenty-one, while still an undergraduate.[64]

Pres Eckert had his quirks. Filled with nervous energy, he would pace the room, bite his nails, leap around, and occasionally stand atop a desk when he was thinking. He wore a watch chain that wasn't connected to a watch, and he would twirl it in his hands as if it were rosary beads. He had a quick temper that would flare and then dissolve into charm. His demand for perfection came from his father, who would walk around construction sites carrying a large pack of crayons with which to scrawl instructions, using different colors to indicate which worker was responsible. "He was sort of a perfectionist and made sure you did it right," his son said. "But he had a lot of charm, really. He got things done most of the time by people wanting to do the stuff." An engineer's engineer, Eckert felt that people like himself were necessary complements to physicists such as Mauchly. "A physicist is one who's concerned with the truth," he later said. "An engineer is one who's concerned with getting the job done."[65]

ENIAC

War mobilizes science. Over the centuries, ever since the ancient Greeks built a catapult and Leonardo da Vinci served as the military engineer for Cesare Borgia, martial needs have propelled advances in technology, and this was especially true in the mid-twentieth century. Many of the paramount technological feats of that era—computers, atomic power, radar, and the Internet—were spawned by the military.

America's entry into World War II in December 1941 provided the impetus to fund the machine that Mauchly and Eckert were devising. The University of Pennsylvania and the Army's Ordnance Department at Aberdeen Proving Ground had been tasked with producing the booklets of firing-angle settings needed for the artillery being shipped to Europe. In order to be aimed properly, the guns required tables that factored in hundreds of conditions, including temperature, humidity, wind speeds, altitude, and gunpowder varieties.

Creating a table for just one category of shell shot by one gun might require calculating three thousand trajectories from a set of differential equations. The work was often done using one of the Differential Analyzers invented at MIT by Vannevar Bush. The machine's calculations were combined with the labor of more than 170 people, most of them women, known as "computers," who tackled equations by punching the keys and cranking the handles of desktop adding machines. Women math majors were recruited from around the nation. But even with all of this effort, it took more than a month to complete just one firing table. By the summer of 1942, it was clear that production was falling further behind every week, rendering some of America's artillery ineffective.

That August, Mauchly wrote a memo that proposed a way to help the Army meet this challenge. It would change the course of computing. Titled "The Use of High Speed Vacuum Tube Devices for Calculating," his memo requested funding for the machine that he and Eckert were hoping to build: a digital electronic computer, using circuits with vacuum tubes, that could solve differential equations and perform other mathematical tasks. "A great gain in the speed of calculation can be obtained if the devices which are used employ electronic means," he argued. He went on to estimate that a missile trajectory could be calculated in "100 seconds."[66]

Mauchly's memo was ignored by Penn's deans, but it was brought to the attention of the Army officer attached to the university, Lieutenant (soon to be Captain) Herman Goldstine, a twenty-nine-year-old who had been a math professor at the University of Michigan. His mission was to speed up the production of firing tables, and he had dispatched his wife, Adele, also a mathematician, on a cross-country tour to recruit more women to join the battalions of human computers at Penn. Mauchly's memo convinced him that there was a better way.

The decision of the U.S. War Department to fund the electronic computer came on April 9, 1943. Mauchly and Eckert stayed up all the night before working on their proposal, but they still hadn't finished it by the time they got into the car for the two-hour ride from Penn to the Aberdeen Proving Ground in Maryland, where officials from the Ordnance Department were gathered. As Lieutenant Goldstine drove, they sat in the backseat writing the remaining sections, and when they arrived in Aberdeen, they continued working in a small room while Goldstine went to the review meeting. It was chaired by Oswald Veblen, the president of the Institute for Advanced Study in Princeton, who was advising the military on mathematical projects. Also present was Colonel Leslie Simon, director of the Army's Ballistic Research Laboratory. Goldstine recalled what happened: "Veblen, after listening for a short while to my presentation and teetering on the back legs of his chair, brought the chair down with a crash, arose, and said, 'Simon, give Goldstine the money.' He thereupon left the room and the meeting ended on this happy note."[67]

Mauchly and Eckert incorporated their memo into a paper they titled "Report on an Electronic Diff. Analyzer." Using the abbreviation *diff.* was cagey; it stood for both *differences*, which reflected the digital nature of the proposed machine, and *differential*, which described the equations it would tackle. Soon it was given a more memorable name: ENIAC, the Electronic Numerical Integrator and Computer. Even though ENIAC was designed primarily for handling differential equations, which were key to calculating missile trajectories, Mauchly wrote that it could have a "programming device" that would allow it to do other tasks, thus making it more of a general-purpose computer.[68]

In June 1943 construction of ENIAC began. Mauchly, who retained his teaching duties, served as a consultant and visionary. Goldstine, as the Army's representative, oversaw the operations and budget. And Eckert, with his passion for detail and perfection, was the chief engineer. Eckert became so dedicated to the project that he would sometimes sleep next to the machine. Once, as a joke, two engineers picked up his cot and gently moved him to an identical room one floor up; when he awoke he briefly feared the machine had been stolen.[69]

Knowing that great conceptions are worth little without precision execution (a lesson Atanasoff learned), Eckert was not shy about micromanaging. He would hover over the other engineers and tell them where to solder a joint or twist a wire. "I took every engineer's work and checked every calculation of every resistor in the machine to make sure that it was done correctly," he asserted. He disdained anyone who dismissed an issue as trivial. "Life is made up of a whole concentration of trivial matters," he once said. "Certainly a computer is nothing but a huge concentration of trivial matters."[70]

Eckert and Mauchly served as counterbalances for each other, which made them typical of so many digital-age leadership duos. Eckert drove people with a passion for precision; Mauchly tended to calm them and make them feel loved. "He was always kidding and joking with people," Eckert recalled. "He was personable." Eckert, whose technical skills came with a nervous energy and scattershot attention span, badly needed an intellectual sounding board, and Mauchly loved being that. Although he was not an engineer, Mauchly did have the ability to connect scientific theories with engineering practicalities in a way that was inspiring. "We got together and did this thing and I don't think either of us would have done it by ourselves," Eckert later conceded.[71]

ENIAC was digital, but instead of a binary system, using just 0s and 1s, it used a decimal system of ten-digit counters. In that regard, it was not like a modern computer. Other than that, it was more advanced than the machines built by Atanasoff, Zuse, Aiken, and Stibitz. Using what was called conditional branching (a capability described by Ada Lovelace a century earlier), it could hop around in a program based on its interim results, and it could repeat blocks of code, known as subroutines, that performed common tasks. "We had the ability to have subroutines and subroutines of subroutines," Eckert explained. When Mauchly proposed this functionality, Eckert recalled, "it was an idea that I instantly recognized as the key to this whole thing."[72]

After a year of building, around the time of D-Day in June 1944, Mauchly and Eckert were able to test the first two components, amounting to about one-sixth of the planned machine. They started with a simple multiplication problem. When it produced the correct answer, they let out a shout. But it took more than another year, until November 1945, for ENIAC to be fully operational. At that point it was able to perform five thousand additions and subtractions in one second, which was more than a hundred times faster than any previous machine. A hundred feet long and eight feet high, filling the space of what could be a modest three-bedroom apartment, it weighed close to thirty tons and had 17,468 vacuum tubes. By contrast, the Atanasoff-Berry computer, then languishing in a basement in Iowa, was the size of a desk, had only three hundred tubes, and could do merely thirty additions or subtractions per second.

BLETCHLEY PARK

Although few outsiders knew it at the time—and would not know for more than three decades—another electronic computer using vacuum tubes had been secretly built at the end of 1943 on the grounds of a redbrick Victorian manor in the town of Bletchley, fifty-four miles northwest of London, where the British had sequestered a team of geniuses and engineers to break the German wartime codes. The computer, known as Colossus, was the first all-electronic, partially programmable computer. Because it was geared for a special task, it was not a general-purpose or "Turing-complete" computer, but it did have Alan Turing's personal fingerprints on it.

Turing had begun to focus on codes and cryptology in the fall of 1936, when he arrived at Princeton just after writing "On Computable Numbers." He explained his interest in a letter to his mother that October:

> I have just discovered a possible application of the kind of thing I am working on at present. It answers the question "What is the most general kind of code or cipher possible," and at the same time (rather naturally) enables me to construct a lot of particular and interesting codes. One of them is pretty well impossible to decode without the key, and very quick to encode. I expect I could sell them to H.M. Government for quite a substantial sum, but am rather doubtful about the morality of such things. What do you think?[73]

Over the ensuing year, as he worried about the possibility of war with Germany, Turing got more interested in cryptology and less interested in trying to make money from it. Working in the machine shop of Princeton's physics building in late 1937, he constructed the first stages of a coding machine that turned letters into binary numbers and, using electromechanical relay switches, multiplied the resulting numerically encoded message by a huge secret number, making it almost impossible to decrypt.

One of Turing's mentors in Princeton was John von Neumann, the brilliant physicist and mathematician who had fled his native

Hungary and was at the Institute for Advanced Study, which for the time being was located in the building that housed the university's Mathematics Department. In the spring of 1938, as Turing was finishing his doctoral thesis, von Neumann offered him a job as his assistant. With the war clouds gathering in Europe, the offer was tempting, but it also felt vaguely unpatriotic. Turing decided to return to his fellowship at Cambridge and shortly thereafter joined the British effort to crack the German military codes.

His Majesty's Government Code and Cypher School was, at the time, located in London and staffed mainly by literary scholars, such as Dillwyn "Dilly" Knox, a classics professor from Cambridge, and Oliver Strachey, a dilettante socialite who played piano and occasionally wrote about India. There were no mathematicians among the eighty staffers until the fall of 1938, when Turing went there. But the following summer, as Britain prepared for war, the department began actively hiring mathematicians, at one point using a contest that involved solving the *Daily Telegraph* crossword puzzle as a recruitment tool, and it relocated to the drab redbrick town of Bletchley, whose main distinction was being at the juncture where the railway line between Oxford and Cambridge intersected with the one from London to Birmingham. A team from the British intelligence service, posing as "Captain Ridley's shooting party," visited the Bletchley Park manor house, a Victorian Gothic monstrosity that its owner wanted to demolish, and discreetly bought it. The code breakers were located in the cottages, stables, and some prefabricated huts that were erected on the grounds.[74]

Turing was assigned to a team working in Hut 8 that was trying to break the German Enigma code, which was generated by a portable machine with mechanical rotors and electrical circuits. It encrypted military messages by using a cipher that, after every keystroke, changed the formula for substituting letters. That made it so tough to decipher that the British despaired of ever doing so. A break came when Polish intelligence officers created a machine based on a captured German coder that was able to crack some of the Enigma codes. By the time the Poles showed the British their machine, however, it had been rendered ineffective because the Germans had added two more rotors and two more plugboard connections to their Enigma machines.

Turing and his team went to work creating a more sophisticated machine, dubbed "the bombe," that could decipher the improved Enigma messages—in particular, naval orders that would reveal the deployment of U-boats that were decimating British supply convoys. The bombe exploited a variety of subtle weaknesses in the coding, including the fact that no letter could be enciphered as itself and that there were certain phrases the Germans used repeatedly. By August 1940 Turing's team had two operating bombes, which were able to break 178 coded messages; by the end of the war they had built close to two hundred.

The Turing-designed bombe was not a notable advance in computer technology. It was an electromechanical device with relay switches and rotors rather than vacuum tubes and electronic circuits. But a subsequent machine produced at Bletchley Park, and Colossus, was a major milestone.

The need for Colossus arose when the Germans started coding important messages, such as orders from Hitler and his high command, with an electronic digital machine that used a binary system and twelve code wheels of unequal size. The electromechanical bombes designed by Turing were powerless to break it. It required an attack using lightning-quick electronic circuits.

The team in charge, based in Hut 11, was known as the Newmanry after its leader, Max Newman, the Cambridge math don who had introduced Turing to Hilbert's problems almost a decade earlier. Newman's engineering partner was the electronics wizard Tommy Flowers, a pioneer of vacuum tubes, who worked at the Post Office Research Station at Dollis Hill, a London suburb.

Turing was not part of Newman's team, but he did come up with a statistical approach, dubbed "Turingery," that detected any departures from a uniform distribution of characters in a stream of ciphered text. A machine was built that could scan two loops of punched paper tapes, using photoelectric heads, in order to compare all possible permutations of the two sequences. The machine was dubbed the

"Heath Robinson," after a British cartoonist who specialized, as did Rube Goldberg in America, in drawing absurdly complex mechanical contraptions.

For almost a decade Flowers had been fascinated by electronic circuits made with vacuum tubes, which he and other Brits called "valves." As an engineer with the Post Office's telephone division, he had created in 1934 an experimental system that used more than three thousand tubes to control connections among a thousand phone lines. He also pioneered the use of vacuum tubes for data storage. Turing enlisted Flowers to help on the bombe machines and then introduced him to Newman.

Flowers realized that the only way to analyze the German encrypted streams quickly enough was to store at least one of them into the internal electronic memory of a machine rather than trying to compare two punched paper tapes. This would require 1,500 vacuum tubes. At first the Bletchley Park managers were skeptical, but Flowers pushed ahead, and by December 1943—after only eleven months—he produced the first Colossus machine. An even bigger version, using 2,400 vacuum tubes, was ready by June 1, 1944. Its first decoded intercepts supported other sources informing General Dwight Eisenhower, who was about to launch the D-Day invasion, that Hitler was not ordering extra troops to Normandy. Within a year, eight more Colossus machines were produced.

This meant that well before ENIAC, which did not become operational until November 1945, the British code breakers had built a fully electronic and digital (indeed binary) computer. The second version, in June 1944, was even capable of some conditional branching. But unlike ENIAC, which had ten times the number of tubes, Colossus was a special-purpose machine geared for code breaking, not a general-purpose computer. With its limited programmability, it could not be instructed to perform all computational tasks, the way that (in theory) ENIAC could.

SO, WHO INVENTED THE COMPUTER?

In assessing how to apportion credit for creating the computer, it's useful to begin by specifying which attributes define the essence of a computer. In the most general sense, the definition of a computer could encompass everything from an abacus to an iPhone. But in chronicling the birth of the Digital Revolution, it makes sense to follow the accepted definitions of what, in modern usage, constitutes a computer. Here are a few:

> "A programmable usually electronic device that can store, retrieve, and process data." (Merriam-Webster Dictionary)

> "An electronic device which is capable of receiving information (data) in a particular form and of performing a sequence of operations in accordance with a predetermined but variable set of procedural instructions (program) to produce a result." (Oxford English Dictionary)

> "A general purpose device that can be programmed to carry out a set of arithmetic or logical operations automatically." (Wikipedia, 2014)

So the ideal computer is a machine that is electronic, general purpose, and programmable. What, then, best qualifies as the first?

George Stibitz's Model K, begun on his kitchen table in November 1937, led to a full-scale model at Bell Labs in January 1940. It was a binary computer and the first such device to be used remotely. But it used electromechanical relays and was thus not fully electronic. It was also a special-purpose computer and not programmable.

Herman Zuse's Z3, completed in May 1941, was the first automatically controlled, programmable, electrical, binary machine. It was designed to do engineering problems rather than be a general-purpose machine. However, it was later shown that, in theory, it could have been used as a Turing-complete machine. Its major difference from modern computers was that it was electromechanical, dependent on clacking and slow relay switches, rather than electronic. Another shortcoming is that it never really went into full-scale service. It was destroyed by the Allied bombing of Berlin in 1943.

The computer designed by John Vincent Atanasoff, which was complete but not fully workable by the time Atanasoff abandoned it to serve in the Navy in September 1942, was the world's first electronic digital computer, but it was only partly electronic. Its add-subtract mechanism used vacuum tubes, but its memory and data retrieval involved mechanical rotating drums. Its other main drawback, in terms of being considered the first modern computer, was that it was not programmable nor general purpose; instead it was hard-wired for the special task of solving linear equations. Also, Atanasoff never got it fully operational, and it disappeared into a basement at Iowa State.

Bletchley Park's Colossus I, completed in December 1943 by Max Newman and Tommy Flowers (with input from Alan Turing), was the first digital computer that was fully electronic, programmable, and operational. It was not, however, a general-purpose or Turing-complete machine; it was geared to the specific purpose of breaking Germany's wartime codes.

Howard Aiken's Harvard Mark I, built with IBM and put into operation in May 1944, was programmable, as we will see in the following chapter, but it was electromechanical rather than electronic.

ENIAC, completed by Presper Eckert and John Mauchly in November 1945, was the first machine to incorporate the full set of traits of a modern computer. It was all-electronic, superfast, and could be programmed by plugging and unplugging the cables connecting its different units. It was capable of changing paths based on interim results, and it qualified as a general-purpose Turing-complete machine, meaning it could in theory tackle any task. Most important, it worked. "That's a big thing with an invention," Eckert later said, contrasting their machine with Atanasoff's. "You have to have a whole system that works."[75] Mauchly and Eckert got their machine to do some very powerful calculations, and it was in constant use for ten years. It became the basis for most subsequent computers.

That last attribute is important. When we ascribe credit for an invention, determining who should be most noted by history, one criterion is looking at whose contributions turned out to have the most influence. Invention implies contributing something to the flow of history and affecting how an innovation developed. Using historic impact as a standard, Eckert and Mauchly are the most noteworthy innovators. Almost all computers of the 1950s trace their roots to ENIAC. The influence of Flowers, Newman, and Turing is somewhat trickier to assess. Their work was kept top-secret, but all three men were involved in the British computers built after the war. Zuse, who was isolated and under bombardment in Berlin, had even less influence on the course of computer development elsewhere. As for Atanasoff, his main influence on the field, perhaps his only influence, came from providing a few inspirations to Mauchly when he visited.

The issue of what inspirations Mauchly gleaned during his four-day visit with Atanasoff in Iowa in June 1941 turned into a protracted legal dispute. That raised another criterion, more legalistic than historical, in assessing credit for invention: Who, if anyone, ended up with the patents? In the case of the first computers, nobody did. But that outcome was due to a controversial legal battle that resulted in the patents of Eckert and Mauchly being nullified.[76]

The saga began in 1947, when Eckert and Mauchly, after leaving Penn, applied for a patent on their ENIAC work, which was finally granted (the patent system being rather slow) in 1964. By then the Eckert-Mauchly company and its patent rights had been sold to Remington Rand, which became Sperry Rand; it began pressuring other companies to pay it licensing fees. IBM and Bell Labs made deals, but Honeywell balked and started looking for a way to challenge the patents. It hired a young lawyer, Charles Call, who had an engineering degree and had worked at Bell Labs. His mission was to upend the Eckert-Mauchly patent by showing that their ideas weren't original.

Pursuing a tip from a Honeywell lawyer who had gone to Iowa State and read about the computer that Atanasoff had built there, Call paid a visit to Atanasoff at his home in Maryland. Atanasoff was charmed by Call's knowledge of his computer and somewhat resentful that he had never gotten much credit for it, so he handed over hundreds of letters and documents that showed how Mauchly had derived some ideas from his visit to Iowa. That evening Call drove to Washington to sit in the back of a lecture Mauchly was giving. In answer to a question about Atanasoff's machine, Mauchly claimed he had barely examined it. Call realized that if he could get Mauchly to say this in a deposition, then he could discredit him at a trial by producing Atanasoff's documents.

When Mauchly found out a few months later that Atanasoff might be helping Honeywell challenge his patents, he made his own visit to Atanasoff's Maryland home, bringing with him a Sperry Rand lawyer. It was an awkward meeting. Mauchly claimed that during his visit to Iowa he hadn't read Atanasoff's paper carefully or examined his computer, and Atanasoff coldly pointed out that this was not true. Mauchly stayed for dinner and tried to ingratiate himself with Atanasoff, but to no avail.

The issue went to trial before a federal judge, Earl Larson, in Minneapolis in June 1971. Mauchly proved a problematic witness. Pleading poor memory, he sounded squirrely about what he had seen during his visit to Iowa, and he repeatedly backtracked from assertions he had made in his earlier deposition, including his claim that he had only seen Atanasoff's computer partly covered and in dim light. Atanasoff, by contrast, was very effective. He described the machine he had built, demonstrated a model, and pointed out which of his ideas Mauchly had borrowed. In all, seventy-seven witnesses were called to testify, another eighty were deposed, and 32,600 exhibits were entered into the record. The trial lasted more than nine months, making it the longest federal trial to that point.

Judge Larson took another nineteen months to write his final decision, which was issued in October 1973. In it he ruled that the Eckert-Mauchly ENIAC patent was invalid: "Eckert and Mauchly did not themselves first invent the automatic electronic digital computer, but instead derived that subject matter from one Dr. John Vincent Atanasoff."[77] Instead of appealing, Sperry settled with Honeywell.*

---

*By then Atanasoff had retired. His career after World War II had been spent in the field of military ordnance and artillery, not computers. He died in 1995. John Mauchly

The judge's opinion, at 248 pages, was thorough, but it disregarded some significant differences between the machines. Mauchly did not derive quite as much from Atanasoff as the judge seemed to think. For example, Atanasoff's electronic circuit used binary logic, whereas Mauchly's was a decimal counter. Had the Eckert-Mauchly patent claims been less sweeping, they probably would have survived.

The case did not determine, even legally, who should get what proportion of the credit for the invention of the modern computer, but it did have two important consequences: it resurrected Atanasoff from the basement of history, and it showed very clearly, though this was not the intent of the judge or either party, that great innovations are usually the result of ideas that flow from a large number of sources. An invention, especially one as complex as the computer, usually comes not from an individual brainstorm but from a collaboratively woven tapestry of creativity. Mauchly had visited and talked to many people. That perhaps made his invention harder to patent, but it did not lessen the impact he had.

Mauchly and Eckert should be at the top of the list of people who deserve credit for inventing the computer, not because the ideas were all their own but because they had the ability to draw ideas from multiple sources, add their own innovations, execute their vision by building a competent team, and have the most influence on the course of subsequent developments. The machine they built was the first general-purpose electronic computer. "Atanasoff may have won a point in court, but he went back to teaching and we went on to build the first real electronic programmable computers," Eckert later pointed out.[78]

A lot of the credit, too, should go to Turing, for developing the concept of a universal computer and then being part of a hands-on team at Bletchley Park. How you rank the historic contributions of the others depends partly on the criteria you value. If you are enticed

---

remained a computer scientist, partly as a consultant with Sperry and as the founding president of the Association for Computing Machinery. He died in 1980. Eckert likewise remained with Sperry much of his career. He died in 1995.

by the romance of lone inventors and care less about who most influenced the progress of the field, you might put Atanasoff and Zuse high. But the main lesson to draw from the birth of computers is that innovation is usually a group effort, involving collaboration between visionaries and engineers, and that creativity comes from drawing on many sources. Only in storybooks do inventions come like a thunderbolt, or a lightbulb popping out of the head of a lone individual in a basement or garret or garage.

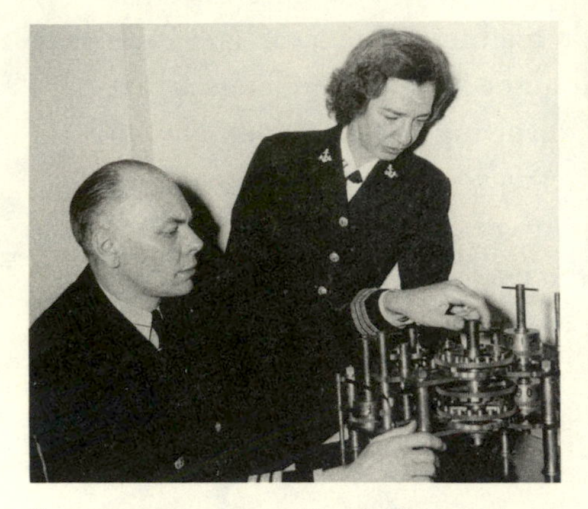

Howard Aiken and Grace Hopper (1906–92) with a part of Babbage's Difference Engine at Harvard in 1946.

Jean Jennings and Frances Bilas with ENIAC.

Jean Jennings (1924–2011) in 1945.

Betty Snyder (1917–2001) in 1944.

CHAPTER THREE

PROGRAMMING

The development of the modern computer required another important step. All of the machines built during the war were conceived, at least initially, with a specific task in mind, such as solving equations or deciphering codes. A *real* computer, like that envisioned by Ada Lovelace and then Alan Turing, should be able to perform, seamlessly and quickly, any logical operation. This required machines whose operations were determined not just by their hardware but by software, the programs they could run. Once again Turing laid out the concept clearly. "We do not need to have an infinity of different machines doing different jobs," he wrote in 1948. "A single one will suffice. The engineering problem of producing various machines for various jobs is replaced by the office work of 'programming' the universal machine to do these jobs."[1]

In theory, machines such as ENIAC could be programmed and even pass for general-purpose machines. But in practice, loading in a new program was a laborious process that often involved replugging by hand the cables that connected different units in the computer. The wartime machines could not switch programs at electronic speeds. This would require the next major step in the creation of the modern computer: figuring out how to store programs inside a machine's electronic memory.

GRACE HOPPER

Starting with Charles Babbage, the men who invented computers focused primarily on the hardware. But the women who became involved during World War II saw early on the importance of programming, just as Ada Lovelace had. They developed ways to code the instructions that told the hardware what operations to perform. In this software lay the magic formulas that could transform the machines in wondrous ways.

The most colorful programming pioneer was a gutsy and spirited, yet also charming and collegial, naval officer named Grace Hopper, who ended up working for Howard Aiken at Harvard and then for Presper Eckert and John Mauchly. Born Grace Brewster Murray in 1906, she was from a prosperous family on the Upper West Side of Manhattan. Her grandfather was a civil engineer who took her around New York on surveying trips, her mother was a mathematician, and her father was an insurance executive. She graduated from Vassar with a degree in math and physics, then went on to Yale, where in 1934 she earned her PhD in math.[2]

Her education wasn't as unusual as you might think. She was the eleventh woman to get a math doctorate from Yale, the first being in 1895.[3] It was not all that uncommon for a woman, especially from a successful family, to get a doctorate in math in the 1930s. In fact, it was more common than it would be a generation later. The number of American women who got doctorates in math during the 1930s was 113, which was 15 percent of the total number of American math doctorates. During the decade of the 1950s, only 106 American women got math doctorates, which was a mere 4 percent of the total. (By the first decade of the 2000s things had more than rebounded, and there were 1,600 women who got math doctorates, 30 percent of the total.)

After marrying a professor of comparative literature, Vincent Hopper, Grace joined the faculty of Vassar. Unlike most math professors, she insisted that her students be able to write well. In her probability course, she began with a lecture on one of her favorite mathematical formulas* and asked her students to write an essay about it. These she would mark for clarity of writing and style. "I'd cover [an essay] up with ink, and I would get a rebellion that they were taking a math course not an English course," she recalled. "Then I would explain, it was no use trying to learn math unless they could communicate it with other people."[4] Throughout her life, she excelled at being able to translate scientific problems—such as those involving trajectories, fluid flows, explosions, and weather patterns—into mathematical equations and then into ordinary English. This talent helped to make her a good programmer.

By 1940 Grace Hopper was bored. She had no children, her marriage was unexciting, and teaching math was not as fulfilling as she had hoped. She took a partial leave from Vassar to study with the noted mathematician Richard Courant at New York University, focusing on methods for solving partial differential equations. She was still studying with Courant when the Japanese attacked Pearl Harbor in December 1941. America's entry into World War II offered her a way to change her life. During the following eighteen months, she quit Vassar, divorced her husband, and at age thirty-six joined the U.S. Navy. She was sent to the Naval Reserve Midshipmen's School at Smith College in Massachusetts, and in June 1944 graduated first in her class as Lieutenant Grace Hopper.

She assumed that she would be assigned to a cryptography and code group, but to her surprise she was ordered to report to Harvard University to work on the Mark I, the behemoth digital computer with clunky electromechanical relays and a motorized turning shaft that, as described above, had been conceived by Howard Aiken in 1937. By the time Hopper was assigned to it, the machine had been commandeered by the Navy; Aiken was still running it, but as a commander in the Navy rather than as a member of the Harvard faculty.

When Hopper reported for duty in July 1944, Aiken gave her a copy of Charles Babbage's memoirs and brought her to see the Mark I. "That is a computing machine," he told her. Hopper just

---

* Stirling's formula, which approximates the value of the factorial of a number.

stared at it silently for a while. "There was this large mass of machinery out there making a lot of racket," she remembered. "It was all bare, all open, and very noisy."[5] Realizing that she would need to understand it fully in order to run it properly, she spent nights analyzing the blueprints. Her strength came from her ability to know how to translate (as she had at Vassar) real-world problems into mathematical equations, and then to communicate those in commands that the machine would understand. "I learned languages of oceanography, of this whole business of minesweeping, of detonators, of proximity fuses, of biomedical stuff," she explained. "We had to learn their vocabularies in order to be able to run their problems. I could switch my vocabulary and speak highly technical for the programmers, and then tell the same things to the managers a few hours later but with a totally different vocabulary." Innovation requires articulation.

Because of her ability to communicate precisely, Aiken assigned her to write what was to become the world's first computer programming manual. "You are going to write a book," he said one day, standing next to her desk.

"I can't write a book," she replied. "I've never written one."

"Well, you're in the Navy now," he declared. "You are going to write one."[6]

The result was a five-hundred-page book that was both a history of the Mark I and a guide to programming it.[7] The first chapter described earlier calculating machines, with an emphasis on those built by Pascal, Leibniz, and Babbage. The frontispiece was a picture of the portion of Babbage's Difference Engine that Aiken had mounted in his office, and Hopper began with an epigraph from Babbage. She understood, as had Ada Lovelace, that Babbage's Analytical Engine had a special quality, one that she and Aiken believed would distinguish the Harvard Mark I from other computers of the time. Like Babbage's unbuilt machine, Aiken's Mark I, which received its marching orders via a punch tape, could be reprogrammed with new instructions.

Every evening Hopper read to Aiken the pages she had written that day, which helped her learn a simple trick of good writers: "He pointed out that if you stumble when you try to read it aloud, you'd better fix that sentence. Every day I had to read five pages of what I had written."[8] Her sentences became simple, crisp, and clear. With their strong partnership, Hopper and Aiken became the modern counterparts, a century later, of Lovelace and Babbage. The more she learned about Ada Lovelace, the more Hopper identified with her. "She wrote the first loop," Hopper said. "I will never forget. None of us ever will."[9]

Hopper's historical sections focused on personalities. In doing so, her book emphasized the role of individuals. In contrast, shortly after Hopper's book was completed, the executives at IBM commissioned their own history of the Mark I that gave primary credit to the IBM teams in Endicott, New York, who had constructed the machine. "IBM interests were best served by replacing individual history with organizational history," the historian Kurt Beyer wrote in a study of Hopper. "The locus of technological innovation, according to IBM, was the corporation. The myth of the lone radical inventor working in the laboratory or basement was replaced by the reality of teams of faceless organizational engineers contributing incremental advancements."[10] In the IBM version of history, the Mark I contained a long list of small innovations, such as the ratchet-type counter and the double-decked card feed, that IBM's book attributed to a bevy of little-known engineers who worked collaboratively in Endicott.*

The difference between Hopper's version of history and IBM's ran deeper than a dispute over who should get the most credit. It showed fundamentally contrasting outlooks on the history of innovation. Some studies of technology and science emphasize, as Hopper did, the role of creative inventors who make innovative leaps. Other studies emphasize the role of teams and institutions, such as the collaborative work done at Bell Labs and IBM's Endicott facility. This latter approach tries to show that what may seem like creative leaps—the Eureka moment—are actually the result of an evolutionary process that occurs when ideas, concepts, technologies, and engineering methods ripen together. Neither way of looking at technological

---

*The display and explanations of the Mark I at Harvard's science center made no mention of Grace Hopper nor pictured any women until 2014, when the display was revised to highlight her role and that of the programmers.

advancement is, on its own, completely satisfying. Most of the great innovations of the digital age sprang from an interplay of creative individuals (Mauchly, Turing, von Neumann, Aiken) with teams that knew how to implement their ideas.

Hopper's partner in operating the Mark I was Richard Bloch, a Harvard math major who had played flute in the university's prank-loving band and served a tour of duty in the Navy. Ensign Bloch began working for Aiken three months before Hopper arrived, and he took her under his wing. "I remember sitting down, long into the night, going over how this machine worked, how to program this thing," he said. He and Hopper alternated twelve-hour shifts tending to the demands of the machine and its equally temperamental commander, Aiken. "Sometimes he would show up at four in the morning," Bloch said, "and his comment was 'are we making numbers?' He was very nervous when the machine stopped."[11]

Hopper's approach to programming was very systematic. She broke down every physics problem or mathematical equation into small arithmetic steps. "You simply step by step told the computer what to do," she explained. "Get this number and add it to that number and put the answer there. Now pick up this number and multiply it by this number and put it there."[12] When the program was punched into a tape and the moment came to test it, the Mark I crew, as a joke that became a ritual, would pull out a prayer rug, face east, and pray that their work would prove acceptable.

Late at night Bloch would sometimes fiddle with the hardware circuits of the Mark I, which would cause problems for the software programs Hopper had written. She had a piss-and-vinegar personality salted with the language of a midshipman, and the ensuing tongue-lashings she gave to the lanky and calmly amused Bloch were a precursor of the mix of confrontation and camaraderie that was to develop between hardware and software engineers. "Every time I got a program running, he'd get in there at night and change the circuits in the computer and the next morning the program would not run," she lamented. "What's more, he was home asleep and couldn't tell me what he had done." As Bloch put it, "all hell broke loose" on such occasions. "Aiken didn't look at these things with great humor."[13]

Such episodes gave Hopper the reputation of being irreverent. That she was. But she also had a software hacker's ability to combine irreverence with a collaborative spirit. This pirate crew camaraderie—something Hopper shared with subsequent generations of coders—actually liberated rather than restrained her. As Beyer wrote, "It was Hopper's collaborative abilities rather than her rebellious nature that created the space for her independent thought and action."[14]

In fact, it was the calm Bloch rather than the spunky Hopper who had the more contentious relationship with Commander Aiken. "Dick was always getting in trouble," Hopper claimed. "I would try to explain to him that Aiken was just like a computer. He's wired a certain way, and if you are going to work with him you must realize how he is wired."[15] Aiken, who initially balked at having a woman on his officer corps, soon made Hopper not only his primary programmer but his top deputy. Years later he would recall fondly the contributions she made to the birth of computer programming. "Grace was a good man," he declared.[16]

Among the programming practices that Hopper perfected at Harvard was the subroutine, those chunks of code for specific tasks that are stored once but can be called upon when needed at different points in the main program. "A subroutine is a clearly defined, easily symbolized, often repeated program," she wrote. "Harvard's Mark I contained subroutines for sine *x*, log10 *x*, and 10*x*, each called for by a single operational code."[17] It was a concept that Ada Lovelace had originally described in her "Notes" on the Analytical Engine. Hopper collected a growing library of these subroutines. She also developed, while programming the Mark I, the concept of a compiler, which would eventually facilitate writing the same program for multiple machines by creating a process for translating source code into the machine language used by different computer processors.

In addition, her crew helped to popularize the terms *bug* and *debugging*. The Mark II version of the Harvard computer was in a building without window screens. One night the machine conked out, and the crew began looking for the problem. They found a moth with a wingspan of four inches that had gotten smashed in one of the electromechanical relays. It was retrieved and pasted into the log book with Scotch tape. "Panel F (moth) in relay," the entry noted. "First actual case of bug being found."[18] From then on, they referred to ferreting out glitches as "debugging the machine."

By 1945, thanks largely to Hopper, the Harvard Mark I was the world's most easily programmable big computer. It could switch tasks simply by getting new instructions via punched paper tape rather than requiring a reconfiguration of its hardware or cables. However, this distinction was largely unnoticed, both then and in history, because the Mark I (and even its 1947 successor, the Mark II) used slow and clackety electromechanical relays rather than electronic components such as vacuum tubes. "By the time anybody knew anything about her," Hopper said of the Mark II, "she was a dead duck, and everybody was going electronic."[19]

Computer innovators, like other pioneers, can find themselves left behind if they get stuck in their ways. The same traits that make them inventive, such as stubbornness and focus, can make them resistant to change when new ideas come along. Steve Jobs was famously stubborn and focused, yet he dazzled and baffled colleagues by suddenly changing his mind when he realized he needed to think different. Aiken lacked that agility. He was not nimble enough to pirouette. He had a naval commander's instinct for centralized authority, so his crew was not as freewheeling as the Mauchly-Eckert team at Penn. Aiken also placed a premium on reliability rather than speed. So he clung to the use of time-tested and dependable electromechanical relays even after it became clear to the people at Penn and Bletchley Park that vacuum tubes were the wave of the future. His Mark I could execute only about three commands per second, while the ENIAC being built at Penn would execute five thousand commands in that time.

When he went to Penn to see ENIAC and attend some lectures, "Aiken was absorbed in his own way of doing things," a report on the meeting noted, "and does not appear to have been aware of the significance of the new electronic machines."[20] The same was true of Hopper when she visited ENIAC in 1945. It seemed to her that the Mark I was superior because it was easily programmable. With ENIAC, she said, "you plugged the pieces and essentially you built a special computer for each job, and we were used to the concept of programming and controlling the computer by our program."[21] The time it took to reprogram ENIAC, which could be a whole day, wiped out the advantage it had in processing speed, unless it was doing the same task over and over.

But unlike Aiken, Hopper was open-minded enough that she soon changed her outlook. Advances were being made that year in ways to reprogram ENIAC more quickly. And the people in the forefront of that programming revolution, to Hopper's delight, were women.

THE WOMEN OF ENIAC

All the engineers who built ENIAC's hardware were men. Less heralded by history was a group of women, six in particular, who turned out to be almost as important in the development of modern computing. As ENIAC was being constructed at Penn in 1945, it was thought that it would perform a specific set of calculations over and over, such as determining a missile's trajectory using different variables. But the end of the war meant that the machine was needed for many other types of calculations—sonic waves, weather patterns, and the explosive power of new types of atom bombs—that would require it to be reprogrammed often.

This entailed switching around by hand ENIAC's rat's nest of cables and resetting its switches. At first the programming seemed to be a routine, perhaps even menial task, which may have been why it was relegated to women, who back then were not encouraged to become engineers. But what the women of ENIAC soon showed, and the men later came to understand, was that the programming of a computer could be just as significant as the design of its hardware.

The tale of Jean Jennings is illustrative of the early women computer programmers.[22] She was born on a farm on the outskirts of Alanthus Grove, Missouri (population: 104), into a family that had almost no money and deeply valued education. Her father taught in a one-room schoolhouse, where Jean became the star pitcher and lone girl on the softball team. Her mother, though she had dropped out of school in eighth grade, helped tutor algebra and geometry. Jean was the sixth of seven children, all of whom went to college. That was back when state governments valued education and realized the economic and social value of making it affordable. She attended Northwest Missouri State Teachers College in Maryville, where the tuition was $76 per year. (In 2013 it was approximately $14,000 per year for in-state residents, a twelve-fold increase *after* adjusting for inflation.) She started out majoring in journalism, but she hated her advisor so switched to math, which she loved.

When she finished in January 1945, her calculus teacher showed her a flyer soliciting women mathematicians to work at the University of Pennsylvania, where women were working as "computers"—humans who performed routinized math tasks—mainly calculating artillery trajectory tables for the Army. As one of the ads put it:

> Wanted: Women With Degrees in Mathematics. . . . Women are being offered scientific and engineering jobs where formerly men were preferred. Now is the time to consider your job in science and engineering. . . . You will find that the slogan there as elsewhere is "WOMEN WANTED!"[23]

Jennings, who had never been out of Missouri, applied. When she received a telegram of acceptance, she boarded the midnight Wabash train heading east and arrived at Penn forty hours later. "Needless to say, they were shocked that I had gotten there so quickly," she recalled.[24]

When Jennings showed up in March 1945, at age twenty, there were approximately seventy women at Penn working on desktop adding machines and scribbling numbers on huge sheets of paper. Captain Herman Goldstine's wife, Adele, was in charge of recruiting and training. "I'll never forget the first time I saw Adele," Jennings said. "She ambled into class with a cigarette dangling from the corner of her mouth, walked over to a table, threw one leg over its corner, and began to lecture in her slightly cleaned up Brooklyn accent." For Jennings, who had grown up as a spirited tomboy bristling at the countless instances of sexism she faced, it was a transforming experience. "I knew I was a long way from Maryville, where women had to sneak down to the greenhouse to grab a smoke."[25]

A few months after she arrived, a memo was circulated among the women about six job openings to work on the mysterious machine that was behind locked doors on the first floor of Penn's Moore School of Engineering. "I had no idea what the job was or what the ENIAC was," Jennings recalled. "All I knew was that I might be getting in on the ground floor of something new, and I believed I could learn and do anything as well as anyone else." She also was looking to do something more exciting than calculating trajectories.

When she got to the meeting, Goldstine asked her what she knew about electricity. "I said that I had had a course in physics and knew that E equaled IR," she recalled, referring to Ohm's Law, which defines how a flow of electrical current is related to voltage and resistance. "No, no," Goldstine replied, "I don't care about that, but are you afraid of it?"[26] The job involved plugging in wires and throwing a lot of switches, he explained. She said that she wasn't afraid. While she was being interviewed, Adele Goldstine came in, looked at her, and nodded. Jennings was selected.

In addition to Jean Jennings (later Bartik), the others were Marlyn Wescoff (later Meltzer), Ruth Lichterman (later Teitelbaum), Betty Snyder (later Holberton), Frances Bilas (later Spence), and Kay McNulty (who later married John Mauchly). They were a typical squad thrown together by the war: Wescoff and Lichterman were Jewish, Snyder a Quaker, McNulty an Irish-born Catholic, and Jennings a lapsed Church of Christ Protestant. "We had a wonderful time with each other, mainly because none of us had ever been in close contact with anyone from one of the others' religions," according to Jennings. "We had some great arguments about religious truths and beliefs. Despite our differences, or perhaps because of them, we really liked one another."[27]

In the summer of 1945, the six women were sent to Aberdeen Proving Ground to learn how to use IBM punch cards and wire up plug boards. "We had great discussions about religion, our families, politics, and our work," McNulty recalled. "We never ran out of things to say to each other."[28] Jennings became a ringleader: "We worked together, lived together, ate together, and sat up until all hours discussing everything."[29] Since they were all single and surrounded by a lot of single soldiers, there were multiple memorable romances nurtured over Tom Collins cocktails in the booths of the officers' club. Wescoff found a Marine who was "tall and quite handsome." Jennings paired up with an Army sergeant named Pete, who was "attractive but not really handsome." He was from Mississippi, and Jennings was outspoken in her opposition to racial segregation: "Pete told me once that he would never take me to Biloxi because I was so outspoken in my views on discrimination that I'd be killed."[30]

After six weeks of training, the six women programmers consigned their boyfriends to memory archives and returned to Penn, where they were given poster-size diagrams and charts describing ENIAC. "Somebody gave us a whole stack of blueprints, and these were the wiring diagrams for all the panels, and they said, 'Here, figure out how the machine works and then figure out how to program it,'" explained McNulty.[31] That required analyzing the differential equations and then determining how to patch the cables to connect to the correct electronic circuits. "The biggest advantage of learning the ENIAC from the diagrams was that we began to understand what it could and could not do," said Jennings. "As a result we could diagnose troubles almost down to the individual vacuum tube." She and Snyder devised a system to figure out which of the eighteen thousand vacuum tubes had burned out. "Since we knew both the application and the machine, we learned to diagnose troubles as well as, if not better than, the engineers. I tell you, those engineers loved it. They could leave the debugging to us."[32]

Snyder described making careful diagrams and charts for each new configuration of cables and switches. "What we were doing then was the beginning of a program," she said, though they did not yet have that word for it. They wrote out each new sequence on paper to protect themselves. "We all felt that we'd be scalped if we ruined the board," said Jennings.[33]

One day Jennings and Snyder were sitting in the second-floor classroom they had commandeered, staring at rolled-out sheets containing the diagrams of ENIAC's many units, when a man came in to inspect some construction. "Hi, my name is John Mauchly," he said. "I was just checking to see if the ceiling's falling in." Neither woman had met the ENIAC visionary before, but they were not the least bit shy or intimidated. "Boy are we glad to see you," Jennings declared. "Tell us how this blasted accumulator works." Mauchly carefully answered the question and then others. When they finished, he told them, "Well, my office is next door. So anytime I'm in my office, you can come in and ask me questions."

Almost every afternoon, they did. "He was a marvelous teacher," according to Jennings. He pushed the women to envision the many things ENIAC might someday do, in addition to calculating artillery trajectories. He knew that in order to make it a true general-purpose computer, it would need to inspire programmers who could coax various tasks out of the hardware. "He used to always try to get us to think of other problems," said Jennings. "He would always want us to invert a matrix or something like that."[34]

Around the same time that Hopper was doing so at Harvard, the women of ENIAC were developing the use of subroutines. They were fretting that the logical circuits did not have enough capacity to compute some trajectories. It was McNulty who pushed a solution. "Oh, I know, I know, I know," she said excitedly one day. "We can use a master programmer to repeat code." They tried it and it worked. "We began to think about how we could have subroutines, and nested subroutines, and all that stuff," recalled Jennings. "It was very practical in terms of doing this trajectory problem, because the idea of not having to repeat a whole program, you could just repeat pieces of it and set up the master programmer to do this. Once you've learned that, you learn how to design your program in modules. Modularizing and developing subroutines were really crucial in learning how to program."[35]

Shortly before she died in 2011, Jean Jennings Bartik reflected proudly on the fact that all the programmers who created the first general-purpose computer were women: "Despite our coming of age in an era when women's career opportunities were generally quite confined, we helped initiate the era of the computer." It happened because a lot of women back then had studied math, and their skills were in demand. There was also an irony involved: the boys with their toys thought that assembling the hardware was the most important task, and thus a man's job. "American science and engineering was even more sexist than it is today," Jennings said. "If the ENIAC's administrators had known how crucial programming would be to the functioning of the electronic computer and how complex it would prove to be, they might have been more hesitant to give such an important role to women."[36]

STORED PROGRAMS

From the beginning, Mauchly and Eckert understood that there were ways to make ENIAC easier to reprogram. But they didn't try to do so because building in that capability would have required them to make the hardware more complicated, and it wasn't necessary for the tasks they originally envisioned. "No attempt has been made to make provision for setting up a problem automatically," they wrote in their year-end 1943 ENIAC progress report. "This is for the sake of simplicity and because it is anticipated that the ENIAC will be used primarily for problems of a type in which one setup will be used many times before another problem is placed on the machine."[37]

But more than a year before ENIAC was finished, indeed as early as the beginning of 1944, Mauchly and Eckert realized that there was a good way to make computers easily reprogrammable: store the programs inside the computer's memory rather than load them in every time. That, they sensed, would be the next great advance in computer development. This "stored-program" architecture would mean that a computer's tasks could be changed almost instantly, without manually reconfiguring cables and switches.[38]

To store a program inside of the machine, they would need to create a large memory capacity. Eckert considered many methods for doing that. "This programming may be of the temporary type set up on alloy discs or of the permanent type on etched discs," he wrote in a January 1944 memo.[39] Because such disks were not yet affordable, he proposed using instead, on the next version of ENIAC, a cheaper storage method, which was called an acoustic delay line. It had been pioneered at Bell Labs by an engineer named William Shockley (about whom there will be a lot more later) and developed at MIT. The acoustic delay line worked by storing data as pulses in a long tube filled with a thick and sluggish liquid, such as mercury. At one end of the tube, an electrical signal carrying a stream of data would be converted by a quartz plug into pulses that would ripple back and forth the length of the tube for a while. The ripples could be refreshed electrically for as long as necessary. When it came time to retrieve the data, the quartz plug would convert it back into an electrical signal. Each tube could handle approximately a thousand bits of data at one-hundredth the cost of using a circuit of vacuum tubes. The next-generation ENIAC successor, Eckert and Mauchly wrote in a memo in the summer of 1944, should have racks of these mercury delay line tubes to store both data and rudimentary programming information in digital form.

JOHN VON NEUMANN

At this point, one of the most interesting characters in the history of computing reenters the tale: John von Neumann, the Hungarian-born mathematician who was a mentor to Turing in Princeton and offered him a job as an assistant. An enthusiastic polymath and urbane intellectual, he made major contributions to statistics, set theory, geometry, quantum mechanics, nuclear weapons design, fluid dynamics, game theory, and computer architecture. He would end up significantly improving upon, getting his name attached to, and reaping most of the credit for the stored-program architecture that Eckert, Mauchly, and their colleagues had begun to consider.[40]

Von Neumann was born to a prosperous Jewish family in Budapest in 1903, during a glittering period after the Austro-Hungarian Empire abolished the restrictive laws against Jews. Emperor Franz Joseph awarded a hereditary title in 1913 to the banker Max Neumann for "meritorious service in the financial field," thus allowing the family to be called margittai Neumann or, in German, von Neumann. János (known as Jancsi and later, in America, as John or Johnny) was the eldest of three brothers, who all converted to Catholicism ("for convenience sake," one admitted) after their father's death.[41]

Von Neumann was another innovator who stood at the intersection of the humanities and sciences. "Father was an amateur poet and he believed that poetry could convey not only emotions but also philosophical ideas," John's brother Nicholas recalled. "He regarded poetry as a language within a language, an idea that might be traced to John's future speculations on the languages of the computer and the brain." Of his mother he wrote, "She believed that music, art, and related aesthetic pleasures had an important place in our lives, that elegance was a quality to be revered."[42]

There is a wealth of stories about young von Neumann's prodigal genius, some of them probably true. At age six, it was later said, he would joke with his father in classical Greek, and he could divide two eight-digit numbers in his head. As a party trick, he would memorize a page of the phone book and recite back the names and numbers, and he could recall verbatim pages from novels or articles he had read, in any of five languages. "If a mentally superhuman race ever develops," the hydrogen bomb developer Edward Teller once said, "its members will resemble Johnny von Neumann."[43]

In addition to school, he had private tutors in math and languages, and by age fifteen he had completely mastered advanced calculus. When the communist Béla Kun briefly took over Hungary in 1919, von Neumann's tutelage was moved to Vienna and a resort on the Adriatic, and he developed a lifelong aversion to communism. He studied chemistry at the Swiss Federal Institute of Technology in Zurich (where Einstein had gone) and mathematics at both Berlin and Budapest, earning his doctorate in 1926. In 1930, he went to Princeton University to teach quantum physics, and he stayed on after being appointed (along with Einstein and Gödel) among the founding faculty of the Institute for Advanced Study.[44]

Von Neumann and Turing, who met in Princeton, would become paired as the grand theorists of the general-purpose computer, but in personality and temperament they were binary opposites. Turing led a spartan existence, lived in boardinghouses and hostels, and generally kept to himself; von Neumann was an elegant bon vivant who hosted sparkling parties with his wife once or twice a week at their huge house in Princeton. Turing was a long-distance runner; there were very few thoughts that could be said never to have crossed von Neumann's mind, but running long distances (or even short distances) was among them. "In dress and habits he tended to be slovenly," Turing's mother once said of her son. Von Neumann, by contrast, wore a three-piece suit at almost all times, including on a donkey ride down the Grand Canyon; even as a student he was so well dressed that, upon first meeting him, the mathematician David Hilbert reportedly had but one question: Who is his tailor?[45]

Von Neumann loved telling jokes and reciting risqué limericks in various languages at his parties, and he ate so heartily that his wife once said he could count anything except calories. He drove cars with an abandon that was reckless but not always wreckless, and he was fond of flashy new Cadillacs. "He bought a new one at least once a year, whether he had wrecked the previous one or not," wrote the science historian George Dyson.[46]

While at the Institute in the late 1930s, von Neumann developed an interest in ways to mathematically model explosive shock waves. This led him to become, in 1943, a member of the Manhattan Project, making frequent trips to the secret facilities in Los Alamos, New Mexico, where atomic weapons were being developed. Because there was not enough uranium-235 to build more than one bomb, the scientists at Los Alamos were also trying to design a device that would use plutonium-239. Von Neumann focused on ways to construct an explosive lens that would compress the plutonium core of the bomb to reach critical mass.*

---

*Von Neumann was successful in this. The plutonium implosion design would result in the first detonation of an atomic device, the Trinity test, in July 1945 near Alamogordo, New Mexico, and it would be used for the bomb that was dropped on Nagasaki on August 9, 1945, three days after the uranium bomb was used on Hiroshima. With his hatred of both the Nazis and the Russian-backed communists, von Neumann became a vocal proponent of atomic weaponry. He attended the Trinity test, as well as later tests on Bikini Atoll in the Pacific, and he argued that a thousand radiation deaths was an acceptable price to pay for the United States attaining a nuclear advantage. He would

Assessing this implosion concept required solving a swarm of equations that could calculate the rate of flow of the compression of air or other material that would occur after an explosion. Thus von Neumann embarked on a mission to understand the potential of high-speed computers.

During the summer of 1944, this quest took him to Bell Labs to study the updated versions of George Stibitz's Complex Number Calculator. The latest one had an innovation that particularly impressed him: the punched tape that fed in the instructions for each task also included the data, all commingled. He also spent time at Harvard trying to determine if Howard Aiken's Mark I could help with the bomb calculations. Throughout the summer and fall of that year, he shuttled by train between Harvard, Princeton, Bell Labs, and Aberdeen, acting as an idea bee, pollinating and cross-pollinating various teams with the notions that had adhered to his mind as he buzzed around. Just as John Mauchly had traveled around picking up the ideas that led to the first working electronic computer, von Neumann wandered around gathering the elements and concepts that became part of stored-program computer architecture.

At Harvard, Grace Hopper and her programming partner, Richard Bloch, set up a place for von Neumann to work in the conference room right next to the Mark I. Von Neumann and Bloch would write equations on the blackboard and feed them into the machine, and Hopper would read off the intermediate results as they spewed out. When the machine was "making numbers," Hopper said, von Neumann would often barge in from the conference room and predict what the results would be. "I'll just never forget that surge out of the backroom, and then surge back again and put them all over the blackboard, and von Neumann predicting what the numbers would turn out to be and, ninety-nine percent of the time, with the greatest of accuracy—fantastic," Hopper enthused. "He just seemed to know how the computation was going, or to feel how it was going."[47]

Von Neumann impressed the Harvard team with how collabora-

---

die twelve years later, at age fifty-three, of bone and pancreatic cancer, which may have been caused by the radiation emitted during those tests.

tive he was. He absorbed their ideas, took credit for some, but also made it clear that nobody should claim ownership of any concept. When the time came to write a report on what they were doing, von Neumann insisted that Bloch's name go first. "I really didn't feel I deserved that; but that's the way it came out, so I prize that," Bloch said.[48] Aiken was similarly open about sharing ideas. "Don't worry about people stealing an idea," he once told a student. "If it's original, you will have to ram it down their throats." Yet even he was struck, and a little discomforted, by von Neumann's cavalier attitude about parsing out who deserved credit for ideas. "He talked about concepts without worrying where they came from," Aiken said.[49]

The problem von Neumann faced at Harvard was that the Mark I, with its electromechanical switches, was excruciatingly slow. It would take months to get his atom bomb calculations done. Even though the paper tape input was useful in reprogramming the computer, it was necessary to switch tapes manually every time there was a call for a subroutine. Von Neumann became convinced that the only solution was to build a computer that worked at electronic speeds and could store and modify programs in an internal memory.

Thus he was primed to be part of the next big advance: the development of a stored-memory computer. So it was fortuitous that, in late August 1944, he had an encounter on the platform of the train station of the Aberdeen Proving Ground.

VON NEUMANN AT PENN

Captain Herman Goldstine, the Army liaison who was working with Mauchly and Eckert on the ENIAC, happened to be on the same platform at Aberdeen waiting for the train north. He had never met von Neumann, but he recognized him instantly. Goldstine tended to be star-struck by brilliant minds, so he was excited by what passed for a celebrity sighting in the world of mathematics. "It was therefore with considerable temerity that I approached this world-famous figure, introduced myself, and started talking," he recalled. "Fortunately for me von Neumann was a warm, friendly person who did his best to make people feel relaxed." The chat intensified when von Neumann discovered what Goldstine was doing. "When it became clear to von Neumann that I was concerned with the development of an electronic computer capable of 333 multiplications per second, the whole atmosphere of our conversation changed from one of relaxed good humor to one more like the oral examination for the doctor's degree in mathematics."[50]

At Goldstine's behest, von Neumann visited Penn a few days later to see ENIAC as it was being built. Presper Eckert was curious to meet the famed mathematician, and he had in mind a test to see if he was "really a genius": if his first question was about the logical structure of the machine. When that indeed was the first question von Neumann asked, he earned Eckert's respect.[51]

ENIAC could solve in less than an hour a partial differential equation that would take Harvard's Mark I close to eighty hours. That impressed von Neumann. However, reprogramming ENIAC to do different tasks could take hours, and von Neumann realized how severe a drawback that was when it came to tackling a bunch of dissimilar problems. Mauchly and Eckert had been wrestling throughout 1944 with ways to store programs inside the machine. The arrival of von Neumann, brimming with ideas from Harvard and Bell Labs and elsewhere, kicked into a higher orbit the thinking about stored-program computers.

Von Neumann, who became a consultant to the ENIAC team, pushed the idea that the computer program should be stored in the same memory as its data, so that the program could be easily modified as it was running. His work began the first week of September 1944, when Mauchly and Eckert explained the machine in detail and shared their thoughts about creating, in their next version, "one storage device with addressable locations" that would serve as a memory for both the data and programming instructions. As Goldstine put it in a letter to his Army superior that week, "We propose a centralized programming device in which the program routine is stored in coded form in the same type storage devices suggested above."[52]

Von Neumann's series of meetings with the ENIAC team, and in particular four formal sessions he held with them in the spring of 1945, took on such significance that minutes were taken under the title "Meetings with von Neumann." Pacing in front of a blackboard and ringleading the discussion with the engagement of a Socratic moderator, he absorbed ideas, refined them, and then wrote them on the board. "He would stand in front of the room like a professor, consulting with us," Jean Jennings recalled. "We would state to him a particular problem that we had, and we were always very careful that the questions represented fundamental problems that we were having and not just mechanical problems."[53]

Von Neumann was open but intellectually intimidating. When he made pronouncements, it was unusual for anyone to push back. But Jennings sometimes did. One day she disputed one of his points, and the men in the room stared at her incredulously. But von Neumann paused, tilted his head, and then accepted her point. Von Neumann could listen well, and he had also mastered the ingratiating art of feigning humility.[54] "He was an amazing combination of a very brilliant man who knows that he's brilliant, but at the same time is very modest and shy about presenting his ideas to other people," according to Jennings. "He was very restless and would march back and forth across the room, yet when he presented his ideas it was almost as though he were apologizing for disagreeing with you or thinking of a better idea."

Von Neumann was especially good at devising the fundamentals of computer programming, which was still an ill-defined craft that had advanced little in the century since Ada Lovelace wrote down the steps for getting the Analytical Engine to generate Bernoulli numbers. Creating an elegant instruction set, he realized, involved both rigorous logic and precise expression. "He was very thorough in explaining why we needed a particular instruction or why we could do without an instruction," Jennings recounted. "It was the first time that I had ever realized the importance of instruction codes, the logic behind them and the ingredients that a whole instruction set must have." It was a manifestation of his broader talent, which was to get to the essence of a new idea. "The thing that Von Neumann had, which I've noticed that other geniuses have, is the ability to pick out, in a particular problem, the one crucial thing that's important."[55]

Von Neumann realized that they were doing more than merely

improving the ENIAC so that it could be reprogram quickly. More significantly, they were fulfilling Ada's vision ing a machine that could perform any logical task on any set bols. "The stored-program computer, as conceived by Alan Tur delivered by John von Neumann, broke the distinction betwee bers that mean things and numbers that do things," George wrote. "Our universe would never be the same."[56]

In addition, von Neumann grasped, more readily than his co leagues, an important attribute of commingling data and program ming instructions in the same stored memory. The memory could be erasable, what we now call read-write memory. This meant that the stored program instructions could be changed not just at the end of a run but anytime the program was running. The computer could modify its own program based on the results it was getting. To facilitate this, von Neumann came up with a variable-address program language that enabled an easy switch to substitute instructions while the program was running.[57]

The team at Penn proposed to the Army that a new and improved ENIAC be built along these lines. It would be binary rather than decimal, use mercury delay lines for memory, and include much, though not all, of what became known as "von Neumann architecture." In the original proposal to the Army, this new machine was called the Electronic Discrete Variable Automatic *Calculator*. Increasingly, however, the team started referring to it as a *computer*, because it would do so much more than merely calculate. Not that it mattered. Everyone simply called it EDVAC.

Over the ensuing years, at patent trials and conferences, in books and dueling historical papers, there would be debates over who deserved the most credit for the ideas developed in 1944 and early 1945 that became part of the stored-program computer. The account above, for example, gives primary credit to Eckert and Mauchly for the stored-program concept and to von Neumann for realizing the importance of the computer's ability to modify its stored program as it ran and for creating a variable-address programming functionality to facilitate this. But more important than parsing provenance of ideas

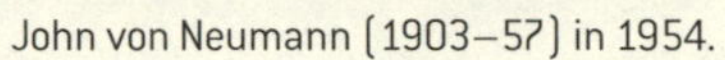

John von Neumann (1903–57) in 1954.

Herman Goldstine (1913–2004) circa 1944.

Presper Eckert (*center*) and CBS's Walter Cronkite (*right*) look at an election prediction from UNIVAC in 1952.

is to appreciate how the innovation at Penn was another example of collaborative creativity. Von Neumann, Eckert, Mauchly, Goldstine, Jennings, and many others batted around ideas collectively and elicited input from engineers, electronics experts, material scientists, and programmers.

Most of us have been involved in group brainstorming sessions that produced creative ideas. Even a few days later, there may be different recollections of who suggested what first, and we realize that the formation of ideas was shaped more by the iterative interplay within the group than by an individual tossing in a wholly original concept. The sparks come from ideas rubbing against each other rather than as bolts out of the blue. This was true at Bell Labs, Los Alamos, Bletchley Park, and Penn. One of von Neumann's great strengths was his talent—questioning, listening, gently floating tentative proposals, articulating, and collating—for being an impresario of such a collaborative creative process.

Von Neumann's propensity to collect and collate ideas, and his lack of concern for pinning down precisely where they came from, was useful in sowing and fertilizing the concepts that became part of EDVAC. But it did sometimes rankle those more concerned about getting credit—or even intellectual property rights—where due. He once proclaimed that it was not possible to attribute the origination of ideas discussed in a group. Upon hearing that, Eckert is said to have responded, "Really?"[58]

The benefits and drawbacks of von Neumann's approach became apparent in June 1945. After ten months of buzzing around the work being done at Penn, he offered to summarize their discussions on paper. And that is what he proceeded to do on a long train ride to Los Alamos.

In his handwritten report, which he mailed back to Goldstine at Penn, von Neumann described in mathematically dense detail the structure and logical control of the proposed stored-program computer and why it was "tempting to treat the entire memory as one organ." When Eckert questioned why von Neumann seemed to be preparing a paper based on the ideas that others had helped to develop, Goldstine reassured him: "He's just trying to get these things clear in his own mind and he's done it by writing me letters so that we can write back if he hasn't understood it properly."[59]

Von Neumann had left blank spaces for inserting references to other people's work, and his text never actually used the acronym EDVAC. But when Goldstine had the paper typed up (it ran to 101 pages), he ascribed sole authorship to his hero. The title page Goldstine composed called it "First Draft of a Report on the EDVAC, by John von Neumann." Goldstine used a mimeograph machine to produce twenty-four copies, which he distributed at the end of June 1945.[60]

The "Draft Report" was an immensely useful document, and it guided the development of subsequent computers for at least a decade. Von Neumann's decision to write it and allow Goldstine to distribute it reflected the openness of academic-oriented scientists, especially mathematicians, who tend to want to publish and disseminate rather than attempt to own intellectual property. "I certainly intend to do my part to keep as much of this field in the public domain (from the patent point of view) as I can," von Neumann explained to a colleague. He had two purposes in writing the report, he later said: "to contribute to clarifying and coordinating the thinking of the group working on the EDVAC" and "to further the development of the art of building high speed computers." He said that he was not trying to assert any ownership of the concepts, and he never applied for a patent on them.[61]

Eckert and Mauchly saw this differently. "You know, we finally regarded von Neumann as a huckster of other people's ideas with Goldstine as his principal mission salesman," Eckert later said. "Von Neumann was stealing ideas and trying to pretend work done at [Penn's] Moore School was work he had done."[62] Jean Jennings agreed, later lamenting that Goldstine "enthusiastically supported von Neumann's wrongful claims and essentially helped the man hijack the work of Eckert, Mauchly, and the others in the Moore School group."[63]

What especially upset Mauchly and Eckert, who tried to patent many of the concepts behind both ENIAC and then EDVAC, was that the distribution of von Neumann's report legally placed those concepts in the public domain. When Mauchly and Eckert tried to patent the architecture of a stored-program computer, they were stymied because (as both the Army's lawyers and the courts eventually ruled) von Neumann's report was deemed to be a "prior publication" of those ideas.

These patent disputes were the forerunner of a major issue of the digital era: Should intellectual property be shared freely and placed whenever possible into the public domain and open-source commons? That course, largely followed by the developers of the Internet and the Web, can spur innovation through the rapid dissemination and crowdsourced improvement of ideas. Or should intellectual property rights be protected and inventors allowed to profit from their proprietary ideas and innovations? That path, largely followed in the computer hardware, electronics, and semiconductor industries, can provide the financial incentives and capital investment that encourages innovation and rewards risks. In the seventy years since von Neumann effectively placed his "Draft Report" on the EDVAC into the public domain, the trend for computers has been, with a few notable exceptions, toward a more proprietary approach. In 2011 a milestone was reached: Apple and Google spent more on lawsuits and payments involving patents than they did on research and development of new products.[64]

THE PUBLIC UNVEILING OF ENIAC

Even as the team at Penn was designing EDVAC, they were still scrambling to get its predecessor, ENIAC, up and running. That occurred in the fall of 1945.

By then the war was over. There was no need to compute artillery trajectories, but ENIAC's first task nevertheless involved weaponry. The secret assignment came from Los Alamos, the atomic weapons lab in New Mexico, where the Hungarian-born theoretical physicist Edward Teller had devised a proposal for a hydrogen bomb, dubbed "the Super," in which a fission atomic device would be used to create a fusion reaction. To determine how this would work, the scientists needed to calculate what the force of the reactions would be at every ten-millionth of a second.

The nature of the problem was highly classified, but the mammoth equations were brought to Penn in October for ENIAC to crunch. It required almost a million punch cards to input the data, and Jennings was summoned to the ENIAC room with some of her colleagues so that Goldstine could direct the process of setting it up. ENIAC solved the equations, and in doing so showed that Teller's design was flawed. The mathematician and Polish refugee Stanislaw Ulam subsequently worked with Teller (and Klaus Fuchs, who turned out to be a Russian spy) to modify the hydrogen bomb concept, based on the ENIAC results, so that it could produce a massive thermonuclear reaction.[65]

Until such classified tasks were completed, ENIAC was kept under wraps. It was not shown to the public until February 15, 1946, when the Army and Penn scheduled a gala presentation with some press previews leading up to it.[66] Captain Goldstine decided that the centerpiece of the unveiling would be a demonstration of a missile trajectory calculation. So two weeks in advance, he invited Jean Jennings and Betty Snyder to his apartment and, as Adele served tea, asked them if they could program ENIAC to do this in time. "We sure could," Jennings pledged. She was excited. It would allow them to get their hands directly on the machine, which was rare.[67] They set to work plugging memory buses into the correct units and setting up program trays.

The men knew that the success of their demonstration was in the hands of these two women. Mauchly came by one Saturday with a bottle of apricot brandy to keep them fortified. "It was delicious," Jennings recalled. "From that day forward, I always kept a bottle of apricot brandy in my cupboard." A few days later, the dean of the engineering school brought them a paper bag containing a fifth of whiskey. "Keep up the good work," he told them. Snyder and Jennings were not big drinkers, but the gifts served their purpose. "It impressed us with the importance of this demonstration," said Jennings.[68]

The night before the demonstration was Valentine's Day, but despite their normally active social lives, Snyder and Jennings did not celebrate. "Instead, we were holed up with that wonderful machine, the ENIAC, busily making the last corrections and checks on the program," Jennings recounted. There was one stubborn glitch they couldn't figure out: the program did a wonderful job spewing out data on the trajectory of artillery shells, but it just didn't know when to stop. Even after the shell would have hit the ground, the program kept calculating its trajectory, "like a hypothetical shell burrowing through the ground at the same rate it had traveled through the air," as Jennings described it. "Unless we solved that problem, we knew the demonstration would be a dud, and the ENIAC's inventors and engineers would be embarrassed."[69]

Jennings and Snyder worked late into the evening before the press briefing trying to fix it, but they couldn't. They finally gave up at midnight, when Snyder needed to catch the last train to her suburban apartment. But after she went to bed, Snyder figured it out: "I woke up in the middle of the night thinking what that error was. . . . I came in, made a special trip on the early train that morning to look at a certain wire." The problem was that there was a setting at the end of a "do loop" that was one digit off. She flipped the requisite switch and the glitch was fixed. "Betty could do more logical reasoning while she was asleep than most people can do awake," Jennings later marveled. "While she slept, her subconscious untangled the knot that her conscious mind had been unable to."[70]

At the demonstration, ENIAC was able to spew out in fifteen seconds a set of missile trajectory calculations that would have taken human computers, even working with a Differential Analyzer, several weeks. It was all very dramatic. Mauchly and Eckert, like good innovators, knew how to put on a show. The tips of the vacuum tubes in the ENIAC accumulators, which were arranged in 10 x 10 grids, poked through holes in the machine's front panel. But the faint light from the neon bulbs, which served as indicator lights, was barely visible. So Eckert got Ping-Pong balls, cut them in half, wrote numbers on them, and placed them over the bulbs. As the computer began processing the data, the lights in the room were turned off so that the audience would be awed by the blinking Ping-Pong balls, a spectacle that became a staple of movies and TV shows. "As the trajectory was being calculated, numbers built up in the accumulators and were transferred from place to place, and the lights started flashing like the bulbs on the marquees in Las Vegas," said Jennings. "We had done what we set out to do. We had programmed the ENIAC."[71] That bears repeating: they had programmed the ENIAC.

The unveiling of ENIAC made the front page of the *New York Times* under the headline "Electronic Computer Flashes Answers, May Speed Engineering." The story began, "One of the war's top secrets, an amazing machine which applies electronic speeds for the first time to mathematical tasks hitherto too difficult and cumbersome for solution, was announced here tonight by the War Department."[72] The report continued inside the *Times* for a full page, with pictures of Mauchly, Eckert, and the room-size ENIAC. Mauchly proclaimed that the machine would lead to better weather predictions (his original passion), airplane design, and "projectiles operating at supersonic speeds." The Associated Press story reported an even grander vision, declaring, "The robot opened the mathematical way to better living for every man."[73] As an example of "better living," Mauchly asserted that computers might one day serve to lower the cost of a loaf of bread. How that would happen he did not explain, but it and millions of other such ramifications did in fact eventually transpire.

Later Jennings complained, in the tradition of Ada Lovelace, that many of the newspaper reports overstated what ENIAC could do by calling it a "giant brain" and implying that it could think. "The ENIAC wasn't a brain in any sense," she insisted. "It couldn't reason, as computers still cannot reason, but it could give people more data to use in reasoning."

Jennings had another complaint that was more personal: "Betty and I were ignored and forgotten following the demonstration. We felt as if we had been playing parts in a fascinating movie that suddenly took a bad turn, in which we had worked like dogs for two weeks to produce something really spectacular and then were written out of the script." That night there was a candle-lit dinner at Penn's venerable Houston Hall. It was filled with scientific luminaries, military brass, and most of the men who had worked on ENIAC. But Jean Jennings and Betty Snyder were not there, nor were any of the other women programmers.[74] "Betty and I weren't invited," Jennings said, "so we were sort of horrified."[75] While the men and various dignitaries celebrated, Jennings and Snyder made their way home alone through a very cold February night.

THE FIRST STORED-PROGRAM COMPUTERS

The desire of Mauchly and Eckert to patent—and profit from—what they had helped to invent caused problems at Penn, which did not yet have a clear policy for divvying up intellectual property rights. They were allowed to apply for patents on ENIAC, but the university then insisted on getting royalty-free licenses as well as the right to sublicense all aspects of the design. Furthermore, the parties couldn't agree on who would have rights to the innovations on EDVAC. The wrangling was complex, but the upshot was that Mauchly and Eckert left Penn at the end of March 1946.[76]

They formed what became the Eckert-Mauchly Computer Corporation, based in Philadelphia, and were pioneers in turning computing from an academic to a commercial endeavor. (In 1950 their company, along with the patents they would be granted, became part of Remington Rand, which morphed into Sperry Rand and then Unisys.) Among the machines they built was UNIVAC, which was purchased by the Census Bureau and other clients, including General Electric.

With its flashing lights and Hollywood aura, UNIVAC became famous when CBS featured it on election night in 1952. Walter Cronkite, the young anchor of the network's coverage, was dubious that the huge machine would be much use compared to the expertise of the network's correspondents, but he agreed that it might provide an amusing spectacle for viewers. Mauchly and Eckert enlisted a Penn statistician, and they worked out a program that compared the early results from some sample precincts to the outcomes in previous elections. By 8:30 p.m. on the East Coast, well before most of the nation's polls had closed, UNIVAC predicted, with 100-to-1 certainty, an easy win for Dwight Eisenhower over Adlai Stevenson. CBS initially withheld UNIVAC's verdict; Cronkite told his audience that the computer had not yet reached a conclusion. Later that night, though, after the vote counting confirmed that Eisenhower had won handily, Cronkite put the correspondent Charles Collingwood on the air to admit that UNIVAC had made the prediction at the beginning of the evening but CBS had not aired it. UNIVAC became a celebrity and a fixture on future election nights.[77]

Eckert and Mauchly did not forget the importance of the women programmers who had worked with them at Penn, even though they had not been invited to the dedication dinner for ENIAC. They hired Betty Snyder, who, under her married name, Betty Holberton, went on to become a pioneer programmer who helped develop the COBOL and Fortran languages, and Jean Jennings, who married an engineer and became Jean Jennings Bartik. Mauchly also wanted to recruit Kay McNulty, but after his wife died in a drowning accident he proposed marriage to her instead. They had five children, and she continued to help on software design for UNIVAC.

Mauchly also hired the dean of them all, Grace Hopper. "He let people try things," Hopper replied when asked why she let him talk her into joining the Eckert-Mauchly Computer Corporation. "He encouraged innovation."[78] By 1952 she had created the world's first workable compiler, known as the A-0 system, which translated symbolic mathematical code into machine language and thus made it easier for ordinary folks to write programs.

Like a salty crew member, Hopper valued an all-hands-on-deck style of collaboration, and she helped develop the open-source method of innovation by sending out her initial versions of the compiler to her friends and acquaintances in the programming world and asking them to make improvements. She used the same open development process when she served as the technical lead in coordinating the creation of COBOL, the first cross-platform standardized business language for computers.[79] Her instinct that programming should be machine-independent was a reflection of her preference for collegiality; even machines, she felt, should work well together. It also showed her early understanding of a defining fact of the computer age: that hardware would become commoditized and that *program-*

ming would be where the true value resided. Until Bill Gates came along, it was an insight that eluded most of the men.*

Von Neumann was disdainful of the Eckert-Mauchly mercenary approach. "Eckert and Mauchly are a commercial group with a commercial patent policy," he complained to a friend. "We cannot work with them directly or indirectly in the same open manner in which we would work with an academic group."[80] But for all of his righteousness, von Neumann was not above making money off his ideas. In 1945 he negotiated a personal consulting contract with IBM, giving the company rights to any inventions he made. It was a perfectly valid arrangement. Nevertheless, it outraged Eckert and Mauchly. "He sold all our ideas through the back door to IBM," Eckert complained. "He spoke with a forked tongue. He said one thing and did something else. He was not to be trusted."[81]

After Mauchly and Eckert left, Penn rapidly lost its role as a center of innovation. Von Neumann also left, to return to the Institute for Advanced Study in Princeton. He took with him Herman and Adele Goldstine, along with key engineers such as Arthur Burks. "Perhaps institutions as well as people can become fatigued," Herman Goldstine later reflected on the demise of Penn as the epicenter of computer development.[82] Computers were considered a tool, not a subject for scholarly study. Few of the faculty realized that computer science would grow into an academic discipline even more important than electrical engineering.

Despite the exodus, Penn was able to play one more critical role in the development of computers. In July 1946 most of the experts in the field—including von Neumann, Goldstine, Eckert, Mauchly, and others who had been feuding—returned for a series of talks and seminars, called the Moore School Lectures, that would disseminate their

---

*In 1967, at age sixty, Hopper was recalled to active duty in the Navy with the mission of standardizing its use of COBOL and validating COBOL compilers. By vote of Congress, she was permitted to extend her tour beyond retirement age. She attained the rank of rear admiral, and finally retired in August 1986 at age seventy-nine as the Navy's oldest serving officer.

knowledge about computing. The eight-week series attracted Howard Aiken, George Stibitz, Douglas Hartree of Manchester University, and Maurice Wilkes of Cambridge. A primary focus was the importance of using stored-program architecture if computers were to fulfill Turing's vision of being universal machines. As a result, the design ideas developed collaboratively by Mauchly, Eckert, von Neumann, and others at Penn became the foundation for most future computers.

The distinction of being the first stored-program computers went to two machines that were completed, almost simultaneously, in the summer of 1948. One of them was an update of the original ENIAC. Von Neumann and Goldstine, along with the engineers Nick Metropolis and Richard Clippinger, worked out a way to use three of ENIAC's function tables to store a rudimentary set of instructions.[83] Those function tables had been used to store data about the drag on an artillery shell, but that memory space could be used for other purposes since the machine was no longer being used to calculate trajectory tables. Once again, the actual programming work was done largely by the women: Adele Goldstine, Klára von Neumann, and Jean Jennings Bartik. "I worked again with Adele when we developed, along with others, the original version of the code required to turn ENIAC into a stored-program computer using the function tables to store the coded instructions," Bartik recalled.[84]

This reconfigured ENIAC, which became operational in April 1948, had a read-only memory, which meant that it was hard to modify programs while they were running. In addition, its mercury delay line memory was sluggish and required precision engineering. Both of these drawbacks were avoided in a small machine at Manchester University in England that was built from scratch to function as a stored-program computer. Dubbed "the Manchester Baby," it became operational in June 1948.

Manchester's computing lab was run by Max Newman, Turing's mentor, and the primary work on the new computer was done by Frederic Calland Williams and Thomas Kilburn. Williams invented a storage mechanism using cathode-ray tubes, which made the machine faster and simpler than ones using mercury delay lines. It worked so well that it led to the more powerful Manchester Mark I, which became operational in April 1949, as well as the EDSAC, completed by Maurice Wilkes and a team at Cambridge that May.[85]

As these machines were being developed, Turing was also trying to develop a stored-program computer. After leaving Bletchley Park, he joined the National Physical Laboratory, a prestigious institute in London, where he designed a computer named the Automatic Computing Engine in homage to Babbage's two engines. But progress on ACE was fitful. By 1948 Turing was fed up with the pace and frustrated that his colleagues had no interest in pushing the bounds of machine learning and artificial intelligence, so he left to join Max Newman at Manchester.[86]

Likewise, von Neumann embarked on developing a stored-program computer as soon as he settled at the Institute for Advanced Study in Princeton in 1946, an endeavor chronicled in George Dyson's *Turing's Cathedral*. The Institute's director, Frank Aydelotte, and its most influential faculty trustee, Oswald Veblen, were staunch supporters of what became known as the IAS Machine, fending off criticism from other faculty that building a computing machine would demean the mission of what was supposed to be a haven for theoretical thinking. "He clearly stunned, or even horrified, some of his mathematical colleagues of the most erudite abstraction, by openly professing his great interest in other mathematical tools than the blackboard and chalk or pencil and paper," von Neumann's wife, Klára, recalled. "His proposal to build an electronic computing machine under the sacred dome of the Institute was not received with applause to say the least."[87]

Von Neumann's team members were stashed in an area that would have been used by the logician Kurt Gödel's secretary, except he didn't want one. Throughout 1946 they published detailed papers about their design, which they sent to the Library of Congress and the U.S. Patent Office, not with applications for patents but with affidavits saying they wanted the work to be in the public domain.

Their machine became fully operational in 1952, but it was slowly abandoned after von Neumann left for Washington to join the Atomic Energy Commission. "The demise of our computer group was a disaster not only for Princeton but for science as a whole," said the physicist Freeman Dyson, a member of the Institute (and George Dyson's father). "It meant that there did not exist at that critical period in the 1950s an academic center where computer people of all kinds could get together at the highest intellectual level."[88] Instead, beginning in the 1950s, innovation in computing shifted to the corporate realm, led by companies such as Ferranti, IBM, Remington Rand, and Honeywell.

That shift takes us back to the issue of patent protections. If von Neumann and his team had continued to pioneer innovations and put them in the public domain, would such an open-source model of development have led to faster improvements in computers? Or did marketplace competition and the financial rewards for creating intellectual property do more to spur innovation? In the cases of the Internet, the Web, and some forms of software, the open model would turn out to work better. But when it came to hardware, such as computers and microchips, a proprietary system provided incentives for a spurt of innovation in the 1950s. The reason the proprietary approach worked well, especially for computers, was that large industrial organizations, which needed to raise working capital, were best at handling the research, development, manufacturing, and marketing for such machines. In addition, until the mid-1990s, patent protection was easier to obtain for hardware than it was for software.* However, there was a downside to the patent protection given to hardware innovation: the proprietary model produced companies that were so entrenched and defensive that they would miss out on the personal computer revolution in the early 1970s.

---

*The U.S. Constitution empowers Congress "to promote the progress of science and useful arts by securing for limited times to authors and inventors the exclusive Right to their respective writings and discoveries." The U.S. Patent and Trademark Office throughout the 1970s generally would not grant patents to innovations whose only departure from existing technology was the use of a new software algorithm. That became murky in the 1980s with conflicting appeals court and Supreme Court rulings. Policies changed in the mid-1990s, when the DC Circuit Court issued a series of rulings permitting patents for software that produces a "useful, concrete and tangible result" and President Bill Clinton appointed as head of the Patent Office a person who had been the chief lobbyist for the Software Publishing Industry.

CAN MACHINES THINK?

As he thought about the development of stored-program computers, Alan Turing turned his attention to the assertion that Ada Lovelace had made a century earlier, in her final "Note" on Babbage's Analytical Engine: that machines could not really *think*. If a machine could modify its own program based on the information it processed, Turing asked, wouldn't that be a form of learning? Might that lead to artificial intelligence?

The issues surrounding artificial intelligence go back to the ancients. So do the related questions involving human consciousness. As with most questions of this sort, Descartes was instrumental in framing them in modern terms. In his 1637 *Discourse on the Method*, which contains his famous assertion "I think, therefore I am," Descartes wrote:

> If there were machines that bore a resemblance to our bodies and imitated our actions as closely as possible for all practical purposes, we should still have two very certain means of recognizing that they were not real humans. The first is that . . . it is not conceivable that such a machine should produce arrangements of words so as to give an appropriately meaningful answer to whatever is said in its presence, as the dullest of men can do. Secondly, even though some machines might do some things as well as we do them, or perhaps even better, they would inevitably fail in others, which would reveal that they are acting not from understanding.

Turing had long been interested in the way computers might replicate the workings of a human brain, and this curiosity was furthered by his work on machines that deciphered coded language. In early 1943, as Colossus was being designed at Bletchley Park, Turing sailed across the Atlantic on a mission to Bell Laboratories in lower Manhattan, where he consulted with the group working on electronic speech encipherment, the technology that could electronically scramble and unscramble telephone conversations.

There he met the colorful genius Claude Shannon, the former

MIT graduate student who wrote the seminal master's thesis in 1937 that showed how Boolean algebra, which rendered logical propositions into equations, could be performed by electronic circuits. Shannon and Turing began meeting for tea and long conversations in the afternoons. Both were interested in brain science, and they realized that their 1937 papers had something fundamental in common: they showed how a machine, operating with simple binary instructions, could tackle not only math problems but all of logic. And since logic was the basis for how human brains reasoned, then a machine could, in theory, replicate human intelligence.

"Shannon wants to feed not just data to [a machine], but cultural things!" Turing told Bell Lab colleagues at lunch one day. "He wants to play music to it!" At another lunch in the Bell Labs dining room, Turing held forth in his high-pitched voice, audible to all the executives in the room: "No, I'm not interested in developing a powerful brain. All I'm after is just a mediocre brain, something like the President of the American Telephone and Telegraph Company."[89]

When Turing returned to Bletchley Park in April 1943, he became friends with a colleague named Donald Michie, and they spent many evenings playing chess in a nearby pub. As they discussed the possibility of creating a chess-playing computer, Turing approached the problem not by thinking of ways to use brute processing power to calculate every possible move; instead he focused on the possibility that a machine might *learn* how to play chess by repeated practice. In other words, it might be able to try new gambits and refine its strategy with every new win or loss. This approach, if successful, would represent a fundamental leap that would have dazzled Ada Lovelace: machines would be able to do more than merely follow the specific instructions given them by humans; they could learn from experience and refine their own instructions.

"It has been said that computing machines can only carry out the purposes that they are instructed to do," he explained in a talk to the London Mathematical Society in February 1947. "But is it necessary that they should always be used in such a manner?" He then discussed the implications of the new stored-program computers that could modify their own instruction tables. "It would be like a pupil who had learnt much from his master, but had added much more by his own work. When this happens I feel that one is obliged to regard the machine as showing intelligence."[90]

When he finished his speech, his audience sat for a moment in silence, stunned by Turing's claims. Likewise, his colleagues at the National Physical Laboratory were flummoxed by Turing's obsession with making thinking machines. The director of the National Physical Laboratory, Sir Charles Darwin (grandson of the evolutionary biologist), wrote to his superiors in 1947 that Turing "wants to extend his work on the machine still further towards the biological side" and to address the question "Could a machine be made that could learn by experience?"[91]

Turing's unsettling notion that machines might someday be able to think like humans provoked furious objections at the time—as it has ever since. There were the expected religious objections and also those that were emotional, both in content and in tone. "Not until a machine can write a sonnet or compose a concerto because of thoughts and emotions felt, and not by the chance fall of symbols, could we agree that machine equals brain," declared a famous brain surgeon, Sir Geoffrey Jefferson, in the prestigious Lister Oration in 1949.[92] Turing's response to a reporter from the London *Times* seemed somewhat flippant, but also subtle: "The comparison is perhaps a little bit unfair because a sonnet written by a machine will be better appreciated by another machine."[93]

The ground was thus laid for Turing's second seminal work, "Computing Machinery and Intelligence," published in the journal *Mind* in October 1950.[94] In it he devised what became known as the Turing Test. He began with a clear declaration: "I propose to consider the question, 'Can machines think?'" With a schoolboy's sense of fun, he then invented a game—one that is still being played and debated—to give empirical meaning to that question. He proposed a purely operational definition of artificial intelligence: If the output of a machine is indistinguishable from that of a human brain, then we have no meaningful reason to insist that the machine is not "thinking."

Turing's test, which he called "the imitation game," is simple: An interrogator sends written questions to a human and a machine in another room and tries to determine from their answers which one is the human. A sample interrogation, he wrote, might be the following:

> Q: Please write me a sonnet on the subject of the Forth Bridge.
> A: Count me out on this one. I never could write poetry.
>
> Q: Add 34957 to 70764.
> A: (Pause about 30 seconds and then give as answer) 105621.
>
> Q: Do you play chess?
> A: Yes.
>
> Q: I have K at my K1, and no other pieces. You have only K at K6 and R at R1. It is your move. What do you play?
> A: (After a pause of 15 seconds) R–R8 mate.

In this sample dialogue, Turing did a few things. Careful scrutiny shows that the respondent, after thirty seconds, made a slight mistake in addition (the correct answer is 105,721). Is that evidence that the respondent was a human? Perhaps. But then again, maybe it was a machine cagily pretending to be human. Turing also flicked away Jefferson's objection that a machine cannot write a sonnet; perhaps the answer above was given by a human who admitted to that inability. Later in the paper, Turing imagined the following interrogation to show the difficulty of using sonnet writing as a criterion of being human:

> Q: In the first line of your sonnet which reads "Shall I compare thee to a summer's day," would not "a spring day" do as well or better?
> A: It wouldn't scan.
>
> Q: How about "a winter's day." That would scan all right.
> A: Yes, but nobody wants to be compared to a winter's day.
>
> Q: Would you say Mr. Pickwick reminded you of Christmas?
> A: In a way.

Q: Yet Christmas is a winter's day, and I do not think Mr. Pickwick would mind the comparison.

A: I don't think you're serious. By a winter's day one means a typical winter's day, rather than a special one like Christmas.

Turing's point was that it might not be possible to tell whether such a respondent was a human or a machine pretending to be a human.

Turing gave his own guess as to whether a computer might be able to win this imitation game: "I believe that in about fifty years' time it will be possible to programme computers . . . to make them play the imitation game so well that an average interrogator will not have more than 70 percent chance of making the right identification after five minutes of questioning."

In his paper Turing tried to rebut the many possible challenges to his definition of thinking. He swatted away the theological objection that God has bestowed a soul and thinking capacity only upon humans, arguing that this "implies a serious restriction of the omnipotence of the Almighty." He asked whether God "has freedom to confer a soul on an elephant if He sees fit." Presumably so. By the same logic, which, coming from the nonbelieving Turing was somewhat sardonic, surely God could confer a soul upon a machine if He so desired.

The most interesting objection, especially for our narrative, is the one that Turing attributed to Ada Lovelace. "The Analytical Engine has no pretensions whatever to *originate* anything," she wrote in 1843. "It can do whatever we know how to order it to perform. It can follow analysis; but it has no power of anticipating any analytical relations or truths." In other words, unlike the human mind, a mechanical contrivance cannot have free will or come up with its own initiatives. It can merely perform as programmed. In his 1950 paper, Turing devoted a section to what he dubbed "Lady Lovelace's Objection."

His most ingenious parry to this objection was his argument that a machine might actually be able to *learn*, thereby growing into its own agent and able to originate new thoughts. "Instead of trying to produce a programme to simulate the adult mind, why not rather try to produce one which simulates the child's?" he asked. "If this were then subjected to an appropriate course of education, one would obtain the adult brain." A machine's learning process would be different from a child's, he admitted. "It will not, for instance, be provided with legs, so that it could not be asked to go out and fill the coal scuttle. Possibly it might not have eyes.... One could not send the creature to school without the other children making excessive fun of it." The baby machine would therefore have to be tutored some other way. Turing proposed a punishment and reward system, which would cause the machine to repeat certain activities and avoid others. Eventually such a machine could develop its own conceptions about how to figure things out.

But even if a machine could mimic thinking, Turing's critics objected, it would not *really* be conscious. When the human player of the Turing Test uses words, he associates those words with real-world meanings, emotions, experiences, sensations, and perceptions. Machines don't. Without such connections, language is just a game divorced from meaning.

This objection led to the most enduring challenge to the Turing Test, which was in a 1980 essay by the philosopher John Searle. He proposed a thought experiment, called the Chinese Room, in which an English speaker with no knowledge of Chinese is given a comprehensive set of rules instructing him on how to respond to any combination of Chinese characters by handing back a specified new combination of Chinese characters. Given a good enough instruction manual, the person might convince an interrogator that he was a real speaker of Chinese. Nevertheless, he would not have understood a single response that he made, nor would he have exhibited any intentionality. In Ada Lovelace's words, he would have no pretensions whatever to originate anything but instead would merely do whatever actions he was ordered to perform. Similarly, the machine in Turing's imitation game, no matter how well it could mimic a human being, would have no understanding or consciousness of what it was saying. It makes no more sense to say that the machine "thinks" than it does to say that the fellow following the massive instruction manual understands Chinese.[95]

One response to the Searle objection is to argue that, even if the man does not really understand Chinese, the entire system incorporated in the room—the man (processing unit), instruction manual (program), and files full of Chinese characters (the data)—as a whole might indeed understand Chinese. There's no conclusive answer. Indeed, the Turing Test and the objections to it remain to this day the most debated topic in cognitive science.

For a few years after he wrote "Computing Machinery and Intelligence," Turing seemed to enjoy engaging in the fray that he provoked. With wry humor, he poked at the pretensions of those who prattled on about sonnets and exalted consciousness. "One day ladies will take their computers for walks in the park and tell each other 'My little computer said such a funny thing this morning!'" he japed in 1951. As his mentor Max Newman later noted, "His comical but brilliantly apt analogies with which he explained his ideas made him a delightful companion."[96]

One topic that came up repeatedly in discussions with Turing, and would soon have a sad resonance, was the role that sexual appetites and emotional desires play in human thinking, unlike in machines. A very public example occurred in a January 1952 televised BBC debate that Turing had with the brain surgeon Sir Geoffrey Jefferson, moderated by Max Newman and the philosopher of science Richard Braithwaite. "A human's interests are determined, by and large, by his appetites, desires, drives, instincts," said Braithwaite, who argued that to create a true thinking machine, "it would seem to be necessary to equip the machine with something corresponding to a set of appetites." Newman chimed in that machines "have rather restricted appetites, and they can't blush when they're embarrassed." Jefferson went even further, repeatedly using "sexual urges" as an example and referring to a human's "emotions and instincts, such as those to do with sex." Man is prey to "sexual urges," he said, and "may make a fool of himself." He spoke so much about how sexual appetites affected human thinking that the BBC editors cut some of it out of the broadcast, including his assertion that he would not believe a machine could think until he saw it touch the leg of a female machine.[97]

Turing, who was still rather discreet about being a homosexual, fell quiet during this part of the discussion. During the weeks leading up to the recording of the broadcast on January 10, 1952, he was engaged in a series of actions that were so very human that a machine would have found them incomprehensible. He had just finished a scientific paper, and he followed it by writing a short story about how he planned to celebrate: "It was quite some time now since he had 'had' anyone, in fact not since he had met that soldier in Paris last summer. Now that his paper was finished he might justifiably consider that he had earned another gay man, and he knew where he might find one who might be suitable."[98]

On Oxford Street in Manchester, he picked up a nineteen-year-old working-class drifter named Arnold Murray and began a relationship. When he returned from taping the BBC show, he invited Murray to move in. One night Turing told young Murray of his fantasy of playing chess against a nefarious computer that he was able to beat by causing it to show anger, then pleasure, then smugness. The relationship became more complex in the ensuing days, until Turing returned home one evening and found that his house had been burglarized. The culprit was a friend of Murray's. When Turing reported the incident to the police, he ended up disclosing to them his sexual relationship with Murray, and they arrested Turing for "gross indecency."[99]

At the trial in March 1952, Turing pled guilty, though he made clear he felt no remorse. Max Newman appeared as a character witness. Convicted and stripped of his security clearance,* Turing was offered a choice: imprisonment or probation contingent on receiving hormone treatments via injections of a synthetic estrogen designed to curb his sexual desires, as if he were a chemically controlled machine. He chose the latter, which he endured for a year.

Turing at first seemed to take it all in stride, but on June 7, 1954, he committed suicide by biting into an apple he had laced with cyanide. His friends noted that he had always been fascinated by the scene in *Snow White* in which the Wicked Queen dips an apple into a poisonous brew. He was found in his bed with froth around his mouth, cyanide in his system, and a half-eaten apple by his side.

Was that something a machine would have done?

---

*At Christmas 2013 Turing was posthumously granted a formal pardon by Queen Elizabeth II.

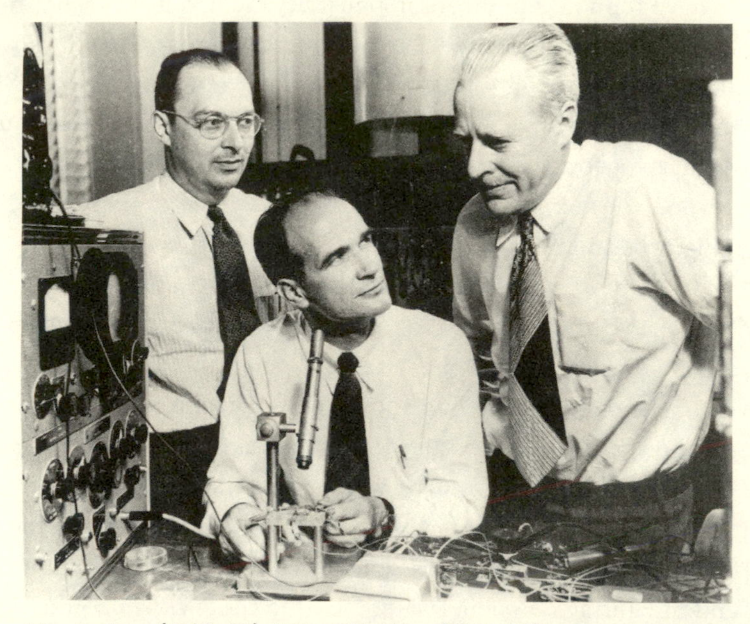

John Bardeen (1908–91), William Shockley (1910–89), and Walter Brattain (1902–87) in a Bell Labs photograph in 1948.

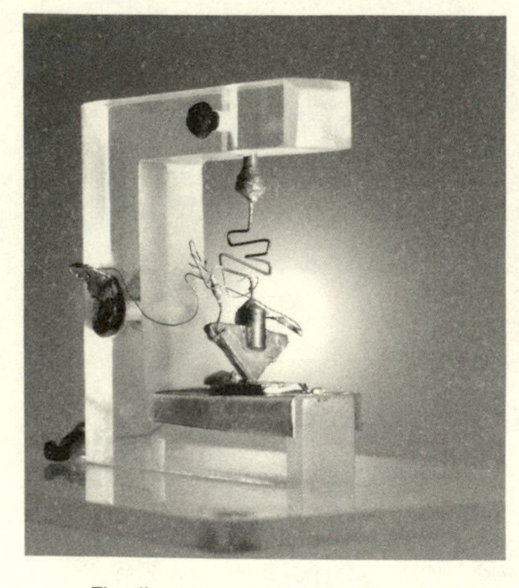

The first transistor at Bell Labs.

William Shockley (*at head of table*) the day he won the Nobel Prize being toasted by colleagues, including Gordon Moore (*seated left*) and Robert Noyce (*standing center with wine glass*) in 1956.

CHAPTER FOUR

THE TRANSISTOR

The invention of computers did not immediately launch a revolution. Because they relied on large, expensive, fragile vacuum tubes that consumed a lot of power, the first computers were costly behemoths that only corporations, research universities, and the military could afford. Instead the true birth of the digital age, the era in which electronic devices became embedded in every aspect of our lives, occurred in Murray Hill, New Jersey, shortly after lunchtime on Tuesday, December 16, 1947. That day two scientists at Bell Labs succeeded in putting together a tiny contraption they had concocted from some strips of gold foil, a chip of semiconducting material, and a bent paper clip. When wiggled just right, it could amplify an electric current and switch it on and off. The transistor, as the device was soon named, became to the digital age what the steam engine was to the Industrial Revolution.

The advent of transistors, and the subsequent innovations that allowed millions of them to be etched onto tiny microchips, meant that the processing power of many thousands of ENIACs could be nestled inside the nose cone of rocket ships, in computers that could sit on your lap, in calculators and music players that could fit in your pocket, and in handheld devices that could exchange information or entertainment with any nook or node of a networked planet.

Three passionate and intense colleagues, whose personalities both complemented and conflicted with one another, would go down in history as the inventors of the transistor: a deft experimentalist named Walter Brattain, a quantum theorist named John Bardeen, and the most passionate and intense of them all—tragically so by the end—a solid-state physics expert named William Shockley.

But there was another player in this drama that was actually as important as any individual: Bell Labs, where these men worked. What made the transistor possible was a mixture of diverse talents rather than just the imaginative leaps of a few geniuses. By its nature, the transistor required a team that threw together theorists who had an intuitive feel for quantum phenomena with material scientists who were adroit at baking impurities into batches of silicon, along with dexterous experimentalists, industrial chemists, manufacturing specialists, and ingenious tinkerers.

BELL LABS

In 1907 the American Telephone and Telegraph Company faced a crisis. The patents of its founder, Alexander Graham Bell, had expired, and it seemed in danger of losing its near-monopoly on phone services. Its board summoned back a retired president, Theodore Vail, who decided to reinvigorate the company by committing to a bold goal: building a system that could connect a call between New York and San Francisco. The challenge required combining feats of engineering with leaps of pure science. Making use of vacuum tubes and other new technologies, AT&T built repeaters and amplifying devices that accomplished the task in January 1915. On the historic first transcontinental call, in addition to Vail and President Woodrow Wilson, was Bell himself, who echoed his famous words from thirty-nine years earlier, "Mr. Watson, come here, I want to see you." This time his former assistant Thomas Watson, who was in San Francisco, replied, "It would take me a week."[1]

Thus was the seed planted for a new industrial organization that became known as Bell Labs. Originally located on the western edge of Manhattan's Greenwich Village overlooking the Hudson River, it brought together theoreticians, materials scientists, metallurgists, engineers, and even AT&T pole climbers. It was where George Stibitz developed a computer using electromagnetic relays and Claude Shannon worked on information theory. Like Xerox PARC and other corporate research satellites that followed, Bell Labs showed how sustained innovation could occur when people with a variety of talents were brought together, preferably in close physical proximity where they could have frequent meetings and serendipitous encounters. That was the upside. The downside was that these were big bureaucracies under corporate thumbs; Bell Labs, like Xerox PARC, showed the limits of industrial organizations when they don't have passionate leaders and rebels who can turn innovations into great products.

The head of Bell Labs' vacuum-tube department was a high-octane Missourian named Mervin Kelly, who had studied to be a metallurgist at the Missouri School of Mines and then got a PhD in physics under Robert Millikan at the University of Chicago. He was able to make vacuum tubes more reliable by devising a water-cooling system, but he realized that tubes would never be an efficient method of amplification or switching. In 1936 he was promoted to research director of Bell Labs, and his first priority was to find an alternative.

Kelly's great insight was that Bell Labs, which had been a bastion of practical engineering, should also focus on basic science and theoretical research, until then the domain of universities. He began a search for the country's brightest young physics PhDs. His mission was to make innovation something that an industrial organization could do on a regular basis rather than ceding that territory to eccentric geniuses holed up in garages and garrets.

"It had become a matter of some consideration at the Labs whether the key to invention was a matter of individual genius or collaboration," Jon Gertner wrote in *The Idea Factory*, a study of Bell Labs.[2] The answer was both. "It takes many men in many fields of science, pooling their various talents, to funnel all the necessary research into the development of one new device," Shockley later explained.[3] He was right. He was also, however, showing a rare flash of feigned humility. More than anyone, he believed in the importance of the individual genius, such as himself. Even Kelly, the proselytizer for collaboration, realized that individual genius also needed to be nurtured. "With all the needed emphasis on leadership, organization and teamwork, the individual has remained supreme—of paramount importance," he once said. "It is in the mind of a single person that creative ideas and concepts are born."[4]

The key to innovation—at Bell Labs and in the digital age in general—was realizing that there was no conflict between nurturing individual geniuses and promoting collaborative teamwork. It was not either-or. Indeed, throughout the digital age, the two approaches went together. Creative geniuses (John Mauchly, William Shockley, Steve Jobs) generated innovative ideas. Practical engineers (Presper Eckert, Walter Brattain, Steve Wozniak) partnered closely with them to turn concepts into contraptions. And collaborative teams of technicians and entrepreneurs worked to turn the invention into a practical product. When part of this ecosystem was lacking, such as for John Atanasoff at Iowa State or Charles Babbage in the shed behind his London home, great concepts ended up being consigned to history's basement. And when great teams lacked passionate visionaries, such as Penn after Mauchly and Eckert left, Princeton after von Neumann, or Bell Labs after Shockley, innovation slowly withered.

The need to combine theorists with engineers was particularly true in a field that was becoming increasingly important at Bell Labs: solid-state physics, which studied how electrons flow through solid materials. In the 1930s, Bell Labs engineers were tinkering with materials such as silicon—after oxygen the most common element in the earth's crust and a key component of sand—in order to juice them into performing electronic tricks. At the same time in the same building, Bell theorists were wrestling with the mind-bending discoveries of quantum mechanics.

Quantum mechanics is based on theories developed by the Danish physicist Niels Bohr and others about what goes on inside an atom. In 1913 Bohr had come up with a model of atomic structure in which electrons orbited around a nucleus at specific levels. They could make a quantum leap from one level to the next, but never be in between. The number of electrons in the outer orbital level helped to determine the chemical and electronic properties of the element, including how well it conducted electricity.

Some elements, such as copper, are good conductors of electricity. Others, such as sulfur, are horrible conductors, and are thus good insulators. And then there are those in between, such as silicon and germanium, which are known as semiconductors. What makes them useful is that they are easy to manipulate into becoming better conductors. For example, if you contaminate silicon with a tiny amount of arsenic or boron, its electrons become more free to move.

The advances in quantum theory came at the same time that metallurgists at Bell Labs were finding ways to create new materials using novel purification techniques, chemical tricks, and recipes for combining rare and ordinary minerals. In seeking to solve some everyday problems, like vacuum-tube filaments that burned out too quickly or telephone-speaker diaphragms that sounded too tinny, they were mixing new alloys and developing methods to heat or cool concoctions until they performed better. By trial and error, like cooks in a kitchen, they were creating a revolution in materials science that would go hand in hand with the theoretical revolution that was occurring in quantum mechanics.

As they experimented with their samples of silicon and germanium, the chemical engineers at Bell Labs stumbled across evidence for much of what the theorists were conjecturing.* It became clear that there was a lot that the theorists, engineers, and metallurgists could learn from one another. So in 1936 a solid-state study group was formed at Bell Labs that included a potent mix of practical and theoretical stars. It met once a week in the late afternoon to share findings, engage in a bit of academic-style trash talk, and then adjourn for informal discussions that lasted late into the night. There

---

*For example, the engineers and theorists discovered that silicon (which has four electrons in its outer orbit) that was doped with phosphorus or arsenic (which have five electrons in their outer orbits) had spare electrons and thus was a negative-charged carrier. The result was called an n-type semiconductor. Silicon that was doped with boron (with three electrons in its outer orbit) had a deficit of electrons—there were "holes" where some electrons would normally be—and thus was positively charged, making it known as a p-type semiconductor.

was value to getting together in person rather than just reading each other's papers: the intense interactions allowed ideas to be kicked into higher orbits and, like electrons, occasionally break loose to spark chain reactions.

Of all the people in the group, one stood out. William Shockley, a theorist who had arrived at Bell Labs right when the study group was being formed, impressed the others, and sometimes frightened them, with both his intellect and his intensity.

WILLIAM SHOCKLEY

William Shockley grew up with a love of both art and science. His father studied mine engineering at MIT, took music courses in New York, and learned seven languages as he wandered through Europe and Asia as an adventurer and mineral speculator. His mother majored in both math and art at Stanford and was one of the first known climbers to succeed in a solo ascent of Mt. Whitney. They met in a tiny Nevada mining village, Tonopah, where he was staking claims and she had gone to do surveying work. After they were married, they moved to London, where their son was born in 1910.

William would be their only child, and for that they were thankful. Even as a baby he had a ferocious temper, with fits of rage so loud and long that his parents kept losing babysitters and apartments. In a journal his father described the boy "screaming at the top of his voice and bending and throwing himself back" and recorded that he "has bitten his mother severely many times."[5] His tenacity was ferocious. In any situation, he simply had to have his way. His parents eventually adopted a policy of surrender. They abandoned any attempt to discipline him, and until he was eight they home-schooled him. By then they had moved to Palo Alto, where his mother's parents lived.

Convinced that their son was a genius, William's parents had him evaluated by Lewis Terman,* who had devised the Stanford–Binet IQ test and was planning a study of gifted children. Young Shockley scored in the high 120s, which was respectable but not enough for

* His son Fred Terman later became the famous dean and provost at Stanford.

Terman to label him a genius. Shockley would become obsessed by IQ tests and use them to assess job applicants and even colleagues, and he developed increasingly virulent theories about race and inherited intelligence that would poison the later years of his life.[6] Perhaps he should have learned from his own life the shortcomings of IQ tests. Despite being certified as a nongenius, he was smart enough to skip middle school and get a degree from Caltech and then a doctorate in solid-state physics from MIT. He was incisive, creative, and ambitious. Even though he loved performing magic tricks and playing practical jokes, he never learned to be easygoing or friendly. He had an intellectual and personal intensity, resonating from his childhood, that made him difficult to deal with, all the more so as he became successful.

When Shockley graduated from MIT in 1936, Mervin Kelly came up from Bell Labs to interview him and offered him a job on the spot. He also gave Shockley a mission: find a way to replace vacuum tubes with a device that was more stable, solid, and cheap. After three years, Shockley became convinced he could find a solution using solid material such as silicon rather than glowing filaments in a bulb. "It has today occurred to me that an amplifier using semiconductors rather than vacuum is in principle possible," he wrote in his lab notebook on December 29, 1939.[7]

Shockley had the ability to visualize quantum theory, how it explained the movement of electrons, the way a choreographer can visualize a dance. His colleagues said that he could look at semiconducting material and see the electrons. However, in order to transform his artist's intuitions into a real invention, Shockley needed a partner who was an adroit experimenter, just as Mauchly needed Eckert. This being Bell Labs, there were many in the building, most notably the merrily cantankerous westerner Walter Brattain, who enjoyed making ingenious devices with semiconducting compounds such as copper oxide. For example, he built electric rectifiers, which turn alternating current into direct current, based on the fact that current flows in only one direction through an interface where a piece of copper meets a layer of copper oxide.

Brattain grew up on an isolated ranch in eastern Washington

State, where as a boy he herded cattle. With his raspy voice and homespun demeanor, he affected the self-deprecating style of a confident cowboy. He was a natural-born tinkerer with deft fingers, and he loved devising experiments. "He could put things together out of sealing wax and paper clips," recalled an engineer he worked with at Bell Labs.[8] But he also had a laid-back cleverness that led him to seek shortcuts rather than plod through repetitious trials.

Shockley had an idea for finding a solid-state replacement for a vacuum tube by putting a grid into a layer of copper oxide. Brattain was skeptical. He laughed and told Shockley that he had tried that approach before, and it never ended up producing an amplifier. But Shockley kept pushing. "It's so damned important," Brattain finally said, "that if you'll tell me how you want it made, we'll try it."[9] But as Brattain predicted, it didn't work.

Before Shockley and Brattain could figure out why it had failed, World War II intervened. Shockley went off to become a research director in the Navy's antisubmarine group, where he developed analyses of bomb detonation depths to improve attacks on German U-boats. He later traveled to Europe and Asia to help B-29 bomber fleets use radar. Brattain likewise left for Washington to work on submarine-detection technologies for the Navy, focusing on airborne magnetic devices.

THE SOLID-STATE TEAM

While Shockley and Brattain were away, the war was transforming Bell Labs. It became part of the triangular relationship that was forged among the government, research universities, and private industry. As the historian Jon Gertner noted, "In the first few years after Pearl Harbor, Bell Labs took on nearly a thousand different projects for the military—everything from tank radio sets to communications systems for pilots wearing oxygen masks to enciphering machines for scrambling secret messages."[10] The staff doubled in size, to nine thousand.

Having outgrown its Manhattan headquarters, most of Bell Labs moved to two hundred rolling acres in Murray Hill, New Jersey. Mervin Kelly and his colleagues wanted their new home to feel like an academic campus, but without the segregation of various disciplines into different buildings. They knew that creativity came through chance encounters. "All buildings have been connected so as to avoid fixed geographical delineation between departments and to encourage free interchange and close contact among them," an executive wrote.[11] The corridors were extremely long, more than the length of two football fields, and designed to promote random meetings among people with different talents and specialties, a strategy that Steve Jobs replicated in designing Apple's new headquarters seventy years later. Anyone walking around Bell Labs might be bombarded with random ideas, soaking them up like a solar cell. Claude Shannon, the eccentric information theorist, would sometimes ride a unicycle up and down the long red terrazzo corridors while juggling three balls and nodding at colleagues.* It was a wacky metaphor for the balls-in-the-air ferment in the halls.

In November 1941 Brattain had made his last journal entry, into his notebook #18194, before leaving Bell Labs in Manhattan for his wartime service. Almost four years later, he picked up that same notebook in his new lab in Murray Hill and began anew with the entry "The war is over." Kelly assigned him and Shockley to a research group that was designed "to achieve a unified approach to the theoretical and experimental work of the solid state area." Its mission was the same as they had before the war: to create a replacement for the vacuum tube using semiconductors.[12]

When Kelly sent around the list of who was going to be on the solid-state research group, Brattain marveled that it included no losers. "By golly! There isn't an s.o.b. in the group," he recalled saying, before pausing to worry, "Maybe I was the s.o.b. in the group." As he later declared, "It was probably one of the greatest research teams ever pulled together."[13]

Shockley was the primary theoretician, but given his duties as the team's supervisor—he was on a different floor—they decided to bring

---

* For a short video of Shannon and his machines juggling, see https://www2.bc.edu/~lewbel/shortsha.mov.

in an additional theorist. They chose a soft-spoken expert in quantum theory, John Bardeen. A child genius who had skipped three grades in school, Bardeen had written his doctoral thesis under Eugene Wigner at Princeton and during his wartime service in the Naval Ordnance Laboratory discussed torpedo design with Einstein. He was one of the world's greatest experts on using quantum theory to understand how materials conduct electricity, and he had, according to colleagues, a "genuine ability to collaborate easily with experimentalist and theorist alike."[14] There was initially no separate office for Bardeen, so he ensconced himself in Brattain's lab space. It was a smart move that showed, once again, the creative energy generated by physical proximity. By sitting together, the theorist and the experimentalist could brainstorm ideas face-to-face, hour after hour.

Unlike Brattain, who was voluble and talkative, Bardeen was so quiet that he was dubbed "Whispering John." To understand his mumbling, people had to lean forward, but they learned that it was worth it. He was also contemplative and cautious, unlike Shockley, who was lightning-quick and impulsively spouted theories and assertions.

Their insights came from interactions with each other. "The close collaboration between experimentalists and theorists extended through all stages of the research, from the conception of the experiment to the analysis of the results," said Bardeen.[15] Their impromptu meetings, usually led by Shockley, occurred almost every day, a quintessential display of finish-each-other's-sentence creativity. "We would meet to discuss important steps almost on the spur of the moment," Brattain said. "Many of us had ideas in these discussion groups, one person's remarks suggesting an idea to another."[16]

These meetings became known as "blackboard sessions" or "chalk talks" because Shockley would stand, chalk in hand, scribbling down ideas. Brattain, ever brash, would pace around the back of the room and shout out objections to some of Shockley's suggestions, sometimes betting a dollar they wouldn't work. Shockley didn't like losing. "I finally found out he was annoyed when he paid me off once in ten dimes," Brattain recalled.[17] The interactions would spill over into their social outings; they often played golf together, went out for beer at a diner called Snuffy's, and joined in bridge matches with their spouses.

THE TRANSISTOR

With his new team at Bell Labs, Shockley resurrected the theory he had been playing with five years earlier for a solid-state replacement for the vacuum tube. If a strong electrical field was placed right next to a slab of semiconducting material, he posited, the field would pull some electrons to the surface and permit a surge of current through the slab. This potentially would allow a semiconductor to use a very small signal to control a much larger signal. A very low-powered current could provide the input, and it could control (or switch on and off) a much higher-powered output current. Thus the semiconductor could be used as an amplifier or an on-off switch, just like a vacuum tube.

There was one small problem with this "field effect": when Shockley tested the theory—his team charged a plate with a thousand volts and put it only a millimeter away from a semiconductor surface—it didn't work. "No observable change in current," he wrote in his lab notebook. It was, he later said, "quite mysterious."

Figuring out why a theory failed can point the way to a better one, so Shockley asked Bardeen to come up with an explanation. The two of them spent hours discussing what are known as "surface states," the electronic properties and quantum-mechanical description of the atom layers closest to the surface of materials. After five months, Bardeen had his insight. He went to the blackboard in the workspace he shared with Brattain and began to write.

Bardeen realized that when a semiconductor is charged, electrons become trapped on its surface. They cannot move about freely. They form a shield, and an electric field, even a strong one a millimeter away, cannot penetrate this barrier. "These added electrons were trapped, immobile, in surface states," Shockley noted. "In effect, the surface states shielded the interior of the semiconductor from the influence of the positively charged control plate."[18]

The team now had a new mission: find a way to break through the shield that formed on the surface of semiconductors. "We concentrated on new experiments related to Bardeen's surface states," Shockley explained. They would have to breach this barrier in order to goose the semiconductor into being able to regulate, switch, and amplify current.[19]

Progress was slow over the next year, but in November 1947 a series of breakthroughs led to what became known as the Miracle Month. Bardeen built on the theory of the "photovoltaic effect," which says that shining light on two dissimilar materials that are in contact with one another will produce an electric voltage. That process, he surmised, might dislodge some of the electrons that created the shield. Brattain, working side by side with Bardeen, devised ingenious experiments to test out ways to do this.

After a while, serendipity proved to be their friend. Brattain conducted some of the experiments in a thermos so he could vary the temperature. But condensation on the silicon kept gunking up the measurements. The best way to solve that would be to put the entire apparatus in a vacuum, but that would have required a lot of work. "I'm essentially a lazy physicist," Brattain admitted. "So I got the idea to immerse the system in a dielectric liquid."[20] He filled the thermos with water, which proved a simple way to avoid the condensation problem. He and Bardeen tried it out on November 17, and it worked beautifully.

That was a Monday. Throughout that week, they bounced through a series of theoretical and experimental ideas. By Friday, Bardeen had come up with a way to eliminate the need to immerse the apparatus in water. Instead, he suggested, they could just use a drop of water, or a little gel, right where a sharp metal point jabbed down into the piece of silicon. "Come on, John," Brattain responded enthusiastically. "Let's go make it." One challenge was that the metal point couldn't be allowed contact with the water drop, but Brattain was an improvisational wizard and solved that with a bit of sealing wax. He found a nice slab of silicon, put a tiny drop of water on it, coated a piece of wire with wax to insulate it, and jabbed the wire through the water drop and into the silicon. It worked. It was able to amplify a current, at least slightly. From this "point-contact" contraption the transistor was born.

Bardeen went into the office the next morning to record the results in his notebook. "These tests show definitely that it is possible to introduce an electrode or grid to control the flow of current in a semiconductor," he concluded.[21] He even went in on Sunday, which he normally reserved for golf. They also decided it was time to call Shockley, who had been immersed for months in other matters. Over the next two weeks he would come down and offer suggestions, but he mainly let his dynamic duo proceed apace.

Sitting side by side at Brattain's lab bench, Bardeen would quietly offer ideas and Brattain would excitedly try them out. Sometimes Bardeen wrote in Brattain's notebook as the experiments were being conducted. Thanksgiving passed with little notice as they tried different designs: germanium instead of silicon, lacquer rather than wax, gold for the contact points.

Usually Bardeen's theories led to Brattain's experiments, but sometimes the process worked in reverse: unexpected results drove new theories. In one of the germanium experiments, the current seemed to flow in the opposite direction from what they expected. But it was amplified by a factor of more than three hundred, far more than they had previously achieved. So they ended up acting out the old physicist joke: they knew that the approach worked in practice, but could they make it work in theory? Bardeen soon found a way to do so. He realized that the negative voltage was driving away electrons, causing an increase in "electron holes," which occur when there is no electron in a position where one could exist. The existence of such holes attracts a flow of electrons.

There was one problem: this new method did not amplify higher frequencies, including audible sounds. That would make it useless for telephones. Bardeen theorized that the water or electrolyte drop was making things sluggish. So he improvised a few other designs. One involved a wire point stuck into the germanium just a tiny distance from a gold plate that was creating a field. It succeeded in amplifying the voltage, at least slightly, and it worked at higher frequencies. Once again Bardeen supplied a theory for the serendipitous results: "The experiment suggested that holes were flowing into the germanium surface from the gold spot."[22]

Like a call-and-response duet sitting together at a piano, Bardeen and Brattain continued their iterative creativity. They realized that the best way to increase the amplification would be to have two point-contacts jabbed into the germanium *really* close together. Bardeen calculated that they should be less than two-thousandths of an inch apart. That was a challenge, even for Brattain. But he came up with a clever method: he glued a piece of gold foil onto a small plastic wedge that looked like an arrowhead, then he used a razor blade to cut a thin slit in the foil at the tip of the wedge, thus forming two gold contact points close together. "That's all I did," Brattain recounted. "I slit carefully with the razor until the circuit opened, and put it on a spring and put it down on the same piece of germanium."[23]

When Brattain and Bardeen tried it on the afternoon of Tuesday, December 16, 1947, something amazing happened: the contraption worked. "I found if I wiggled it just right," Brattain recalled, "that I had an amplifier with the order of magnitude of one hundred amplification, clear up to the audio range."[24] On his way home that evening, the voluble and talkative Brattain told the others in his carpool he had just done "the most important experiment that I'd ever do in my life." He then made them pledge not to say anything.[25] Bardeen, as was his wont, was less talkative. When he got home that night, however, he did something unusual: he told his wife about something that happened at the office. It was only a sentence. As she was peeling carrots at the kitchen sink, he mumbled quietly, "We discovered something important today."[26]

Indeed, the transistor was one of the most important discoveries of the twentieth century. It came from the partnership of a theorist and an experimentalist working side by side, in a symbiotic relationship, bouncing theories and results back and forth in real time. It also came from embedding them in an environment where they could walk down a long corridor and bump into experts who could manipulate the impurities in germanium, or be in a study group populated by people who understood the quantum-mechanical explanations of surface states, or sit in a cafeteria with engineers who knew all the tricks for transmitting phone signals over long distances.

Shockley convened a demonstration for the rest of the semiconductor group and a few Bell Labs supervisors on the following Tuesday, December 23. The executives put on earphones and took turns speaking into a microphone so that they could hear for themselves the actual amplification of a human voice using a simple, solid-state device. It was a moment that should have resonated like Alexander Graham Bell's first words barked on a telephone, but no one later could recall the words spoken into the device on that momentous afternoon. Instead the event was memorialized for history by understated entries made into lab notebooks. "By switching the device in and out, a distinct gain in speech level could be heard," Brattain wrote.[27] Bardeen's entry was even more matter-of-fact: "Voltage amplification was obtained with use of two gold electrodes on a specifically prepared germanium surface."[28]

SHOCKLEY'S ONE-UPMANSHIP

Shockley signed Bardeen's historic notebook entry as a witness, but he did not make any entries of his own that day. He was clearly rattled. The pride he should have felt in the success of his team was overshadowed by his intense and dark competitive drive. "My emotions were somewhat conflicted," he later admitted. "My elation with the group's success was tempered by not being one of the inventors. I experienced some frustration that my personal efforts, started more than eight years before, had not resulted in a significant inventive contribution of my own."[29] There were demons that increasingly gnawed away deep in his psyche. He would never again be friends with Bardeen and Brattain. Instead he started working feverishly to claim equal credit for the invention and to create, on his own, an even better version.

Shortly after Christmas, Shockley took the train to Chicago to attend two conferences, but he spent most of his time in his room at the Bismarck Hotel devising a revised method for creating the device. On New Year's Eve, as partygoers danced in the ballroom below, he wrote seven pages of notes on lined graph paper. When he woke up on New Year's Day of 1948, he wrote thirteen more. These he sent by airmail back to a colleague at Bell Labs who glued them into Shockley's lab notebook and asked Bardeen to sign them as a witness.

By then Mervin Kelly had assigned one of the Bell Lab attorneys to produce, as fast as possible, a set of patent applications for the new device. This was not Iowa State, where there was no one on staff to handle such a task. When Shockley returned from Chicago, he discovered that Bardeen and Brattain had already been consulted, and he was upset. He called them into his office separately and explained why he should get the primary—perhaps even sole—credit. "He thought," Brattain recalled, "that he could write a patent, starting with the field effect, on the whole damn thing." Bardeen was characteristically silent, though he did mutter bitterly once it was over. Brattain, as was his wont, was blunt. "Oh hell, Shockley," he yelled. "There's enough glory in this for everybody."[30]

Shockley pressed Bell's lawyers to apply for a very broad patent based on his own initial insight about how a field effect could influence current in a semiconductor. But in their research the lawyers discovered that a patent had been granted in 1930 to a little-known physicist named Julius Lilienfeld, who had proposed (but never built or understood) a device using the field effect. So they decided to pursue a patent for the more narrow invention of a point-contact method of making a semiconductor device, and the only names on that particular application would be Bardeen and Brattain. The attorneys questioned the two of them separately, and both said that it had been a joint effort in which each of them contributed equally. Shockley was furious that he was being left off the most important of the patent applications. Bell executives tried to paper over the rift by requiring that all publicity photos and press releases include all three men.

Over the next few weeks, Shockley became increasingly disconcerted, so much so that he had trouble sleeping.[31] His "will to think," as he called it, was driven by "my own motivation to play a more significant personal, rather than managerial, role in what was obviously a development of enormous potential importance."[32] At odd hours of the night, he would pace around searching for better ways to make the device. Early on the morning of January 23, 1948, a month after the demonstration of the Bardeen-Brattain invention, Shockley woke up with an insight that pulled together the thinking he had done on his trip to Chicago. Sitting at his kitchen table, he began writing furiously.

Shockley's idea involved a way to make a semiconductor amplifier that was less rickety than the contrivance that Bardeen and Brattain had rigged up. Instead of jamming gold points into a slab of germanium, Shockley envisioned a simpler "junction" approach that looked like a sandwich. It would have a top and bottom layer of germanium that had been doped with impurities so that they had an excess of electrons, and sandwiched between them would be a thin slice of germanium that had holes or a deficit of electrons. The layers with an excess of electrons were called "n-type" germanium, for *negative*, and the layer with a deficit or holes where electrons could be was called "p-type," for *positive.* Each of the layers would be attached to a wire that allowed its voltage to be tweaked. The middle layer would be an adjustable barrier that, depending how turned on it was by voltage, regulated the current of electrons that flowed between the top and bottom layers. Applying a small positive voltage to this barrier would, Shockley wrote, "increase the flow of electrons over the barrier exponentially." The stronger the charge on this inside p-type layer, the more it would suck electrons from one outside n-type layer to the other. In other words, it could amplify or switch off the current going through the semiconductor—and do so in mere billionths of a second.

Shockley put some notes in his lab book, but he kept his idea secret for almost a month. "I had a competitive urge to make some important transistor inventions on my own," he later conceded.[33] He did not tell his colleagues until mid-February, when they were at a presentation of some related work by a Bell Labs scientist. Shockley recalled being "startled" when the scientist presented some findings that supported the theoretical basis for a junction device, and he realized that someone in the audience, most likely Bardeen, might take the logical next steps. "From that point on," he asserted, "the concept of using p-n junctions rather than metal point contacts would have been but a small step and the junction transistor would have been invented." So before Bardeen or anyone else could suggest such a device, Shockley leaped up and took the stage to reveal the design he had been working on. "I did not want to be left behind on this one," he later wrote.[34]

Bardeen and Brattain were taken aback. The fact that Shockley had been so secretive about his new idea—thus violating the code of sharing that was part of the Bell culture—upset them. Yet they could not help but be impressed by the simple beauty of Shockley's approach.

After patent applications for both methods had been filed, the Bell Labs brass decided it was time to make the new device public. But first they needed a name for it. Internally it had been called a "semiconductor triode" and a "surface-state amplifier," but those were not catchy enough names for an invention that, they correctly believed, would revolutionize the world. One day a colleague named John Pierce wandered into Brattain's office. In addition to being a good engineer, he was a clever wordsmith who wrote science fiction under the pseudonym J. J. Coupling. Among his many quips were "Nature abhors a vacuum tube" and "After growing wildly for years, the field of computing appears to be reaching its infancy." Brattain declared, "You're just the man I want to see." He posed the naming question, and after just a moment Pierce came up with a suggestion. Since the device had the property of transresistance and should have a name similar to devices such as the thermistor and varistor, Pierce proposed *transistor*. Exclaimed Brattain, "That's it!" The naming process still had to go through a formal poll of all the other engineers, but *transistor* easily won the election over five other options.[35]

On June 30, 1948, the press gathered in the auditorium of Bell Labs' old building on West Street in Manhattan. The event featured Shockley, Bardeen, and Brattain as a group, and it was moderated by the director of research, Ralph Bown, dressed in a somber suit and colorful bow tie. He emphasized that the invention sprang from a combination of collaborative teamwork and individual brilliance: "Scientific research is coming more and more to be recognized as a group or teamwork job. . . . What we have for you today represents a fine example of teamwork, of brilliant individual contributions, and of the value of basic research in an industrial framework."[36] That precisely described the mix that had become the formula for innovation in the digital age.

The *New York Times* buried the story on page 46 as the last item in its "News of Radio" column, after a note about an upcoming broadcast of an organ concert. But *Time* made it the lead story of its science section, with the headline "Little Brain Cell." Bell Labs enforced the rule that Shockley be in every publicity photo along with Bardeen and Brattain. The most famous one shows the three of them in Brattain's lab. Just as it was about to be taken, Shockley sat down in Brattain's chair, as if it were his desk and microscope, and became the focal point of the photo. Years later Bardeen would describe Brattain's lingering dismay and his resentment of Shockley: "Boy, Walter hates this picture. . . . That's Walter's equipment and our experiment, and Bill didn't have anything to do with it."[37]

TRANSISTOR RADIOS

Bell Labs was a cauldron of innovation. In addition to the transistor, it pioneered computer circuitry, laser technology, and cellular telephony. It was, however, less good at capitalizing on its inventions. As part of a regulated company that had a monopoly on most phone services, it was not hungry for new products, and it was legally restrained from leveraging its monopoly to enter other markets. In order to stave off public criticism and antitrust actions, it liberally licensed its patents to other companies. For the transistor, it set a remarkably low fee, $25,000, for any company that wanted to make them, and it even offered seminars explaining manufacturing techniques.

Despite these promiscuous policies, one fledgling firm had trouble wrangling a license: a Dallas-based oil exploration company that had reoriented and renamed itself Texas Instruments. Its executive vice president, Pat Haggerty, who would later take over the firm, had served in the Navy's Bureau of Aeronautics and become convinced that electronics were about to transform almost all aspects of life. When he heard about transistors, he decided that Texas Instruments would find a way to exploit them. Unlike many established companies, it was bold enough to reinvent itself. But the folks at Bell Labs were, Haggerty recalled, "visibly amused at the effrontery of our conviction that we could develop the competence to compete in the field." At least initially, Bell resisted selling Texas Instruments a license. "This business is not for you," the firm was told. "We don't think you can do it."[38]

In the spring of 1952, Haggerty was finally able to convince Bell Labs to let Texas Instruments buy a license to manufacture transistors. He also hired away Gordon Teal, a chemical researcher who worked on one of Bell Labs' long corridors near the semiconductor team. Teal was an expert at manipulating germanium, but by the time he joined Texas Instruments he had shifted his interest to silicon, a more plentiful element that could perform better at high temperatures. By May 1954 he was able to fabricate a silicon transistor that used the n-p-n junction architecture developed by Shockley.

Speaking at a conference that month, near the end of reading a thirty-one-page paper that almost put listeners to sleep, Teal shocked the audience by declaring, "Contrary to what my colleagues have told you about the bleak prospects for silicon transistors, I happen to have a few of them here in my pocket." He proceeded to dunk a germanium transistor connected to a record player into a beaker of hot oil, causing it to die, and then did the same with one of his silicon transistors, during which Artie Shaw's "Summit Ridge Drive" continued to blare undiminished. "Before the session ended," Teal later said, "the astounded audience was scrambling for copies of the talk, which we just happened to bring along."[39]

Innovation happens in stages. In the case of the transistor, first there was the invention, led by Shockley, Bardeen, and Brattain. Next came the production, led by engineers such as Teal. Finally, and equally important, there were the entrepreneurs who figured out how to conjure up new markets. Teal's plucky boss Pat Haggerty was a colorful case study of this third step in the innovation process.

Like Steve Jobs, Haggerty was able to project a reality distortion field that he used to push people to accomplish things they thought impossible. Transistors were being sold in 1954 to the military for about $16 apiece. But in order to break into the consumer market, Haggerty insisted that his engineers find a way to make them so that they could be sold for less than $3. They did. He also developed a Jobs-like knack, which would serve him then and in the future, for conjuring up devices that consumers did not yet know they needed but would soon find indispensable. In the case of the transistor, Haggerty came up with the idea of a small pocket radio. When he tried to convince RCA and other big firms that made tabletop radios to become a partner in the venture, they pointed out (rightly) that consumers were not demanding a pocket radio. But Haggerty understood the importance of spawning new markets rather than merely chasing old ones. He convinced a small Indianapolis company that built TV antenna boosters to join forces on what would be called the Regency TR-1 radio. Haggerty made the deal in June 1954 and, typically, insisted that the device be on the market by that November. It was.

The Regency radio, the size of a pack of index cards, used four transistors and sold for $49.95. It was initially marketed partly as a security item, now that the Russians had the atom bomb. "In event of an enemy attack, your Regency TR-1 will become one of your most valued possessions," the first owner's manual declared. But it quickly became an object of consumer desire and teenage obsession. Its plastic case came, iPod-like, in four colors: black, ivory, Mandarin Red, and Cloud Gray. Within a year, 100,000 had been sold, making it one of the most popular new products in history.[40]

Suddenly everyone in America knew what a transistor was. IBM's chief Thomas Watson Jr. bought a hundred Regency radios and gave them to his top executives, telling them to get to work using transistors in computers.[41]

More fundamentally, the transistor radio became the first major example of a defining theme of the digital age: technology making devices personal. The radio was no longer a living-room appliance to be shared; it was a personal device that allowed you to listen to your own music where and when you wanted—even if it was music that your parents wanted to ban.

Indeed, there was a symbiotic relationship between the advent of the transistor radio and the rise of rock and roll. Elvis Presley's first commercial recording, "That's All Right," came out at the same time as the Regency radio. The rebellious new music made every kid want a radio. And the fact that the radios could be taken to the beach or the basement, away from the disapproving ears and dial-controlling fingers of parents, allowed the music to flourish. "The only regret I have about the transistor is its use for rock and roll," its coinventor Walter Brattain often lamented, presumably half in jest. Roger McGuinn, who became the lead singer of the Byrds, got a transistor radio for his thirteenth birthday, in 1955. "I heard Elvis," he recalled. "It was a game changer for me."[42]

The seeds were planted for a shift in perception of electronic technology, especially among the young. It would no longer be the province only of big corporations and the military. It could also empower individuality, personal freedom, creativity, and even a bit of a rebellious spirit.

SETTING THE WORLD ON FIRE

One problem with successful teams, particularly intense ones, is that sometimes they break up. It takes a special type of leader—inspiring yet also nurturing, competitive yet collaborative—to hold such teams together. Shockley was not such a leader. Just the opposite. As he had shown when he went off on his own to devise the junction transistor, he could be competitive and secretive with his own coworkers. Another skill of great team leaders is the ability to instill a nonhierarchical esprit de corps. Shockley was bad at that as well. He was autocratic, often snuffing out spirit by quashing initiative. The great triumph of Brattain and Bardeen had come when Shockley was offering up a few suggestions but not micromanaging or bossing them. After that he became more overbearing.

During weekend golf games, Bardeen and Brattain would share their dismay about Shockley. At one point Brattain decided that Mervin Kelly, the president of Bell Labs, needed to be clued in. "Do you want to call him, or do you want me to?" he asked Bardeen. The task fell, not surprisingly, to the more voluble Brattain.

He and Kelly met one afternoon in the wood-paneled study of Kelly's home in the nearby suburb of Short Hills. Brattain laid out their grievances, describing how ham-fisted Shockley was as a manager and colleague. Kelly batted away the complaints. "So finally, without thinking of its impact, I inadvertently said to him that John Bardeen and I knew when Shockley invented the PNP [junction] transistor," Brattain recalled. In other words, he had let slip a veiled threat that some of the concepts in the patent application for the junction transistor, which listed Shockley as the inventor, had actually arisen from the work that Brattain and Bardeen had done. "Kelly realized that neither Bardeen nor I, if we ever went on the stand in a patent fight, would lie about what we knew. This changed his whole attitude. And after that, my position in the Laboratories was a little bit more satisfactory."[43] Bardeen and Brattain no longer had to report to Shockley.

The new arrangement turned out not to be enough to satisfy Bardeen, who turned his focus away from semiconductors and began working on superconductivity theory. He took a job at the University of Illinois. "My difficulties stem from the invention of the transistor," he wrote in a resignation letter to Kelly. "Before that there was an excellent research atmosphere here. . . . After the invention Shockley at first refused to allow anyone else in the group to work on the problem. In short, he used the group largely to exploit his own ideas."[44]

Bardeen's resignation and Brattain's complaints did not help Shockley's standing at Bell Labs. His prickly personality meant that he was passed over for promotions. He appealed to Kelly and even the president of AT&T, but to no avail. "The hell with that," he told a colleague. "I'll go set up my own business, I'll make a million dollars that way. And by the way, I'll do it out in California." When he heard of Shockley's plans, Kelly did not try to dissuade him. Quite the contrary: "I told him that if he thinks he can earn a million dollars, go ahead!" Kelly even called Laurence Rockefeller to recommend that he help finance Shockley's proposed venture.[45]

As he grappled with his situation in 1954, Shockley went through a midlife crisis. After helping his wife fight ovarian cancer, he left her while she was in remission and found himself a girlfriend, whom he would later marry. He took a leave from Bell Labs. And this being a classic midlife crisis, he even bought a sports car, a green Jaguar XK120 two-seat convertible.

Shockley spent a semester as a visiting professor at Caltech and took a gig consulting with the Army's Weapons Systems Evaluation Group in Washington, but much of the time he traveled the country trying to figure out his new venture, visiting technology companies, and meeting with successful entrepreneurs such as William Hewlett and Edwin Land. "Think I shall try to raise some capital and start on my own," he wrote his girlfriend. "After all, it is obvious I am smarter, more energetic, and understand folks better than most of these other folks." His journals for 1954 show him struggling to make sense of his quest. "Lack of appreciation by bosses, means what?" he wrote at one point. As happens in many biographies, there was also the theme of living up to a late father. Contemplating his plan to create a company that would make transistors ubiquitous, he wrote, "Idea of setting world on fire, father proud."[46]

Setting the world on fire. Despite the fact that he would never turn out to be successful in business, Shockley would accomplish that. The company that he was about to found would transform a valley known for its apricot orchards into one famed for turning silicon into gold.

SHOCKLEY SEMICONDUCTOR

At the February 1955 annual gala of the Los Angeles Chamber of Commerce, two pioneers of electronics were honored: Lee de Forest, who had invented the vacuum tube, and Shockley, an inventor of its replacement. Shockley sat with a distinguished industrialist, Arnold Beckman, the chamber's vice chairman. Like Shockley, Beckman had worked for Bell Labs, where he developed techniques for making vacuum tubes. As a professor at Caltech he had invented a variety of measuring instruments, including one that measured the acidity of lemons, and he used his invention as the foundation for building a large manufacturing company.

That August, Shockley invited Beckman to serve on the board of his proposed transistor company. "I asked him a little bit more about who else was going to be on the board," Beckman recalled, "and it turned out that he was going to have a board composed of almost everyone who was in the instrument business, all of whom would be his competitors." Beckman realized how "unbelievably naïve" Shockley was, so in order to help him devise a more sensible approach, he invited him to spend a week in Newport Beach, where Beckman kept his sailboat.[47]

Shockley's plan was to make transistors by using gas diffusion to dope silicon with impurities. By adjusting the time, pressure, and temperature, he could precisely control the process, thus allowing different varieties of transistors to be mass-manufactured. Impressed by the idea, Beckman convinced Shockley not to launch his own company and instead to lead a new division of Beckman Instruments, which Beckman would fund.

Beckman wanted it located in the Los Angeles area, where most of his other divisions were. But Shockley insisted that it be located in Palo Alto, where he had been raised, so that he could be near his aging mother. They doted on each other intensely, which some found weird but which had the historic significance of helping to create Silicon Valley.

Palo Alto was still, as it had been in Shockley's childhood, a small college town surrounded by orchards. But during the 1950s its population would double, to fifty-two thousand, and twelve new elementary schools would be built. The influx was partly due to the boom in the cold war defense industry. Canisters of film dropped from America's U-2 spy planes were sent to the NASA Ames Research Center in nearby Sunnyvale. Defense contractors took root in the surrounding areas, such as the Lockheed Missiles and Space Division, which built submarine-launched ballistic missiles, and Westinghouse, which produced tubes and transformers for the missile systems. Neighborhoods of tract houses sprang up to accommodate young engineers and Stanford junior professors. "You had all these military companies on the cutting edge," recalled Steve Jobs, who was born in 1955 and grew up in the area. "It was mysterious and high-tech and made living there very exciting."[48]

Sprouting alongside the defense contractors were companies that made electrical measuring instruments and other technological devices. The sector's roots stretched back to 1938, when the electronics entrepreneur Dave Packard and his new wife moved into a home in Palo Alto that had a shed where his friend Bill Hewlett was soon ensconced. The house also had a garage—an appendage that would prove both useful and iconic in the valley—in which they tinkered around until they had their first product, an audio oscillator. By the 1950s Hewlett-Packard had become the pace horse for the region's tech startups.[49]

Fortunately there was a place for entrepreneurs who had outgrown their garages. Fred Terman, a doctoral student of Vannevar Bush's at MIT who became Stanford University's dean of engineering, created an industrial park in 1953 on seven hundred acres of undeveloped university property, where tech companies could lease land inexpensively and build new offices. It helped transform the area. Hewlett and Packard had been Terman's students, and he had persuaded them to stay in Palo Alto when they founded their company rather than move east, as most of Stanford's top graduates had been doing. They became one of the first tenants in the Stanford Research Park. Throughout the 1950s Terman, who went on to become Stanford's provost, grew the industrial park by encouraging its occupants to have a symbiotic relationship with Stanford; employees and executives could study or teach part-time at the university, and its professors were given leeway to advise new businesses. Stanford's office park would end up nurturing hundreds of companies, from Varian to Facebook.

When Terman learned that Shockley was thinking of locating his new enterprise in Palo Alto, he wrote a courtship letter that described all the incentives that proximity to Stanford would offer. "I believe that its location here would be mutually advantageous," he concluded. Shockley agreed. While its new Palo Alto headquarters was being constructed, Shockley Semiconductor Laboratory, a division of Beckman Instruments, set up temporarily in a Quonset shed that had served as a storage space for apricots. Silicon had come to the valley.

ROBERT NOYCE AND GORDON MOORE

Shockley tried to recruit some of the researchers he had worked with at Bell Labs, but they knew him too well. So he set about compiling a list of the best semiconductor engineers in the country and calling them cold. The most notable of them all, destined to be a momentous choice, was Robert Noyce, a charismatic Iowa golden boy with a doctorate from MIT, who was at the time a twenty-eight-year-old research manager at Philco in Philadelphia. In January 1956 Noyce picked up the phone and heard the words "Shockley here." He knew immediately who it was. "It was like picking up the phone and talking to God," Noyce declared.[50] He later joked, "When he came out here to organize Shockley Labs, he whistled and I came."[51]

Noyce, the third of four sons of a Congregationalist minister, grew up in a series of tiny Iowa farm towns—Burlington, Atlantic, Decorah, Webster City—where his father found himself called. Both of Noyce's grandfathers were also ministers of the Congregationalist Church, a Nonconformist Protestant movement that was a product of the Puritan Reformation. Even though he didn't inherit their religious faith, Noyce did absorb their denomination's aversion to hierarchy, centralized authority, and autocratic leadership.[52]

When Noyce was twelve, his family finally settled down in Grinnell (population 5,200 at the time), about fifty miles east of Des Moines, where his father got an administrative job with the Church. The centerpiece of the town was Grinnell College, founded in 1846 by a group of Congregationalists from New England. Noyce, with an infectious grin and a taut graceful body, flourished in the town's high school as a scholar, athlete, and heartthrob. "The quick lopsided smile, the good manners and fine family, the wavy hair high on his forehead, the dash of rapscallion—it made for an appealing combination," wrote his biographer Leslie Berlin. Said his high school girlfriend, "He was probably the most physically graceful man I've ever met."[53]

Years later the literary journalist Tom Wolfe wrote a glimmering profile of Noyce for *Esquire*, in which he came close to canonizing him:

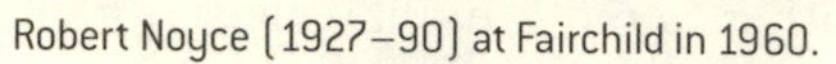

Robert Noyce (1927–90) at Fairchild in 1960.

Gordon Moore (1929–) at Intel in 1970.

Gordon Moore (*far left*), Robert Noyce (*front center*), and the other "traitorous eight" who in 1957 left Shockley to form Fairchild Semiconductor.

> Bob had a certain way of listening and staring. He would lower his head slightly and look up with a gaze that seemed to be about one hundred amperes. While he looked at you he never blinked and never swallowed. He absorbed everything you said and then answered very levelly in a soft baritone voice and often with a smile that showed off his terrific set of teeth. The stare, the voice, the smile; it was all a bit like the movie persona of the most famous of all Grinnell College's alumni, Gary Cooper. With his strong face, his athlete's build, and the Gary Cooper manner, Bob Noyce projected what psychologists call the halo effect. People with the halo effect seem to know exactly what they're doing and, moreover, make you want to admire them for it. They make you see the halos over their heads.[54]

As a kid, Noyce benefited from a situation that was common back then: "Dad always managed to have some sort of workshop in the basement." Young Noyce loved to make things, including a vacuum-tube radio, a sled with a propeller, and a headlight to use on his early-morning paper route. Most famously, he built a hang glider that he flew by hitching it to the back of a fast-moving car or by leaping with it off a barn roof. "I grew up in small town America, so we had to be self-sufficient. If something was broke you fix it yourself."[55]

Like his brothers, Noyce was at the top of his class academically. He mowed the lawn of Grant Gale, a beloved professor who taught physics at Grinnell College. With the help of his mother, who knew the Gales from church, he wrangled permission to take Gale's college course during his senior year of high school. Gale became Noyce's intellectual mentor, which continued the following year, when he enrolled at Grinnell as an undergraduate.

There he pursued a double major in math and physics, starring in all endeavors, academic and extracurricular, with a grace worn lightly. He made a point of deriving every formula in physics class from scratch, became the Midwest conference champion diver on the swim team, played oboe in the band, sang in the chorus, designed circuits for the model airplane club, had the lead in a radio soap opera, and helped his math professor teach a calculus class on complex numbers. Most amazingly, he was, despite all this, well liked.

His scampish affability sometimes got him into trouble. When his dormitory decided to throw a spring luau in his junior year, Noyce and a friend volunteered to procure the pig that would be roasted. After a few drinks, they snuck into a farm nearby and, combining strength with agility, kidnapped a twenty-five-pound suckling. After they butchered the squealing pig with knives in an upstairs shower in the dorm, they roasted it. There followed much cheering, applause, eating, and drinking. The next morning brought a moral hangover. Noyce went with his friend to the farmer and confessed, offering to pay for what they had taken. In a storybook he would have been given the George Washington cherry tree award. But in the struggling farm country of Iowa, the larceny he had committed was neither funny nor forgivable. The farm was owned by the dour mayor of the town, and he threatened to press charges. Eventually Professor Gale helped broker a compromise: Noyce would pay for the pig and be suspended for one semester, but not expelled. Noyce took it in stride.[56]

When Noyce returned in February 1949, Gale did him what may have been an even bigger favor. The professor had been a college friend of John Bardeen, and when he read about the transistor that Bardeen had coinvented at Bell Labs he wrote and asked for a sample. He also contacted the president of Bell Labs, who was a Grinnell alum and the father of two current students. A batch of technical monographs arrived followed by a transistor. "Grant Gale got hold of one of the first point contact transistors that was ever made," Noyce recalled. "That was during my junior year there. I suppose that was one of the things that influenced me to get involved in transistors." In a later interview, Noyce described his excitement more vividly: "The concept hit me like the atom bomb. It was simply astonishing. Just the whole concept, that you could get amplification without a vacuum. It was one of those ideas that just jolts you out of the rut, gets you thinking in a different way."[57]

Upon graduation, Noyce received what was, for someone with his style and charm, the college's highest honor, awarded by a vote of his classmates: the Brown Derby Prize, given to "the senior man who earned the best grades with the least amount of work." But when he arrived at MIT to pursue a doctorate, he realized that he was going to have to apply himself more diligently. He was deemed deficient in theoretical physics and had to take an introductory course in the topic. By his second year he had regained his stride and won an academic fellowship. His dissertation investigated how the photoelectric effect was manifest in the surface state of insulators. Although it was not a triumph of lab work or analysis, it did familiarize him with Shockley's research in that field.

Thus when he got the summons from Shockley, he was eager to accept. But there was one strange hoop he had to leap through. Shockley, who had failed to triumph on an IQ test as a kid and was starting to show the creepy paranoia that would mar his later career, insisted that his new hires go through a battery of psychological and intelligence exams. So Noyce spent an entire day at a testing firm in Manhattan reacting to ink blots, opining on weird drawings, and filling out aptitude quizzes. He was judged to be an introvert and not a good potential manager, which revealed a lot more about the weaknesses of the tests than of Noyce.[58]

Shockley's other great hire, also judged by the psychology firm to be a poor potential manager, was the soft-spoken chemist Gordon Moore, who also got a phone call from Shockley out of the blue. Shockley was carefully assembling a team with different scientific talents that could be mixed together to catalyze innovation. "He knew chemists had been useful to him at Bell Laboratories, so he thought he needed one in his new operation, and he got my name and gave me a call," Moore said. "Fortunately, I recognized who it was. I picked up the phone, he says, 'Hello, this is Shockley.'"[59]

With his self-effacing and genial manner cloaking a precision-guided mind, Gordon Moore would become one of the most revered and beloved figures in Silicon Valley. He had grown up near Palo Alto, in Redwood City, where his father was a deputy sheriff. When he was eleven, the kid next door got a chemistry set. "In those days there was really neat stuff in chemistry sets," Moore recalled, lamenting that government regulations and parental fears have since neutered such kits and probably deprived the nation of some needed scientists. He was able to turn out a small quantity of nitroglycerin, which he made into dynamite. "A couple of ounces of dynamite makes an absolutely fantastic firecracker," he gleefully recounted in an interview, wiggling all ten of his fingers to show that they had survived such childhood foolery.[60] His fun with chemistry sets, he said, helped set him on a path to a chemistry degree from Berkeley and a doctorate from Caltech.

From his birth until he finished his doctorate, Moore never ventured farther east than Pasadena. He was a true-bred Californian, easygoing and affable. For a brief period after getting his PhD, he went to work at a Navy physics laboratory in Maryland. But he and his beloved wife, Betty, also a native of northern California, were restless to get home, so he was receptive when the call from Shockley came.

When Moore went for his interview, he was twenty-seven, a year younger than Noyce, and was already balding in a distinguished manner. Shockley peppered him with questions and brainteasers, holding a stopwatch to time his answers. Moore did so well that Shockley took him to dinner at Rickeys Hyatt House, the local hangout, and did his magic trick of bending a spoon without seeming to apply any physical force.[61]

The dozen engineers recruited by Shockley, almost all under thirty, considered him a bit bizarre but absolutely brilliant. "He just showed up in my lab at MIT one day, and I thought, my God, I've never met anybody this brilliant," remarked the physicist Jay Last. "I changed my whole career plans and said, I wanna go to California and work with this man." Among the others were Jean Hoerni, a Swiss-born physicist, and Eugene Kleiner, who later became a great venture capitalist. By April 1956 there were enough new employees to throw a welcome party. Noyce drove across the country from Philadelphia, rushing to make it in time. He arrived at 10 p.m., while Shockley was doing a solo tango with a rose in his mouth. One of the engineers described Noyce's arrival to his biographer Berlin: "He hadn't shaved, he looked like he'd been living in his suit for a week—and he was thirsty. There was a big goddamn bowl of martinis on the table there. Noyce picks up the goddamn bowl, and starts drinking [from] it. Then he passes out. I said to myself, 'this is going to be a whole lot of fun.'"[62]

SHOCKLEY UNRAVELS

Some leaders are able to be willful and demanding while still inspiring loyalty. They celebrate audaciousness in a way that makes them charismatic. Steve Jobs, for example; his personal manifesto, dressed in the guise of a TV ad, began, "Here's to the crazy ones. The misfits. The rebels. The trouble-makers. The round pegs in the square holes." Amazon's founder Jeff Bezos has that same ability to inspire. The knack is to get people to follow you, even to places they may not think they can go, by motivating them to share your sense of mission. Shockley did not have this talent. Because of his aura, he was able to recruit brilliant employees, but soon after they began working together, they were rankling under his ham-fisted management, just as Brattain and Bardeen had.

One useful leadership talent is knowing when to push ahead against doubters and when to heed them. Shockley had trouble striking this balance. One case arose when he devised a four-layer diode that he thought would be faster and more versatile than a three-layer transistor. In some ways, it was the first step toward an integrated circuit, because the new device would perform tasks that would require four or five transistors on a circuit board. But it was difficult to manufacture (the paper-thin silicon had to be doped differently on either side), and most of the ones that came off the line proved useless. Noyce tried to get Shockley to abandon the diode, but to no avail.

Many transformative innovators have been similarly stubborn about pushing a new idea, but Shockley crossed the line from being visionary to being hallucinatory, turning him into a case study in bad leadership. In his pursuit of the four-layer diode, he was secretive, rigid, authoritarian, and paranoid. He formed private teams and refused to share information with Noyce, Moore, and others. "He couldn't face up to the fact that he'd made a bad decision so he started blaming everybody around him," recalled Jay Last, one engineer who resisted him. "He was very abusive. I went from being his fair-haired boy to being one of the causes of all his problems."[63]

His paranoia, already diffusing into his personality layers, was manifest in disruptive incidents. For example, when a secretary at the firm cut her finger opening a door, Shockley became convinced it was a sabotage scheme. He ordered everyone in the firm to take a lie detector test. Most refused, and Shockley had to back down. It was later discovered that the cut was caused by the remains of a thumbtack that had been used to post a notice on the door. "I don't think 'tyrant' begins to encapsulate Shockley," Moore said. "He was a complex person. He was very competitive and even competed with the people that worked for him. My amateur diagnosis is he was also paranoid."[64]

Worse yet, Shockley's infatuation with the four-layer diode turned out to be misplaced. Sometimes the difference between geniuses and jerks hinges on whether their ideas turn out to be right. If Shockley's diode had proved practical, or if he had evolved it into an integrated circuit, he may have again been regarded as a visionary. But that didn't happen.

The situation became even worse after Shockley, along with his erstwhile partners Bardeen and Brattain, won the Nobel Prize. When Shockley got the call early on the morning of November 1, 1956, his first reaction was that it was a Halloween trick. Later he would become darkly suspicious that there were people who had tried to deny him the prize, and he would write the Nobel committee seeking information about those who wrote in opposition to him, a request that was denied. But for that day, at least, there was a respite in the tension and a chance to celebrate. A champagne lunch was held at Rickeys.

Shockley was still estranged from Bardeen and Brattain, but the atmosphere was cordial when they convened with their families in Stockholm for the awards ceremony. The chair of the Nobel committee used his speech to highlight the combination of individual genius and teamwork that was involved in the invention of the transistor. He called it a "supreme effort of foresight, ingenuity and perseverance, exercised individually and as a team." Late that night, Bardeen and Brattain were having drinks at the bar of the Grand Hotel when, shortly after midnight, Shockley walked in. They had barely spoken to him for six years, but they put their differences aside and invited him to join their table.

When Shockley returned from Stockholm, his head was swelled but his insecurities were undiminished. In a talk to coworkers, he noted that it was "about time" that his contributions were recognized. The atmosphere at the firm "deteriorated very rapidly," Last observed, until it began to resemble "a big psychiatric institute." Noyce told Shockley of the "general feeling of resentment" that was accumulating, but his warning had little effect.[65]

Shockley's unwillingness to share credit made it hard for him to create a spirit of collaboration. When some of his employees wrote papers to be presented at the American Physical Society in December 1956, the month after he had received his Nobel, Shockley required that his name be listed on them all as a coauthor. The same was true on most patent applications coming out of his firm. Yet he insisted, somewhat contradictorily, that there was truly only one real inventor of any device, because "there's only one light bulb to go on in somebody's head." Any other people involved, he added, were "mere helpers."[66] His own experience with the team that invented the transistor should have disabused him of such a notion.

Shockley's ego caused him to clash not only with subordinates but also with his nominal boss and owner, Arnold Beckman. When Beckman flew up for a meeting about the need to control costs, Shockley surprised everyone by declaring in front of the entire senior staff, "Arnold, if you don't like what we're doing up here I can take this group and get support any place else." He then stormed out of the room, leaving his owner humiliated in front of the staff.

Thus Beckman was attentive when he was called in May 1957 by Gordon Moore, who had been tapped by other restless colleagues to present their grievances. "Things aren't going well up there, are they?" Beckman asked.

"No, they really are not," replied Moore, who assured Beckman that the top staff would stay if Shockley quit.[67] The reverse was also true, Moore warned; if Shockley was not replaced by a competent manager, the staff would likely leave.

Moore and his colleagues had recently seen *The Caine Mutiny*, and they started plotting against their own Captain Queeg.[68] Over the next few weeks, in a series of secret meetings and dinners with Beckman and seven disgruntled top staffers led by Moore, a deal was hammered out to move Shockley into a senior consulting role with no management duties. Beckman took Shockley to dinner and informed him of the change.

At first Shockley acquiesced. He would allow Noyce to manage the lab and confine his own duties to offering ideas and strategic advice. But then he changed his mind. It was not in Shockley's nature to cede control. Plus, he had qualms about Noyce's executive ability. He told Beckman that Noyce would not be an "aggressive leader" or decisive enough, and there was some merit to that criticism. Shockley may have been too driven and decisive, but Noyce, who was naturally congenial and accommodating, could have benefited from a dose of toughness. A key challenge for managers is how to strike a balance between being decisive and being collegial, and neither Shockley nor Noyce got the calibration precise.

When forced to choose between Shockley and the staff, Beckman got cold feet. "With one of my misdirected feelings of loyalty, I felt I owed Shockley and should give him enough of a chance to prove himself," Beckman later explained. "If I had known what I know now, I would have said goodbye to Shockley."[69] Beckman stunned Moore and his supporters with his decision. "Beckman essentially told us, 'Shockley's the boss, take it or leave it,'" Moore recalled. "We discovered a group of young PhDs couldn't push aside a new Nobel Prize winner very easily." A revolt became inevitable. "We were just completely sandbagged, and we realized then we had to leave," said Last.[70]

Abandoning an established enterprise to start a rival was rather unusual back then, so it took some courage. "The business culture that existed in this country was that you go to work for a company, and you stay with that company, and you retire with that company," observed Regis McKenna, who became a marketing maven for technology firms. "This was what traditional East Coast—and even Midwestern—American values were." That's no longer true, of course, and the Shockley rebels contributed to the cultural shift. "It looks easy nowadays because we have a tradition—largely set in motion by those guys—where it's accepted in this town," said Michael Malone, a historian of Silicon Valley. "You're better off to go out and start your own company and fail than it is to stick at one company for thirty years. But that wasn't true in the 1950s. It must've been scary as hell."[71]

Moore rallied the rebel contingent. There were seven of them at first—Noyce had not yet enlisted—and they decided to form their own company. But that required funding. So one of them, Eugene Kleiner, wrote a letter to his father's stockbroker at the venerable Wall Street brokerage firm Hayden, Stone & Co. After describing their credentials, he declared, "We believe that we could get a company into the semiconductor business within three months." The letter ended up on the desk of Arthur Rock, a thirty-year-old analyst who had been succeeding with risky investments since his days at Harvard Business School. Rock convinced his boss, Bud Coyle, that it was worth a trip west to investigate.[72]

When Rock and Coyle met with the seven in San Francisco's Clift Hotel, they found one thing missing: a leader. So they urged the rebels to recruit Noyce, who was resisting because of his feeling of commitment to Shockley. Moore was finally able to persuade him to come to the next meeting. Rock was impressed: "As soon as I saw Noyce I was struck by his charisma, and I could tell he was their natural leader. They deferred to him."[73] At that meeting, the group, including Noyce, made a pact that they would all leave together to form a new firm. Coyle pulled out some crisp new dollar bills, which they signed as a symbolic contract with each other.

It was hard to get money, especially from established corporations, to start a completely independent company. The idea of seed funding for startups was not yet well established; that important innovation would have to wait, as we shall see, until the next time Noyce and Moore leaped into a new venture. So they searched for a corporate sponsor that might set them up as a semiautonomous division, just as Beckman had done with Shockley. Over the next few days, the cabal pored over the *Wall Street Journal* and came up with a list of thirty-five firms that might adopt them. Rock started making calls when he got back to New York, but to no avail. "None of them were willing to take on a separate company division," he recalled. "They felt that their own employees would have problems with it. We had a couple of months of doing this and were about to give up, when someone suggested that I see Sherman Fairchild."[74]

It was a fine match. Fairchild, the owner of Fairchild Camera and Instrument, was an inventor, playboy, entrepreneur, and the largest single stockholder in IBM, which his father had cofounded. A great tinkerer, as a Harvard freshman he invented the first synchronized camera and flash. He went on to develop aerial photography, radar cameras, specialized airplanes, methods to illuminate tennis courts, high-speed tape recorders, lithotypes for printing newspapers, color engraving machines, and a wind-resistant match. In the process, he added a second fortune to his inheritance, and he was as joyful spending it as he had been making it. He frequented the 21 Club and the El Morocco nightclub wearing (in the words of *Fortune*) "a fresh pretty girl every few days like a new boutonniere," and he designed for himself a futuristic house on Manhattan's Upper East Side with glass walls and ramps overlooking an atrium garden with green ceramic-clad rocks.[75]

Fairchild readily put up $1.5 million to start the new company—about twice what the eight founders had originally thought necessary—in return for an option deal. If the company turned out to be successful, he would be able to buy it outright for $3 million.

Dubbed "the traitorous eight," Noyce and his posse set up shop just down the road from Shockley on the outskirts of Palo Alto. Shockley Semiconductor never recovered. Six years later, Shockley gave up and joined the faculty of Stanford. His paranoia deepened, and he became obsessed with his notion that blacks were genetically inferior in terms of IQ and should be discouraged from having children. The genius who conceptualized the transistor and brought people to the promised land of Silicon Valley became a pariah who could not give a lecture without facing hecklers.

The traitorous eight who formed Fairchild Semiconductor, by contrast, turned out to be the right people at the right place at the right time. The demand for transistors was growing because of the pocket radios that Pat Haggerty had launched at Texas Instruments, and it was about to skyrocket even higher; on October 4, 1957, just three days after Fairchild Semiconductor was formed, the Russians launched the *Sputnik* satellite and set off a space race with the United States. The civilian space program, along with the military program to build ballistic missiles, propelled the demand for both computers and transistors. It also helped assure that the development of these two technologies became linked. Because computers had to be made small enough to fit into a rocket's nose cone, it was imperative to find ways to cram hundreds and then thousands of transistors into tiny devices.

Jack Kilby (1923–2005)
at Texas Instruments in 1965.

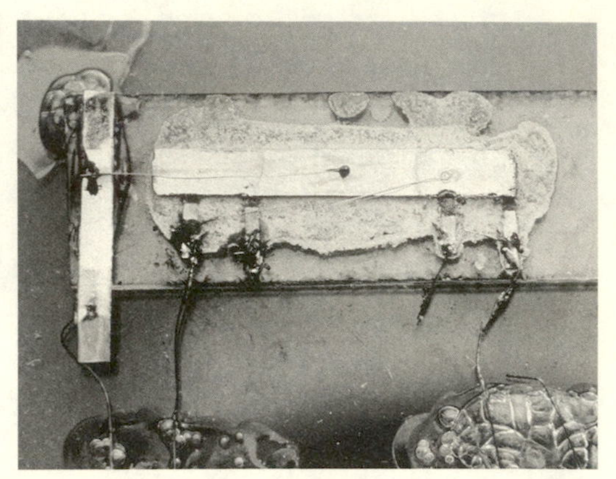

Kilby's microchip.

Arthur Rock (1926–) in 1997.

Andy Grove (1936–)
with Noyce and Moore at Intel in 1978.

CHAPTER FIVE

THE MICROCHIP

In a paper written to celebrate the tenth anniversary of the transistor, published in 1957 just when Fairchild Semiconductor was formed and *Sputnik* launched, a Bell Labs executive identified a problem that he dubbed "the tyranny of numbers." As the number of components in a circuit increased, the number of connections increased way faster. If a system had, for example, ten thousand components, that might require 100,000 or more little wire links on the circuit boards, most often soldered by hand. This was not a recipe for reliability.

It was, instead, part of a recipe for an innovation. The need to solve this growing problem coincided with hundreds of small advances in ways to manufacture semiconductors. This combination produced an invention that occurred independently in two different places, Texas Instruments and Fairchild Semiconductor. The result was an integrated circuit, also known as a microchip.

JACK KILBY

Jack Kilby was another of those boys from the rural Midwest who tinkered in the workshop with his dad and built ham radios.[1] "I grew up among the industrious descendants of the western settlers of the American Great Plains," he declared when he won a Nobel Prize.[2] He was raised in Great Bend, in the middle of Kansas, where his father ran a local utility company. In the summer they would drive in the family Buick to far-flung generating plants and, when something had gone wrong, crawl through them together looking for the problem. During one bad blizzard they used a ham radio to keep in touch with areas where customers had lost phone service, and young Kilby became fascinated by the importance of such technologies. "It was during an ice storm in my teens," he told the *Washington Post*'s T. R. Reid, "that I first saw how radio and, by extension, electronics, could really impact people's lives by keeping them informed and connected, and giving them hope."[3] He studied to get a ham operator's license and kept upgrading his radio using parts that he scrounged.

After being turned down by MIT, he went to the University of Illinois, interrupting his studies after Pearl Harbor to join the Navy. Deployed to a radio repair facility in India, he made runs to Calcutta to buy parts on the black market, using them to build better receivers and transmitters in a pup-tent lab. He was a gentle guy with a wide smile and an easygoing, taciturn manner. What made him special was his insatiable curiosity about inventions. He began to read every new patent issued. "You read everything—that's part of the job," he said. "You accumulate all this trivia, and you hope that someday maybe a millionth of it will be useful."[4]

His first job was at Centralab, a Milwaukee firm that made electronic parts. It experimented with ways of combining the components used to make hearing aids onto a single ceramic base, a rough precursor of the idea for a microchip. In 1952 Centralab was one of the companies that paid $25,000 for a license to make transistors, and it was the beneficiary of Bell's willingness to share its knowledge. Kilby attended a two-week Bell Labs seminar—staying with dozens of others at a Manhattan hotel and being loaded every morning onto a bus for Murray Hill—that included in-depth sessions on transistor design, hands-on experience in the labs, and visits to a manufacturing plant. Bell sent all attendees three volumes of technical papers. With its extraordinary willingness to license its patents cheaply and share its knowledge, Bell Labs laid the foundations for the Digital Revolution, even though it didn't fully capitalize on it.

In order to be at the forefront of transistor development, Kilby realized that he needed to work at a bigger company. Weighing a variety of offers, he decided in the summer of 1958 to join Texas Instruments, where he would get to work with Pat Haggerty and his brilliant transistor research team led by Willis Adcock.

The policy at Texas Instruments was for everyone to take off the same two weeks in July. So when Kilby arrived in Dallas with no accrued vacation time, he was one of the very few people in the semiconductor lab. This gave him time to think about what could be done with silicon other than fabricate it into transistors.

He knew that if you created a bit of silicon without any impurities, it would act as a simple resistor. There was also a way, he realized, to make a p-n junction in a piece of silicon act as a capacitor, meaning it could store a small electrical charge. In fact, you could make any electronic component out of differently treated silicon. From that he came up with what became known as the "monolithic idea": you could make all of these components in one monolithic piece of silicon, thus eliminating the need to solder together different components on a circuit board. In July 1958, six months before Noyce wrote down a similar idea, Kilby described it in his lab notebook in a sentence that would later be quoted in his Nobel Prize citation: "The following circuit elements could be made on a single slice: resistors, capacitor, distributed capacitor, transistor." Then he drew a few crude sketches of how to construct these components by configuring sections of silicon that had been doped with impurities to have different properties on a single slab.

When his boss, Willis Adcock, returned from vacation he was not fully persuaded that this would be practical. There were other things for the lab to do that seemed more pressing. But he made Kilby a deal: if he could make a working capacitor and resistor, Adcock would authorize an effort to do a complete circuit on a single chip.

All went as planned, and in September 1958 Kilby prepared a demonstration that was similar in drama to the one Bardeen and Brattain had done for their superiors at Bell Labs eleven years earlier. On a silicon chip the size of a short toothpick, Kilby assembled the components that would, in theory, make an oscillator. Under the gaze of a group of executives, including the chairman of the company, a nervous Kilby hooked up the tiny chip to an oscilloscope. He looked at Adcock, who shrugged as if to say, Here goes nothing. When he pushed a button, the line on the oscilloscope screen undulated in waves, just as it should. "Everybody broke into broad smiles," Reid reported. "A new era in electronics had begun."[5]

It was not the most elegant device. In the models that Kilby built that fall of 1958, there were a lot of tiny gold wires connecting some of the components within the chip. It looked like expensive cobwebs sticking out of a silicon twig. Not only was it ugly; it was also impractical. There would be no way to manufacture it in large quantities. Nevertheless, it was the first microchip.

In March 1959, a few weeks after filing for a patent, Texas Instruments announced its new invention, which it dubbed a "solid circuit." It also put a few prototypes on display, with much fanfare, at the Institute of Radio Engineers annual conference in New York City. The company's president declared that the invention would be the most important one since the transistor. It seemed like hyperbole, but it was an understatement.

The Texas Instruments announcement struck like a thunderbolt at Fairchild. Noyce, who had jotted down his own version of the concept two months earlier, was disappointed at being scooped and fearful of the competitive advantage it might give Texas Instruments.

NOYCE'S VERSION

There are often different paths to the same innovation. Noyce and his Fairchild colleagues had been pursuing the possibility of a microchip from another direction. It began when they found themselves hit with a messy problem: their transistors were not working very well. Too many of them failed. A tiny piece of dust or even exposure to some gases could cause them to fizzle. So, too, might a sharp tap or bump.

Jean Hoerni, a Fairchild physicist who was one of the traitorous eight, came up with an ingenious fix. On the surface of a silicon transistor, he would place a thin layer of silicon oxide, like icing atop a layer cake, that would protect the silicon below. "The building up of an oxide layer . . . on the surface of the transistor," he wrote in his notebook, "will protect the otherwise exposed junctions from contamination."[6]

The method was dubbed "the planar process" because of the flat plane of oxide that sat on top of the silicon. In January 1959 (after Kilby had come up with his ideas but before they were patented or announced), Hoerni had another "epiphany" while showering one morning: tiny windows could be engraved in this protective oxide layer to allow impurities to be diffused at precise spots in order to create the desired semiconductor properties. Noyce loved this idea of "building a transistor inside a cocoon," and he compared it to "setting up your jungle operating room—you put the patient inside a plastic bag and you operate inside of that, and you don't have all the flies of the jungle sitting on the wound."[7]

The role of patent lawyers is to protect good ideas, but sometimes they also stimulate them. The planar process became an example of this. Noyce called in John Ralls, Fairchild's patent lawyer, to prepare an application. So Ralls began grilling Hoerni, Noyce, and their coworkers: What practical things could be done with this planar process? Ralls was probing to obtain the widest range of possible uses to put in the patent application. Recalled Noyce, "The challenge from Ralls was, 'What else can we do with these ideas in terms of patent protection?'"[8]

At the time, Hoerni's idea was merely designed to build a reliable transistor. It had not yet occurred to them that the planar process with its tiny windows could be used to permit many types of transistors and other components to be etched onto a single piece of silicon. But Ralls's persistent questioning got Noyce thinking, and he spent time that January batting around ideas with Moore, scribbling them on a blackboard and jotting them into his notebook.

Noyce's first realization was that the planar process could eliminate the tiny wires that stuck out of each layer of the transistor. In their place, little copper lines could be printed on top of the oxide layer. That would make manufacturing the transistors faster and more reliable. This led to Noyce's next insight: if you used these printed copper lines to connect the regions of a transistor, you could also use them to connect two or more transistors that were on the same piece of silicon. The planar process with its window technique would allow you to diffuse impurities so that multiple transistors could be placed on the same silicon chip, and the printed copper wires could connect them into a circuit. He walked into Moore's office and drew the idea on the blackboard for him.

Noyce was a talkative bundle of energy and Moore was a taciturn yet insightful sounding board, and they played off each other well. The next leap was easy: the same chip could also contain various components, such as resistors and capacitors. Noyce scribbled on Moore's blackboard to show how a small section of pure silicon could serve as a resistor, and a few days later he sketched out how to make a silicon capacitor. The little metal lines printed on the oxide surface could integrate all of these components into a circuit. "I don't remember any time when a light bulb went off and the whole thing was there," conceded Noyce. "It was more like, every day, you would say, 'Well, if I could do this, then maybe I could do that, and that would let me do this,' and eventually you had the concept."[9] After this flurry of activity he wrote an entry in his notebook, in January 1959: "It would be desirable to make multiple devices on a single piece of silicon."[10]

Noyce had come up with the concept of a microchip independently of (and a few months later than) Kilby, and they had gotten there in different ways. Kilby was trying to solve the problem of how to overcome the tyranny of numbers by creating circuits with many components that did not have to be soldered together. Noyce was mainly motivated by trying to figure out all the neat tricks that could come from Hoerni's planar process. There was one other, more practical difference: Noyce's version didn't have a messy spider's nest of wires protruding from it.

PROTECTING DISCOVERIES

Patents present an inevitable source of tension in the history of invention, especially so in the digital age. Innovations tend to proceed through collaboration and building on the work of others, so it is difficult to ascribe with precision the ownership of ideas or intellectual property rights. Occasionally this is made gloriously irrelevant when a group of innovators agrees to engage in an open-source process that allows the fruits of their creativity to be in the public domain. More often, however, an innovator wants credit. Sometimes this is for ego reasons, as was the case when Shockley maneuvered to be listed on the patents for the transistor. At other times it is for financial reasons, especially when it involves companies such as Fairchild and Texas Instruments that need to reward investors in order to have the working capital necessary to keep inventing things.

In January 1959 the lawyers and executives at Texas Instruments began scrambling to file a patent application for Kilby's idea of an integrated circuit—not because they knew what Noyce was jotting in his notebook but because of rumors that RCA had come up with the same idea. They decided to make the application sweeping and broad. That strategy carried a risk because the claims might be easier to dispute, as happened with Mauchly and Eckert's broad claims for their computer patent. But if granted it would serve as an offensive weapon against anyone who tried to make a product that was similar. Kilby's invention, the patent application declared, was "a new and totally different concept for miniaturization." Although the application described only two circuits that Kilby had devised, it asserted, "There is no limit upon the complexity or configuration of circuits that can be made in this manner."

In the rush, however, there wasn't time to produce pictures of the various methods that might work for wiring together the components on the proposed microchip. The only example available was Kilby's spidery demonstration model with a snarl of tiny gold wires threading through it. The Texas Instruments team decided to use this "flying wire picture," as it was later derisively called, as the depiction. Kilby had already figured out that there could be a simpler version using printed metal connections, so at the last moment he told his lawyers to add a passage to the application claiming rights to that concept as well. "Instead of using the gold wires in making electrical connections," it noted, "connections may be provided in other ways. For example . . . silicon oxide may be evaporated onto the semiconductor circuit wafer. . . . Material such as gold may then be laid down on the insulating material to make the necessary electrical connections." It was filed in February 1959.[11]

When Texas Instruments made its public announcement the next month, Noyce and his team at Fairchild hastened to file a competing patent application. Because they were seeking a shield against Texas Instruments' sweeping claim, the Fairchild lawyers focused very specifically on what was special about Noyce's version. They emphasized that the planar process, which Fairchild had already filed to patent, permitted a printed-circuit method "for making electrical connections to the various semiconductor regions" and "to make unitary circuit structures more compact and more easily fabricated." Unlike circuits in which "electrical connection had to be made by fastening wires," declared the Fairchild application, Noyce's method meant that "the leads can be deposited at the same time and in the same manner as the contacts themselves." Even if Texas Instruments should be awarded a patent for putting multiple components on a single chip, Fairchild hoped to be awarded a patent for making the connections through printed metal lines instead of wires. Because this would be necessary for mass-producing microchips, Fairchild knew it would give them some parity in patent protection and force Texas Instruments to enter into a cross-licensing deal. The Fairchild application was filed in July 1959.[12]

As happened with the patent dispute over the computer, the legal system took years grappling with the issue of who deserved what patents on the integrated circuit, and it never quite resolved the question. The rival applications from Texas Instruments and Fairchild were assigned to two different examiners, who each seemed unaware of the other. Although filed second, the Noyce patent application was ruled on first; in April 1961 it was granted. Noyce was declared the inventor of the microchip.

The Texas Instruments lawyers filed a "priority contest," claiming that Kilby had the idea first. That led to the case of *Kilby v. Noyce*, run by the Board of Patent Interferences. Part of the case involved looking at the respective notebooks and other testimony to see who had come up with the general concept first; there was broad agreement, even from Noyce, that Kilby's ideas had come a few months earlier. But there was also a dispute over whether the Kilby application really covered the key technological process of printing metal lines on top of an oxide layer, rather than using a lot of little wires, to make a microchip. This involved many conflicting arguments about the phrase Kilby had inserted at the end of the application, that "material such as gold may then be laid down" on the oxide layer. Was that a specific process he had discovered or merely a casual speculation he had tossed in?[13]

As the wrangling dragged on, the patent office confused things a bit further by ruling, in June 1964, on Kilby's original application—and granting it. That made the priority contest all the more important. It was not until February 1967 that the verdict finally came, in Kilby's favor. It had been eight years since he had filed for his patent, and now he and Texas Instruments were declared the inventors of the microchip. Except that didn't end things. Fairchild appealed, and the Court of Customs and Patent Appeals, after hearing all the arguments and testimony, ruled in November 1969 the other way. "Kilby has not demonstrated," the appeals court declared, "that the term 'laid down' had . . . or has since acquired a meaning in electronic or semiconductor arts which necessarily connotes adherence."[14] Kilby's lawyer tried to appeal to the U.S. Supreme Court, which declined to take the case.

Noyce's victory, after a decade of back-and-forth and more than a million dollars in legal fees, turned out to mean little. The subhead on the small story in *Electronic News* was "Patent Reversal Won't Change Much." By this point the legal proceedings had become almost irrelevant. The market for microchips had exploded so rapidly that the businesslike folks at Fairchild and Texas Instruments realized that the stakes were too high to leave to the legal system. In the summer of 1966, three years before the final legal resolution, Noyce and his Fairchild lawyers met with the president and counsel of Texas Instruments and hammered out a peace treaty. Each company granted that the other had some intellectual property rights to the microchip, and they agreed to cross-license to each other whatever rights they had. Other companies would have to make licensing deals with both, usually paying a royalty fee that totaled about 4 percent of their profit.[15]

So who invented the microchip? As with the question of who invented the computer, the answer cannot be settled simply by reference to legal rulings. The nearly simultaneous advances made by Kilby and Noyce showed that the atmosphere of the time was primed for such an invention. Indeed, many people around the country and world, including Werner Jacobi at Siemens in Germany and Geoffrey Dummer of the Royal Radar Establishment in Britain, had earlier proposed the possibility of an integrated circuit. What Noyce and Kilby did, in collaboration with teams at their companies, was figure out practical methods to produce such a device. Although Kilby was a few months earlier in coming up with a way to integrate components on a chip, Noyce did something more: he devised the right way to connect these components. His design could be mass-produced efficiently, and it became the general model for future microchips.

There is an inspiring lesson in how Kilby and Noyce personally handled the question of who invented the microchip. They were both decent people; they came from tight-knit small communities in the Midwest and were well grounded. Unlike Shockley, they did not suffer from a toxic mix of ego and insecurity. Whenever the topic of credit for the invention came up, each was generous in praising the contributions of the other. It soon became accepted to give them joint credit and refer to them as coinventors. In one early oral history, Kilby gently grumbled, "It doesn't fit with what I understand to be co-invention, but that's become accepted."[16] But he, too, eventually embraced the idea and was ever afterward gracious about it. When Craig Matsumoto of *Electronic Engineering Times* asked him about the controversy many years later, "Kilby heaped praise on Noyce and said the semiconductor revolution came from the work of thousands, not from one patent."[17]

When Kilby was told that he had won the Nobel Prize in 2000, ten years after Noyce had died,* among the first things he did was praise Noyce. "I'm sorry he's not still alive," he told reporters. "If he were, I suspect we'd share this prize." When a Swedish physicist introduced him at the ceremony by saying that his invention had launched the global Digital Revolution, Kilby displayed his aw-

---

*Only living people can be selected for a Nobel.

shucks humility. "When I hear that kind of thing," he responded, "it reminds me of what the beaver told the rabbit as they stood at the base of Hoover Dam: 'No, I didn't build it myself, but it's based on an idea of mine.'"[18]

MICROCHIPS BLAST OFF

The first major market for microchips was the military. In 1962 the Strategic Air Command designed a new land-based missile, the Minuteman II, that would each require two thousand microchips just for its onboard guidance system. Texas Instruments won the right to be the primary supplier. By 1965 seven Minutemen were being built each week, and the Navy was also buying microchips for its submarine-launched missile, the Polaris. With a coordinated astuteness not often found among military procurement bureaucracies, the designs of the microchips were standardized. Westinghouse and RCA began supplying them as well. So the price soon plummeted, until microchips were cost-effective for consumer products and not just missiles.

Fairchild also sold chips to weapons makers, but it was more cautious than its competitors about working with the military. In the traditional military relationship, a contractor worked hand in glove with uniformed officers, who not only managed procurement but also dictated and fiddled with design. Noyce believed such partnerships stifled innovation: "The direction of the research was being determined by people less competent in seeing where it ought to go."[19] He insisted that Fairchild fund the development of its chips using its own money so that it kept control of the process. If the product was good, he believed, military contractors would buy it. And they did.

America's civilian space program was the next big booster for microchip production. In May 1961 President John F. Kennedy declared, "I believe that this nation should commit itself to achieving the goal, before this decade is out, of landing a man on the moon and returning him safely to the earth." The Apollo program, as it became known, needed a guidance computer that could fit into a nose cone. So it was designed from scratch to use the most powerful microchips that could be made. The seventy-five Apollo Guidance Computers that were built ended up containing five thousand microchips apiece, all identical, and Fairchild landed the contract to supply them. The program beat Kennedy's deadline by just a few months; in July 1969 Neil Armstrong set foot on the moon. By that time the Apollo program had bought more than a million microchips.

These massive and predictable sources of demand from the government caused the price of each microchip to fall rapidly. The first prototype chip for the Apollo Guidance Computer cost $1,000. By the time they were being put into regular production, each cost $20. The average price for each microchip in the Minuteman missile was $50 in 1962; by 1968 it was $2. Thus was launched the market for putting microchips in devices for ordinary consumers.[20]

The first consumer devices to use microchips were hearing aids because they needed to be very small and would sell even if they were rather expensive. But the demand for them was limited. So Pat Haggerty, the president of Texas Instruments, repeated a gambit that had served him in the past. One aspect of innovation is inventing new devices; another is inventing popular ways to use these devices. Haggerty and his company were good at both. Eleven years after he had created a huge market for inexpensive transistors by pushing pocket radios, he looked for a way to do the same for microchips. The idea he hit upon was pocket calculators.

On a plane ride with Jack Kilby, Haggerty sketched out his idea and handed Kilby his marching orders: Build a handheld calculator that can do the same tasks as the thousand-dollar clunkers that sit on office desks. Make it efficient enough to run on batteries, small enough to put into a shirt pocket, and cheap enough to buy on impulse. In 1967 Kilby and his team produced almost what Haggerty envisioned. It could do only four tasks (add, subtract, multiply, and divide) and was a bit heavy (more than two pounds) and not very cheap ($150).[21] But it was a huge success. A new market had been created for a device people had not known they needed. And following the inevitable trajectory, it kept getting smaller, more powerful, and cheaper. By 1972 the price of a pocket calculator had dropped to $100, and 5 million units were sold. By 1975 the price was down to

$25, and sales were doubling every year. In 2014 a Texas Instruments pocket calculator cost $3.62 at Walmart.

MOORE'S LAW

That became the pattern for electronic devices. Every year things got smaller, cheaper, faster, more powerful. This was especially true—and important—because two industries were growing up simultaneously, and they were intertwined: the computer and the microchip. "The synergy between a new component and a new application generated an explosive growth for both," Noyce later wrote.[22] The same synergy had happened a half century earlier when the oil industry grew in tandem with the auto industry. There was a key lesson for innovation: Understand which industries are symbiotic so that you can capitalize on how they will spur each other on.

If someone could provide a pithy and accurate rule for predicting the trend lines, it would help entrepreneurs and venture capitalists to apply this lesson. Fortunately, Gordon Moore stepped forward at that moment to do so. Just as the microchip sales were starting to skyrocket, he was asked to forecast the future market. His paper, titled "Cramming More Components onto Integrated Circuits," was published in the April 1965 issue of *Electronics* magazine.

Moore began with a glimpse of the digital future. "Integrated circuits will lead to such wonders as home computers—or at least terminals connected to a central computer—automatic controls for automobiles, and personal portable communications equipment," he wrote. Then he produced an even more prescient prediction that was destined to make him famous. "The complexity for minimum component costs has increased at a rate of roughly a factor of two per year," he noted. "There is no reason to believe it will not remain nearly constant for at least ten years."[23]

Roughly translated, he was saying that the number of transistors that could be crammed, cost-effectively, onto a microchip had been doubling every year, and he expected it to do so for at least the next ten years. One of his friends, a professor at Caltech, publicly dubbed this "Moore's Law." In 1975, when the ten years had passed, Moore was proved right. He then modified his law by cutting the predicted rate of increase by half, prophesying that the future numbers of transistors crammed onto a chip would show "a doubling every two years, rather than every year." A colleague, David House, offered a further modification, now sometimes used, which said chip "performance" would double every eighteen months because of the increased power as well as the increased numbers of transistors that would be put onto a microchip. Moore's formulation and its variations proved to be useful at least through the subsequent half century, and it helped chart the course for one of the greatest bursts of innovation and wealth creation in human history.

Moore's Law became more than just a prediction. It was also a goal for the industry, which made it partly self-fulfilling. The first such example occurred in 1964, as Moore was formulating his law. Noyce decided that Fairchild would sell its simplest microchips for less than they cost to make. Moore called the strategy "Bob's unheralded contribution to the semiconductor industry." Noyce knew that the low price would cause device makers to incorporate microchips into their new products. He also knew that the low price would stimulate demand, high-volume production, and economies of scale, which would turn Moore's Law into a reality.[24]

Fairchild Camera and Instrument decided, not surprisingly, to exercise its right to buy out Fairchild Semiconductor in 1959. That made the eight founders rich but sowed seeds of discord. The corporation's East Coast executives refused to give Noyce the right to hand out stock options to new and valued engineers, and they sucked up the semiconductor division profits to fund less successful investments in more mundane realms, such as home movie cameras and stamp machines.

There were also internal problems in Palo Alto. Engineers began defecting, thus seeding the valley with what became known as Fairchildren: companies that sprouted from spores emanating from Fairchild. The most notable came in 1961, when Jean Hoerni and three of the other eight defectors from Shockley left Fairchild to join a startup, funded by Arthur Rock, that became Teledyne. Others followed, and by 1968 Noyce himself was ready to leave. He had been passed over for the top corporate job at Fairchild, which ticked him off, but he also realized that he did not really want it. Fairchild, the corporation as a whole and even the semiconductor division in Palo Alto, had become too big and bureaucratic. Noyce yearned to shed some managerial duties and return to being close to the lab.

"How about starting a new company?" he asked Moore one day.

"I like it here," Moore replied.[25] They had helped to create the culture of the California tech world, in which people left established companies to form new ones. But now, as they were both hitting forty, Moore no longer had the urge to jump off the roof in a hang glider. Noyce kept pressing. Finally, as the summer of 1968 approached, he simply told Moore he was leaving. "He had a way of making you want to take a leap with him," Moore said many years later, laughing. "So finally I said, 'Okay, let's go.'"[26]

"As [the company] has grown larger and larger, I have enjoyed my daily work less and less," Noyce wrote in his letter of resignation to Sherman Fairchild. "Perhaps this is partly because I grew up in a small town, enjoying all the personal relationships of a small town. Now we employ twice the total population of my largest 'home town.'" His desire, he said, was to "get close to advanced technology again."[27]

When Noyce called Arthur Rock, who had put together the financing deal that launched Fairchild Semiconductor, Rock immediately asked, "What took you so long?"[28]

ARTHUR ROCK AND VENTURE CAPITAL

In the eleven years since he had assembled the deal for the traitorous eight to form Fairchild Semiconductor, Arthur Rock had helped to build something that was destined to be almost as important to the digital age as the microchip: venture capital.

For much of the twentieth century, venture capital and private equity investing in new companies had been mainly the purview of a few wealthy families, such as the Vanderbilts, Rockefellers, Whitneys, Phippses, and Warburgs. After World War II, many of these clans set up firms to institutionalize the business. John Hay "Jock" Whitney, an heir to multiple family fortunes, hired Benno Schmidt Sr. to form J. H. Whitney & Co., which specialized in what they originally called "adventure capital" to fund entrepreneurs with interesting ideas who could not get bank loans. The six sons and one daughter of John D. Rockefeller Jr., led by Laurence Rockefeller, started a similar firm, which eventually became Venrock Associates. That same year, 1946, also saw the birth of the most influential entry, one that was based on business acumen rather than family wealth: the American Research and Development Corporation (ARDC). It was founded by Georges Doriot, a former dean of the Harvard Business School, in partnership with a former MIT president, Karl Compton. ARDC scored big by doing a startup investment in Digital Equipment Corporation in 1957, which was worth five hundred times as much when the company went public eleven years later.[29]

Arthur Rock took this concept west, ushering in the silicon age of venture capital. When he put together Noyce's traitorous eight with Fairchild Camera, Rock and his company took a stake in the deal. After that, he realized that he could raise a fund of money and do similar deals without relying on one corporate patron. He had a background in business research, a love of technology, an intuitive feel for business leadership, and a lot of East Coast investors he had made happy. "The money was on the East Coast but the exciting companies were in California, so I decided to move west knowing that I could connect the two," he said.[30]

Rock grew up the son of Russian Jewish immigrants in Rochester, New York, where he worked as a soda jerk in his father's candy store and developed a good feel for personalities. One of his key investment maxims was to bet primarily on the people rather than the idea. In addition to going over business plans, he conducted incisive personal interviews with those who sought funding. "I believe so strongly in people that I think talking to the individual is much more important than finding out too much about what they want to do," he explained. On the surface, he wore the cloak of the curmudgeon, with a gruff and taciturn style. But those who looked at his face closely enough could tell from the light in his eyes and the hints of a smile that he enjoyed people and had a warm sense of humor.

When he got to San Francisco, he was introduced to Tommy Davis, a talkative deal maker who was investing the money of the Kern County Land Co., a cattle and oil empire flush with cash. They went into business together as Davis and Rock, raised $5 million from Rock's East Coast investors (as well as some of the Fairchild founders), and started funding new companies in return for a chunk of the equity. Stanford's provost Fred Terman, still seeking to build his university's ties to the growing tech boom, encouraged his engineering professors to spend time advising Rock, who took a night course in electronics at the university. Two of his first bets were on Teledyne and Scientific Data Systems, which both paid off handsomely. By the time Noyce called him about finding an exit strategy from Fairchild in 1968, Rock's partnership with Davis had amiably dissolved (their investments had shot up thirtyfold in seven years) and he was on his own.

"If I wanted to start a company," Noyce asked, "could you find me the money?" Rock assured him it would be easy. What could better fit his theory that you place your money on the jockeys—that you invest based on your assessment of the people running the company—than an enterprise that would be led by Robert Noyce and Gordon Moore? He barely asked what they were going to make, and at first he didn't even think they needed to do a business plan or description. "It was the only investment that I've ever made that I was 100 percent sure would succeed," he later claimed.[31]

When he had sought a home for the traitorous eight in 1957, he pulled out a single piece of legal-pad paper, wrote a numbered list of names, and methodically phoned each one, crossing off names as he went down the list. Now, eleven years later, he took another sheet of paper and listed people who would be invited to invest and how many of the 500,000 shares* available at $5 apiece he would offer to

*The vehicle he used was convertible debentures, which were loans that could be converted into common stock if the company became successful but were worthless (at the end of the line of creditors) if it failed.

each. This time around, he would cross out only one name. ("Johnson at Fidelity"* didn't come in.) Rock needed a second sheet to revise the allocations because most people wanted to invest more than he offered them. It took him less than two days to raise the money. The lucky investors included Rock himself, Noyce, Moore, Grinnell College (Noyce wanted to make it rich, and he did), Laurence Rockefeller, Rock's Harvard classmate Fayez Sarofim, Max Palevsky of Scientific Data Systems, and Rock's old investment firm, Hayden, Stone. Most notably, the other six members of the traitorous eight, many of them now working at firms that would have to compete with this new one, were given a chance to invest. All did.

Just in case someone desired a prospectus, Rock himself typed up a three-and-a-half-page sketch of the proposed company. It opened by describing Noyce and Moore and then gave a perfunctory three-sentence overview of the "transistor technologies" the company would develop. "Lawyers later screwed up venture investing by forcing us to write prospectus books that were so long and complex and carefully vetted that it's a joke," Rock complained later, pulling the pages out of his file cabinet. "All I had to tell people was that it was Noyce and Moore. They didn't need to know much else."[32]

The first name that Noyce and Moore chose for their new company was NM Electronics, their initials. That was not very exciting. After many clunky suggestions—Electronic Solid State Computer Technology Corp. was one—they finally decided on Integrated Electronics Corp. That wasn't very thrilling, either, but it had the virtue that it could be abridged—as Intel. That had a nice ring to it. It was smart and knowing, in many different ways.

THE INTEL WAY

Innovations come in a variety of guises. Most of those featured in this book are physical devices, such as the computer and the transistor,

---

*Edward "Ned" Johnson III, then running the Fidelity Magellan Fund. In 2013 Rock still had these two sheets, along with the older one seeking the patron for what became Fairchild, tucked in a filing cabinet in his office overlooking San Francisco Bay.

and related processes, such as programming, software, and networking. Also important are the innovations that produce new services, such as venture capital, and those that create organizational structures for research and development, such as Bell Labs. But this section is about a different type of creation. There arose at Intel an innovation that had almost as much of an impact on the digital age as any of these. It was the invention of a corporate culture and management style that was the antithesis of the hierarchical organization of East Coast companies.

The roots of this style, like much of what happened in Silicon Valley, were at Hewlett-Packard. During World War II, while Bill Hewlett was in the military, Dave Packard slept on a cot at the office many nights and managed three shifts of workers, many of them women. He realized, partly out of necessity, that it helped to give his workers flexible hours and plenty of leeway in determining how to accomplish their objectives. The management hierarchy was flattened. During the 1950s this approach merged with the casual lifestyle of California to create a culture that included Friday beer bashes, flexible hours, and stock options.[33]

Robert Noyce took this culture to the next level. To understand him as a manager, it's useful to recall that he was born and bred a Congregationalist. His father and both grandfathers were ministers of the dissenting denomination that had as its core creed the rejection of hierarchy and all of its trappings. The Puritans had purified the church of all pomp and levels of authority, even going as far as eliminating elevated pulpits, and those who spread this Nonconformist doctrine to the Great Plains, including the Congregationalists, were just as averse to hierarchical distinctions.

It also helps to remember that, from his early days as a student, Noyce loved madrigal singing. Every Wednesday evening he attended rehearsals of his twelve-voice group. Madrigals don't rely on lead singers and soloists; the polyphonic songs weave multiple voices and melodies together, none of them dominant. "Your part depends on [the others' and] it always supports the others," Noyce once explained.[34]

Gordon Moore was similarly unpretentious, nonauthoritarian,

averse to confrontation, and uninterested in the trappings of power. They complemented each other well. Noyce was Mr. Outside; he could dazzle a client with the halo effect that had followed him since childhood. Moore, always temperate and thoughtful, liked being in the lab, and he knew how to lead engineers with subtle questions or (the sharpest arrow in his quiver) a studied silence. Noyce was great at strategic vision and seeing the big picture; Moore understood the details, particularly of the technology and engineering.

So they were perfect partners, except in one way: with their shared aversion to hierarchy and unwillingness to be bossy, neither was a decisive manager. Because of their desire to be liked, they were reluctant to be tough. They guided people but didn't drive them. If there was a problem or, heaven forbid, a disagreement, they did not like to confront it. So they wouldn't.

That's where Andy Grove came in.

Grove, born András Gróf in Budapest, did not come from a madrigal-singing Congregationalist background. He grew up Jewish in Central Europe as fascism was rising, learning brutal lessons about authority and power. When he was eight, the Nazis took over Hungary; his father was sent to a concentration camp, and András and his mother were forced to move into a special cramped apartment for Jews. When he went outside, he had to wear a yellow Star of David. One day when he got sick, his mother was able to convince a non-Jewish friend to bring some ingredients for soup, which led to the arrest of both his mother and the friend. After she was released, she and András assumed false identities while friends sheltered them. The family was reunited after the war, but then the communists took over. Grove decided, at age twenty, to flee across the border to Austria. As he wrote in his memoir, *Swimming Across*, "By the time I was twenty, I had lived through a Hungarian Fascist dictatorship, German military occupation, the Nazis' Final Solution, the siege of Budapest by the Soviet Red Army, a period of chaotic democracy in the years immediately after the war, a variety of repressive Communist regimes, and a popular uprising that was put down at gunpoint."[35] It wasn't like mowing lawns and singing in a small-town Iowa choir, and it did not instill genial mellowness.

Grove arrived in the United States a year later and, as he taught himself English, was able to graduate first in his class at City College of New York and then earn a PhD in chemical engineering from Berkeley. He joined Fairchild in 1963 right out of Berkeley, and in his spare time wrote a college textbook titled *Physics and Technology of Semiconductor Devices.*

When Moore told him of his plans to leave Fairchild, Grove volunteered to come along. In fact, he almost forced himself on Moore. "I really respected him and wanted to go wherever he went," Grove declared. He became the third person at Intel, serving as the director of engineering.

Grove had deep admiration for Moore's technical skills but not his management style. That was understandable, given Moore's aversion to confrontation and almost any aspect of management beyond proffering gentle advice. If there was a conflict, he would watch quietly from afar. "He is either constitutionally unable or simply unwilling to do what a manager has to do," Grove said of Moore.[36] The feisty Grove, by contrast, felt that honest confrontation was not only a managerial duty but one of life's invigorating spices, which as a hardened Hungarian he relished.

Grove was even more appalled by the management style of Noyce. At Fairchild he had simmered with fury when Noyce ignored the incompetence of one of his division heads, who showed up late and drunk at meetings. Thus he groaned when Moore said that his new venture would be in partnership with Noyce. "I told him that Bob was a better leader than Andy gave him credit for," Moore said. "They just had different styles."[37]

Noyce and Grove got along socially better than they did professionally. They went with their families to Aspen, where Noyce helped Grove learn to ski and even buckled his boots for him. Nevertheless, Grove detected a detachment in Noyce that could be disconcerting: "He was the only person I can think of who was both aloof and charming."[38] In addition, despite their weekend friendship, Grove found himself irritated and sometimes appalled by Noyce at the office. "I had nothing but unpleasant, discouraging dealings with him as I watched Bob manage a troubled company," he recalled. "If two people argued and we all looked to him for a decision, he would put a pained look on his face and said something like, 'Maybe you should work that out.' More often he didn't say that, he just changed the subject."[39]

What Grove did not realize at the time, but came to understand later, was that effective management need not always come from having one strong leader. It can come from having the right combination of different talents at the top. Like a metallic alloy, if you get the right mix of elements the result can be strong. Years later, after Grove had learned to appreciate this, he read Peter Drucker's *The Practice of Management*, which described the ideal chief executive as an outside person, an inside person, and a person of action. Grove realized that instead of being embodied in one person, such traits could exist in a leadership team. That was the case at Intel, Grove said, and he made copies of the chapter for Noyce and Moore. Noyce was the outside guy, Moore the inside, and Grove was the man of action.[40]

Arthur Rock, who put together the funding for the trio and initially served as their board chair, understood the virtue of creating an executive team whose members complemented each other. He also noted a corollary: it was important that the trifecta become CEO in the order that they did. Noyce he described as "a visionary who knew how to inspire people and sell the company to others when it was getting off the ground." Once that was done, Intel needed to be led by someone who could make it a pioneer in each new wave of technology, "and Gordon was such a brilliant scientist he knew how to drive the technology." Then, when there were dozens of other companies competing, "we needed a hard-charging, no-nonsense manager who could focus on driving us as a business." That was Grove.[41]

The Intel culture, which would permeate the culture of Silicon Valley, was a product of all three men. As might be expected in a congregation where Noyce was the minister, it was devoid of the trappings of hierarchy. There were no reserved parking places. Everyone, including Noyce and Moore, worked in similar cubicles. Michael Malone, a reporter, described visiting Intel to do an interview: "I couldn't find Noyce. A secretary had to come out and lead me to his cubicle, because his cubicle was almost indistinguishable from all the other cubicles in this vast prairie dog town of cubicles."[42]

When one early employee wanted to see the company's organization chart, Noyce made an X in the center of a page and then drew a bunch of other Xs around it, with lines leading to each. The employee was at the center, and the others were people he would be dealing with.[43] Noyce noticed that at East Coast companies the clerks and secretaries got little metal desks while those of top executives were expansive ones made of mahogany. So Noyce decided that he would work at a small gray aluminum desk, even as newly hired support staffers were given bigger wooden ones. His dented and scratched desk was near the center of the room, in open view, for everyone to see. It prevented anyone else from demanding some vestment of power. "There were no privileges anywhere," recalled Ann Bowers, who was the personnel director and later married Noyce.* "We started a form of company culture that was completely different than anything had been before. It was a culture of meritocracy."[44]

It was also a culture of innovation. Noyce had a theory that he developed after bridling under the rigid hierarchy at Philco. The more open and unstructured a workplace, he believed, the faster new ideas would be sparked, disseminated, refined, and applied. "The idea is people should not have to go up through a chain of command," said one of Intel's engineers, Ted Hoff. "If you need to talk to a particular manager you go talk to him."[45] As Tom Wolfe put it in his profile, "Noyce realized how much he detested the eastern corporate system of class and status with its endless gradations, topped off by the CEOs and vice-presidents who conducted their daily lives as if they were a corporate court and aristocracy."

By avoiding a chain of command, both at Fairchild Semiconductor and then at Intel, Noyce empowered employees and forced them to be entrepreneurial. Even though Grove cringed when disputes went unresolved at meetings, Noyce was comfortable letting junior employees resolve problems rather than bucking them up to a higher layer of management that would tell them what to do. Responsibil-

---

*After she married Noyce she had to leave Intel, and she moved to the fledgling Apple Computer, where she became Steve Jobs's first director of human resources and also a calming maternal influence on him.

ity was thrust on young engineers, who found themselves having to be innovators. Every now and then, a staffer might be unnerved by a tough problem. "He would go to Noyce and hyperventilate and ask him what to do," Wolfe reported. "And Noyce would lower his head, turn on his 100 ampere eyes, listen, and say: 'Look, here are your guidelines. You've got to consider A, you've got to consider B, and you've got to consider C.' Then he would turn on the Gary Cooper smile: 'But if you think I'm going to make your decision for you, you're mistaken. Hey . . . it's your ass.'"

Instead of proposing plans to top management, Intel's business units were entrusted to act as if they were their own little and agile company. Whenever there was a decision that required buy-in from other units, such as a new marketing plan or a change in a product strategy, the issue would not be bucked up to bosses for a decision. Instead an impromptu meeting would be convened to hash it out, or try to. Noyce liked meetings, and there were rooms set aside for whenever anyone felt the need to call one. At these meetings everyone was treated as an equal and could challenge the prevailing wisdom. Noyce was there not as a boss but as a pastor guiding them to make their own decisions. "This wasn't a corporation," Wolfe concluded. "It was a congregation."[46]

Noyce was a great leader because he was inspiring and smart, but he was not a great manager. "Bob operated on the principle that if you suggested to people what the right thing to do would be, they would be smart enough to pick it up and do it," said Moore. "You didn't have to worry about following up."[47] Moore admitted that he was not much better: "I was never very eager to exert authority or be the boss either, which might mean we were too much alike."[48]

Such a management style needed someone to impose discipline. Early on at Intel, well before it was his turn in the lineup to become CEO, Grove helped institute some management techniques. He created a place where people were held accountable for sloppiness. Failures had consequences. "Andy would fire his own mother if she got in the way," said one engineer. Another colleague explained that this was necessary in an organization headed by Noyce: "Bob really has to be a nice guy. It's important for him to be liked. So somebody has to kick ass and take names. And Andy happens to be very good at that."[49]

Grove began to study and absorb the art of management as if it were the science of circuitry. He would later become a best-selling author of books with titles such as *Only the Paranoid Survive* and *High Output Management*. He did not try to impose a hierarchal command on what Noyce had wrought. Instead he helped to instill a culture that was driven, focused, and detail-aware, traits that would not naturally have arisen from Noyce's laid-back, nonconfrontational style. His meetings were crisp and decisive, unlike those run by Noyce, where people tended to hang around as long as possible knowing that he was likely to tacitly assent to the last person who had his ear.

What saved Grove from seeming like a tyrant was that he was so irrepressible, which made him hard not to like. When he smiled, his eyes lit up. He had a pixielike charisma. With his Hungarian accent and goofy grin, he was by far the most colorful engineer in the valley. He succumbed to the dubious fashions of the early 1970s by attempting, in an immigrant geek manner worthy of a *Saturday Night Live* skit, to be *groovy*. He grew his sideburns long and his mustache droopy and wore open shirts with gold chains dangling over his chest hair. None of which hid the fact that he was a real engineer, one who had been a pioneer of the metal-oxide semiconductor transistor that became the workhorse of modern microchips.

Grove nurtured Noyce's egalitarian approach—he worked in an exposed cubicle his entire career, and loved it—but he added an overlay of what he called "constructive confrontation." He never put on airs, but he never let down his guard. In contrast to Noyce's sweet gentility, Grove had a blunt, no-bullshit style. It was the same approach Steve Jobs would later use: brutal honesty, clear focus, and a demanding drive for excellence. "Andy was the guy who made sure the trains all ran on time," recalled Ann Bowers. "He was a taskmaster. He had very strong views about what you should do and what you shouldn't do and he was very direct about that."[50]

Despite their different styles, there was one thing that Noyce and Moore and Grove shared: an unwavering goal of making sure that innovation, experimentation, and entrepreneurship flourished at Intel. Grove's mantra was "Success breeds complacency. Complacency breeds failure. Only the paranoid survive." Noyce and Moore may not have been paranoid, but they were never complacent.

THE MICROPROCESSOR

Inventions sometimes occur when people are confronted with a problem and scramble to solve it. At other times, they happen when people embrace a visionary goal. The tale of how Ted Hoff and his team at Intel invented the microprocessor is a case of both.

Hoff, who had been a young teacher at Stanford, became the twelfth employee at Intel, where he was assigned to work on chip design. He realized that it was wasteful and inelegant to design many types of microchips that each had a different function, which Intel was doing. A company would come in and ask it to build a microchip designed to do a specific task. Hoff envisioned, as did Noyce and others, an alternative approach: creating a general-purpose chip that could be instructed, or programmed, to do a variety of different applications as desired. In other words, a general-purpose computer on a chip.[51]

This vision coincided with a problem that was dumped in Hoff's lap in the summer of 1969. A Japanese company named Busicom was planning a powerful new desktop calculator, and it had drawn up specifications for twelve special-purpose microchips (different ones to handle display, calculations, memory, etc.) that it wanted Intel to build. Intel agreed, and a price was set. Noyce asked Hoff to oversee the project. Soon a challenge arose. "The more I learned about this design, the more concerned I became that Intel may have undertaken more than it was prepared to deliver," Hoff recalled. "The number of chips and their complexity was much greater than I had expected." There was no way Intel could build them at the agreed price. Making matters worse, the growing popularity of Jack Kilby's pocket calculator was forcing Busicom to cut its price even further.

"Well, if there's anything you can think of to simplify the design, why don't you pursue it," Noyce suggested.[52]

Hoff proposed that Intel design a single logic chip that could perform almost all of the tasks that Busicom wanted. "I know this can be done," he said of the general-purpose chip. "It can be made to emulate a computer." Noyce told him to try it.

Before they could sell the idea to Busicom, Noyce realized he had to convince someone who might be even more resistant: Andy Grove, who nominally worked for him. Part of what Grove saw as his mandate was keeping Intel focused. Noyce would say yes to almost anything; Grove's job was to say no. When Noyce sauntered over to Grove's workspace and sat on the corner of his desk, Grove was immediately on guard. He knew that Noyce's effort to appear nonchalant was a sign that something was afoot. "We're starting another project," Noyce said, affecting a laugh.[53] Grove's first reaction was to tell Noyce he was crazy. Intel was a fledgling company still struggling to manufacture its memory chips, and it didn't need any distractions. But after he heard Noyce describe Hoff's idea, Grove realized that resistance was probably wrong and definitely futile.

By September 1969 Hoff and his colleague Stan Mazor had sketched out the architecture of a general-purpose logic chip that could follow programming instructions. It would be able to do the work of nine of the twelve chips that Busicom had requested. Noyce and Hoff presented the option to Busicom executives, who agreed that it was the better approach.

When it came time to renegotiate the price, Hoff made a critical recommendation to Noyce, one that helped create a huge market for general-purpose chips and assured that Intel would remain a driver of the digital age. It was a deal point that Bill Gates and Microsoft would emulate with IBM a decade later. In return for giving Busicom a good price, Noyce insisted that Intel retain the rights to the new chip and be allowed to license it to other companies for purposes other than making a calculator. He realized that a chip that could be programmed to perform any logical function would become a standard component in electronic devices, the way two-by-four pieces of lumber were a standard component in the construction of houses. It would replace custom chips, which meant it could be manufactured in bulk and thus continually decline in price. It would also usher in a more subtle shift in the electronics industry: the importance of hardware engineers, who designed the placement of the components on a circuit board, began to be supplanted by a new breed, software engineers, whose job it was to program a set of instructions into the system.

Because it was essentially a computer processor on a chip, the new device was dubbed a *microprocessor*. In November 1971 Intel unveiled the product, the Intel 4004, to the public. It took out ads in trade magazines announcing "a new era of integrated electronics—a micro-programmable computer on a chip!" It was priced at $200, and orders, as well as thousands of requests for the manual, began pouring in. Noyce was attending a computer show in Las Vegas on the day of the announcement and was thrilled to watch potential customers cramming into the Intel suite.

Noyce became an apostle of the microprocessor. At a reunion in San Francisco he hosted for his extended family in 1972, he stood up in the bus he had chartered and waved a wafer over his head. "This is going to change the world," he told them. "It's going to revolutionize your home. In your own house, you'll all have computers. You will have access to all sorts of information." His relatives passed the wafer around the bus like an object of veneration. "You won't need money anymore," he prophesied. "Everything will happen electronically."[54]

He was exaggerating only slightly. Microprocessors began showing up in smart traffic lights and car brakes, coffeemakers and refrigerators, elevators and medical devices, and thousands of other gizmos. But the foremost success of the microprocessor was making possible smaller computers, most notably *personal* computers that you could have on your desk and in your home. And if Moore's Law continued to hold true (as it would), a personal computer industry would grow up symbiotically with a microprocessor industry.

That is what happened in the 1970s. The microprocessor spawned hundreds of new companies making hardware and software for personal computers. Intel not only developed the leading-edge chips; it also created the culture that inspired venture-funded startups to transform the economy and uproot the apricot orchards of Santa

Clara Valley, the forty-mile stretch of flat land from south San Francisco through Palo Alto to San Jose.

The valley's main artery, a bustling highway named El Camino Real, was once the royal road that connected California's twenty-one mission churches. By the early 1970s—thanks to Hewlett-Packard, Fred Terman's Stanford Industrial Park, William Shockley, Fairchild and its Fairchildren—it connected a bustling corridor of tech companies. In 1971 the region got a new moniker. Don Hoefler, a columnist for the weekly trade paper *Electronic News*, began writing a series of columns entitled "Silicon Valley USA," and the name stuck.[55]

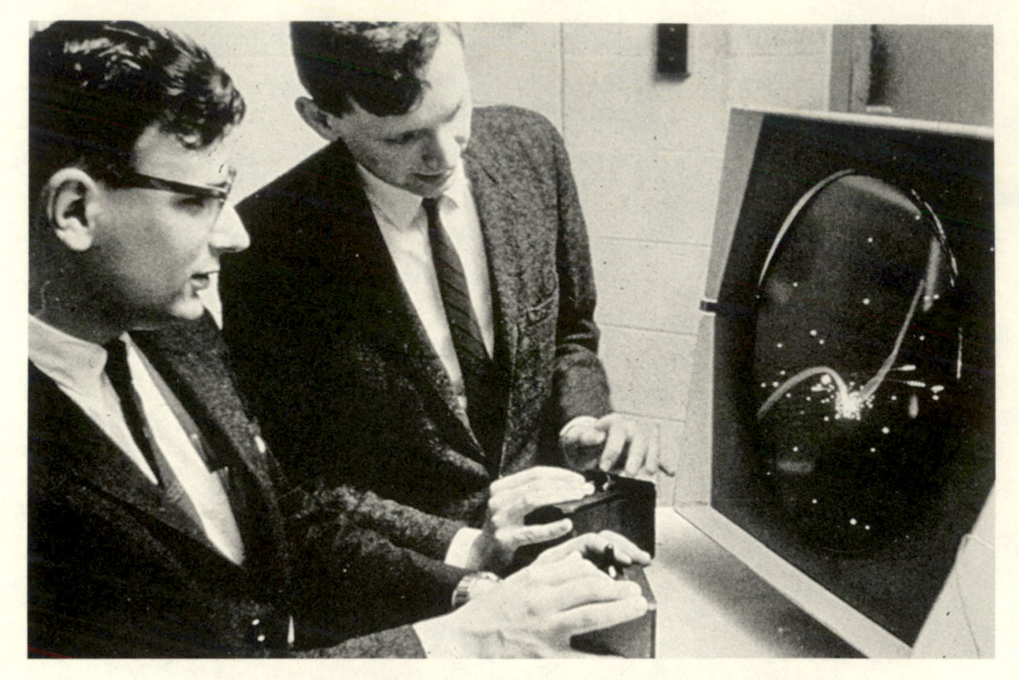

Dan Edwards and Peter Samson in 1962 playing *Spacewar* at MIT.

Nolan Bushnell (1943–).

CHAPTER SIX

VIDEO GAMES

The evolution of microchips led to devices that were, as Moore's Law forecast, smaller and more powerful each year. But there was another impetus that would drive the computer revolution and, eventually, the demand for personal computers: the belief that computers weren't merely for number-crunching. They could and should be fun for people to use.

Two cultures contributed to the idea that computers should be things that we interact and play with. There were the hard-core hackers who believed in "the hands-on imperative" and loved pranks, clever programming tricks, toys, and games.[1] And there were the rebel entrepreneurs eager to break into the amusement games industry, which was dominated by syndicates of pinball distributors and ripe for a digital disruption. Thus was born the video game, which turned out to be not merely an amusing sideshow but an integral part of the lineage that led to today's personal computer. It also helped to propagate the idea that computers should interact with people in real time, have intuitive interfaces, and feature delightful graphic displays.

STEVE RUSSELL AND *SPACEWAR*

The hacker subculture, as well as the seminal video game *Spacewar*, emanated from MIT's Tech Model Railroad Club, a geeky student organization founded in 1946 that met in the bowels of a building where radar had been developed. Its bunker was almost completely filled by a model train board with dozens of tracks, switches, trolleys, lights, and towns, all compulsively crafted and historically accurate. Most of its members obsessed over fashioning picture-perfect pieces to display on the layout. But there was a subset of the club that was more interested in what was underneath the sprawling chest-high board. The members of the "Signals and Power Subcommittee" tended to the relays, wires, circuits, and crossbar switches, which were rigged together on the underside of the board to provide a complex hierarchy of controllers for the numerous trains. In this tangled web they saw beauty. "There were neat regimental lines of switches, and achingly regular rows of dull bronze relays, and a long, rambling tangle of red, blue, and yellow wires—twisting and twirling like a rainbow-colored explosion of Einstein's hair," Steven Levy wrote in *Hackers*, which begins with a colorful depiction of the club.[2]

Members of the Signals and Power Subcommittee embraced the term *hacker* with pride. It connoted both technical virtuosity and playfulness, not (as in more recent usage) lawless intrusions into a network. The intricate pranks devised by MIT students—putting a live cow on the roof of a dorm, a plastic cow on the Great Dome of the main building, or causing a huge balloon to emerge midfield during the Harvard-Yale game—were known as hacks. "We at TMRC use the term 'hacker' only in its original meaning, someone who applies ingenuity to create a clever result, called a 'hack,'" the club proclaimed. "The essence of a 'hack' is that it is done quickly, and is usually inelegant."[3]

Some of the early hackers had been infused with the aspiration of creating machines that could think. Many were students at MIT's Artificial Intelligence Lab, founded in 1959 by two professors who would become fabled: John McCarthy, a Santa Claus lookalike who coined the term *artificial intelligence*, and Marvin Minsky, who was so clever that he seemed a refutation of his own belief that computers would someday surpass human intelligence. The prevailing doctrine of the lab was that, given enough processing power, machines could replicate neural networks like those of the human brain and be able to interact intelligently with users. Minsky, a puckish man with twinkling eyes, had built a learning machine designed to model the brain, which he named SNARC (Stochastic Neural Analog Reinforcement Calculator), hinting that he was serious but might also be joking a bit. He had a theory that intelligence could be a product of the interaction of nonintelligent components, such as small computers connected by giant networks.

A seminal moment for the hackers of the Tech Model Railroad Club came in September 1961, when the Digital Equipment Corporation (DEC) donated the prototype of its PDP-1 computer to MIT. About the size of three refrigerators, the PDP-1 was the first computer to be designed for direct interaction with the user. It could connect to a keyboard and a monitor that displayed graphics, and it could be operated easily by a single person. Like moths to a flame, a handful of hard-core hackers began to circle this new computer, and they formed a cabal to conjure up something fun to do with it. Many of the discussions took place in a rundown apartment on Hingham Street in Cambridge, so the members dubbed themselves the Hingham Institute. The high-minded name was ironic. Their goal was not to come up with some elevated use for the PDP-1 but instead to do something clever.

Previous hackers had created a few rudimentary games for earlier computers. One at MIT had a dot on a screen that represented a mouse trying to navigate a maze to find a wedge of cheese (or, in later versions, a martini); another, at the Brookhaven National Lab on Long Island, used an oscilloscope on an analog computer to simulate a tennis match. But the members of the Hingham Institute knew that with the PDP-1 they had the chance to create the first real computer video game.

The best programmer in their group was Steve Russell, who was helping Professor McCarthy create the language LISP, which was designed to facilitate artificial intelligence research. Russell was a consummate geek, brimming with passions and intellectual obsessions that ranged from steam trains to thinking machines. Short and excitable, he had thick glasses and curly hair. When he spoke, he sounded like someone had punched his fast-forward button. Although he was intense and energetic, he was prone to procrastination, earning him the nickname "Slug."

Like most of his hacker friends, Russell was an avid fan of bad movies and pulp science fiction. His favorite author was E. E. "Doc" Smith, a failed food engineer (an expert on the bleaching of flour, he concocted doughnut mixes) who specialized in a trashy sci-fi subgenre known as space opera. It featured melodramatic adventures filled with battles against evil, interstellar travel, and clichéd romance. Doc Smith "wrote with the grace and refinement of a pneumatic drill," according to Martin Graetz, a member of the Tech Model Railroad Club and the Hingham Institute, who wrote a reminiscence about the creation of *Spacewar*. Graetz recalled a typical Doc Smith tale:

> After some preliminary foofaraw to get everyone's name right, a bunch of overdeveloped Hardy Boys go trekking off through the universe to punch out the latest gang of galactic goons, blow up a few planets, kill all sorts of nasty life forms, and just have a heck of a good time. In a pinch, which is where they usually were, our heroes could be counted on to come up with a complete scientific theory, invent the technology to implement it, and produce the weapons to blow away the baddies, all while being chased in their spaceship hither and thither through the trackless wastes of the galaxy.*

---

*A sample of Doc Smith's prose, from his novel *Triplanetary* (1948): "Nerado's vessel was completely ready for any emergency. And, unlike her sister-ship, she was manned by scientists well-versed in the fundamental theory of the weapons with which they fought. Beams, rods and lances of energy flamed and flared; planes and pencils cut, slashed and stabbed; defensive screens glowed redly or flashed suddenly into intensely brilliant, coruscating incandescence. Crimson opacity struggled sullenly against violet curtains of annihilation. Material projectiles and torpedoes were launched under full-beam control; only to be exploded harmlessly in mid-space, to be blasted into nothingness or to disappear innocuously against impenetrable polycyclic screens."

Afflicted by their passion for such space operas, it's not surprising that Russell, Graetz, and their friends decided to concoct a space-war game for the PDP-1. "I had just finished reading Doc Smith's *Lensman* series," Russell recalled. "His heroes had a strong tendency to get pursued by the villain across the galaxy and have to invent their way out of their problem while they were being pursued. That sort of action was the thing that suggested *Spacewar*."[4] Proudly nerdy, they reconstituted themselves into the Hingham Institute Study Group on Space Warfare, and Slug Russell proceeded to code.[5]

Except that, true to his nickname, he didn't. He knew what the starting point of his game program would be. Professor Minsky had stumbled upon an algorithm that drew a circle on the PDP-1 and was able to modify it so that it would display three dots on the screen that interacted with each other, weaving beautiful little patterns. Minsky called his hack the Tri-Pos, but his students dubbed it "the Minskytron." That was a good foundation for creating a game featuring interacting spaceships and missiles. Russell spent weeks mesmerized by the Minskytron and grokking its ability to make patterns. But he bogged down when it came time to write the sine-cosine routines that would determine the motion of his spaceships.

When Russell explained this obstacle, a fellow club member named Alan Kotok knew how to solve it. He drove out to the suburban Boston headquarters of DEC, which made the PDP-1, and found a sympathetic engineer who had the routines necessary to make the calculations. "Alright, here are the sine-cosine routines," Kotok told Russell. "Now what's your excuse?" Russell later admitted, "I looked around and I didn't find an excuse, so I had to settle down and do some figuring."[6]

Throughout the Christmas vacation of 1961 Russell hacked away, and within weeks he had produced a method to maneuver dots on the screen by using the toggle switches of the control panel to make them speed up, slow down, and turn. Then he converted the dots into two cartoonish spaceships, one of them fat and bulging like a cigar and the other thin and straight like a pencil. Another subroutine allowed each spaceship to shoot a dot out of its nose, mimicking a missile. When the position of the missile dot coincided with that of a spaceship, the latter would "explode" into randomly moving dots. By February 1962 the basics had been completed.

At that point *Spacewar* became an open-source project. Russell put his program tape in the box that held other PDP-1 programs, and his friends began to make improvements. One of them, Dan Edwards, decided it would be cool to introduce a gravitational force, so he programmed in a big sun that exerted a tug on the ships. If you didn't pay attention, it could suck you in and destroy you, but good players learned to whip close to the sun and use its gravitational pull to gain momentum and swing around at higher speeds.

Another friend, Peter Samson, "thought my stars were random and unrealistic," Russell recalled.[7] Samson decided the game needed "the real thing," meaning astronomically correct constellations rather than miscellaneous dots. So he created a programming addition he called "Expensive Planetarium." Using information from the *American Ephemeris and Nautical Almanac*, he encoded a routine that showed all the stars in the night sky down to the fifth magnitude. By specifying how many times a display point on the screen fired, he was even able to replicate each star's relative brightness. As the spaceships sped along, the constellations slowly scrolled past.

This open-source collaboration produced many more clever contributions. Martin Graetz came up with what he called "the ultimate panic button," which was the ability to get out of a jam by toggling a switch and disappearing temporarily into another dimension of hyperspace. "The idea was that when everything else failed you could jump into the fourth dimension and disappear," he explained. He had read about something similar, called a "hyper-spatial tube," in one of Doc Smith's novels. There were, however, some limits: you could toggle into hyperspace only three times in a game; your disappearance gave your opponent a breather; and you never knew where your spaceship would reappear. It might end up in the sun or right in the sights of your opponent. "It was something you could use, but not something you wanted to use," Russell explained. Graetz added an homage to Professor Minsky: a ship disappearing into hyperspace left behind one of the signature patterns of the Minskytron.[8]

One lasting contribution came from two active members of the Tech Model Railroad Club, Alan Kotok and Bob Sanders. They realized that players crammed in front of a PDP-1 console jostling elbows and frantically grabbing at the computer's switches was both awkward and dangerous. So they rummaged around under the train set in the clubroom and commandeered some of the toggles and relays. These they pieced together inside two plastic boxes to make remote controls, complete with all the necessary function switches and the hyperspace panic button.

The game quickly spread to other computer centers and became a staple of hacker culture. DEC began shipping the game preloaded into its computers, and programmers created new versions for other systems. Hackers around the world added more features, such as cloaking powers, exploding space mines, and ways to shift into a first-person perspective from the view of one of the pilots. As Alan Kay, one of the pioneers of the personal computer, said, "The game of *Spacewar* blossoms spontaneously wherever there is a graphics display connected to a computer."[9]

Spacewar highlighted three aspects of the hacker culture that became themes of the digital age. First, it was created collaboratively. "We were able to build it together, working as a team, which is how we liked to do things," Russell said. Second, it was free and open-source software. "People asked for copies of the source code, and of course we gave them out." *Of course*—that was in a time and place when software yearned to be free. Third, it was based on the belief that computers should be personal and interactive. "It allowed us to get our hands on a computer and make it respond to us in real time," said Russell.[10]

NOLAN BUSHNELL AND ATARI

Like many computer science students in the 1960s, Nolan Bushnell was a *Spacewar* fanatic. "The game was seminal to anyone who loved computers, and for me it was transforming," he recalled. "Steve Russell was like a god to me." What set Bushnell apart from other computer bums who got their kicks by maneuvering blips on a screen was that he was also enthralled by amusement parks. He worked in one to help pay for college. In addition, he had the boisterous temperament of an entrepreneur, relishing the mix of thrill-seeking and risk-taking. Thus it was that Nolan Bushnell became one of those innovators who turned an invention into an industry.[11]

When Bushnell was fifteen, his father died. He had been a construction contractor in a growing exurb of Salt Lake City, and he left behind several unfinished jobs for which he hadn't been paid. Young Bushnell, already big and boisterous, finished them off, adding to his natural bravado. "When you do something like that as a 15-year-old, you begin to believe you can do anything," he said.[12] Not surprisingly, he became a poker player, and as good luck would have it he lost, fortuitously forcing him to take a job on the midway at the Lagoon Amusement Park while studying at the University of Utah. "I learned all the various tricks for getting people to put up their quarters, and that sure served me well."[13] He was soon promoted to the pinball and game arcade, where animated driving games such as Speedway, made by Chicago Coin Machine Manufacturing Company, were the new rage.

He was fortunate as well in landing at the University of Utah. It had the best computer graphics program in the country, run by professors Ivan Sutherland and David Evans, and became one of the first four nodes on the ARPANET, the precursor to the Internet. (Other students included Jim Clark, who founded Netscape; John Warnock, who cofounded Adobe; Ed Catmull, who cofounded Pixar; and Alan Kay, about whom more later.) The university had a PDP-1, complete with a *Spacewar* game, and Bushnell combined his love of the game with his understanding of the economics of arcades. "I realized you could make a whole lot of quarters if you could put a computer with a game in an arcade," he said. "And then I did the division and realized that even a whole lot of quarters coming in every day would never add up to the million-dollar cost of a computer. You divide twenty-five cents into a million dollars and you give up."[14] And so he did, for the moment.

When he graduated in 1968 ("last in his class," he often bragged), Bushnell went to work for Ampex, which made recording equipment. He and a colleague there, Ted Dabney, continued to concoct schemes for turning a computer into an arcade video game. They considered ways to adapt the Data General Nova, a $4,000 refrigerator-size minicomputer that came out in 1969. But no matter how they juggled the numbers, it was neither cheap enough nor powerful enough.

In his attempts to push the Nova to support *Spacewar*, Bushnell looked for elements of the game, such as the background of stars, that could be generated by the hardware circuits rather than by the processing power of the computer. "Then I had a great epiphany," he recalled. "Why not do it all with hardware?" In other words, he could design circuits to perform each of the tasks that the program would have done. That made it cheaper. It also meant that the game had to be a lot simpler. So he turned *Spacewar* into a game that had only one user-controlled spaceship, which fought against two simple saucers generated by the hardware. Eliminated, too, were the sun's gravity and the panic button to disappear into hyperspace. But it was still a fun game, and it could be built at a reasonable cost.

Bushnell sold the idea to Bill Nutting, who had formed a company to make an arcade game called *Computer Quiz*. In keeping with that name, they dubbed Bushnell's game *Computer Space*. He and Nutting hit it off so well that Bushnell quit Ampex in 1971 to join Nutting Associates.

As they were working on the first *Computer Space* consoles, Bushnell heard that he had competition. A Stanford grad named Bill Pitts and his buddy Hugh Tuck from California Polytechnic had become addicted to *Spacewar*, and they decided to use a PDP-11 minicomputer to turn it into an arcade game. When Bushnell heard this, he invited Pitts and Tuck to visit. They were appalled at the sacrifices—indeed sacrileges—Bushnell was perpetrating in stripping down *Spacewar* so that it could be produced inexpensively. "Nolan's thing was a totally bastardized version," Pitts fumed.[15] For his part, Bushnell was contemptuous of their plan to spend $20,000 on equipment, including a PDP-11 that would be in another room and connected by yards of cable to the console, and then charge ten cents a game. "I was surprised at how clueless they were about the business model," he said. "Surprised and relieved. As soon as I saw what they were doing, I knew they'd be no competition."

Galaxy Game by Pitts and Tuck debuted at Stanford's Tresidder student union coffeehouse in the fall of 1971. Students gathered around each night like cultists in front of a shrine. But no matter how many lined up their coins to play, there was no way the machine could pay for itself, and the venture eventually folded. "Hugh and I were both engineers and we didn't pay attention to business issues at all," conceded Pitts.[16] Innovation can be sparked by engineering talent, but it must be combined with business skills to set the world afire.

Bushnell was able to produce his game, *Computer Space*, for only $1,000. It made its debut a few weeks after *Galaxy Game* at the Dutch Goose bar in Menlo Park near Palo Alto and went on to sell a respectable 1,500 units. Bushnell was the consummate entrepreneur: inventive, good at engineering, and savvy about business and consumer demand. He also was a great salesman. One reporter remembered running into him at a Chicago trade show: "Bushnell was about the most excited person I've ever seen over the age of six when it came to describing a new game."[17]

Computer Space turned out to be less popular in beer halls than it was in student hangouts, so it was not as successful as most pinball games. But it did acquire a cult following. More important, it launched an industry. Arcade games, once the domain of pinball companies based in Chicago, would soon be transformed by engineers based in Silicon Valley.

Unimpressed by his experience with Nutting Associates, Bushnell decided to form his own company for his next video game. "Working for Nutting was a great learning experience, because I discovered that I couldn't screw things up any worse than they did," he recalled.[18] He decided to name the new company Syzygy, a barely pronounceable term for when three celestial bodies are in a line. Fortunately, that name was not available because a hippie candle-making commune had registered it. So Bushnell decided to call his new venture Atari, adopting a term from the Japanese board game Go.

PONG

On the day that Atari was incorporated, June 27, 1972, Nolan Bushnell hired his first engineer. Al Alcorn was a high school football player from a rough neighborhood of San Francisco who taught himself television repair through an RCA correspondence course. At Berkeley he participated in a work-study program that brought him to Ampex, where he worked under Bushnell. He graduated just as Bushnell was forming Atari.

Many of the key partnerships in the digital age paired people with different skills and personalities, such as John Mauchly and Presper Eckert, John Bardeen and Walter Brattain, Steve Jobs and Steve Wozniak. But occasionally the partnerships worked because the personalities and enthusiasms were similar, as was the case of Bushnell and Alcorn. Both were burly and fun-loving and irreverent. "Al is one of my favorite people in world," Bushnell asserted more than forty years later. "He was the perfect engineer and funny, so he was well-suited to video games."[19]

At the time, Bushnell had a contract to make a new video game for the Chicago firm Bally Midway. The plan was to do a car racing game, which seemed likely to be more appealing than spaceship navigation to beer drinkers in workingmen's bars. But before tossing the task to Alcorn, Bushnell decided to give him a warm-up exercise.

At a trade show, Bushnell had checked out the Magnavox Odyssey, a primitive console for playing games on home television sets. One of the offerings was a version of Ping-Pong. "I thought it was kind of crappy," Bushnell said years later, after he had been sued for stealing its idea. "It had no sound, no score, and the balls were square. But I noticed some people were having some fun with it." When he arrived back at Atari's little rented office in Santa Clara, he described the game to Alcorn, sketched out some circuits, and asked him to build an arcade version of it. He told Alcorn he had signed a contract with GE to make the game, which was untrue. Like many entrepreneurs, Bushnell had no shame about distorting reality in order to motivate people. "I thought it would be a great training program for Al."[20]

Alcorn got a prototype wired up in a few weeks, completing it at the beginning of September 1972. With his childlike sense of fun, he came up with enhancements that turned the monotonous blip bouncing between paddles into something amusing. The lines he created had eight regions so that when the ball hit smack in the center of a paddle it bounced back straight, but as it hit closer to the paddle's edges it would fly off at angles. That made the game more challenging and tactical. He also created a scoreboard. And in a stroke of simple genius, he added just the right "thonk" sound from the sync generator to sweeten the experience. Using a $75 Hitachi black-and-white TV set, Alcorn hard-wired the components together inside a four-foot-tall wooden cabinet. Like *Computer Space*, the game did not use a microprocessor or run a line of computer code; it was all done in hardware with the type of digital logic design used by television engineers. Then he slapped on a coin box taken from an old pinball machine, and a star was born.[21] Bushnell dubbed it *Pong*.

One of *Pong*'s most ingenious features was its simplicity. *Computer Space* had required complex instructions; there were enough directives on its opening screen (among them, for example, "There is no gravity in space; rocket speed can only be changed by engine thrust") to baffle a computer engineer. *Pong*, by contrast, was simple enough that a beer-sloshed barfly or stoned sophomore could figure it out after midnight. There was only one instruction: "Avoid missing ball for high score." Consciously or not, Atari had hit upon one of the most important engineering challenges of the computer age: creating user interfaces that were radically simple and intuitive.

Bushnell was so pleased by Alcorn's creation that he decided it should be more than a training exercise: "My mind changed the minute it got really fun, when we found ourselves playing it for an hour or two after work every night."[22] He flew to Chicago to persuade Bally Midway to accept *Pong* as a fulfillment of their contract rather than push for a car racing game. But the company declined to take it. It was wary of games that required two players.

This turned out to be a lucky break. To test out *Pong*, Bushnell and Alcorn installed the prototype at Andy Capp's, a beer bar in the working-class town of Sunnyvale that had peanut shells on the floor and guys playing pinball in the back. After a day or so, Alcorn got a call from the bar's manager complaining that the machine had stopped working. He should come fix it right away, because it had been surprisingly popular. So Alcorn hurried over. As soon as he opened the machine, he discovered the problem: the coin box was so filled with quarters that it was jammed. The money gushed onto the floor.[23]

Bushnell and Alcorn knew they had a hit on their hands. An average machine made $10 a day; *Pong* was taking in $40. Suddenly Bally's decision to decline it seemed like a blessing. The true entrepreneur in Bushnell came out: he decided that Atari would manufacture the game on its own, even though it had no financing or equipment.

He took the gamble of deciding to bootstrap the whole operation; he would fund as much as possible from the cash flow he made on sales. He looked at how much money he had in the bank, divided it by the $280 cost of making each machine, and figured that he could build thirteen of them initially. "But that was an unlucky number," he recalled, "so we decided to build twelve."[24]

Bushnell made a small model of the console shell he desired out of clay, then took it to a boat manufacturer who began producing them in fiberglass. It took just a week to build each complete game and another few days to sell it for $900, so with the $620 profit he had a positive cash flow to keep things going. Some of the early proceeds were spent on a sales brochure, which featured a beautiful young woman in a slinky sheer nightgown draping her arm over the game machine. "We hired her from the topless bar down the street," Bushnell recounted forty years later to an audience of earnest high school students, who seemed somewhat baffled by the tale and unsure what a topless bar was.[25]

Venture capital, a realm that had just begun in Silicon Valley with Arthur Rock's financing of Intel, was not available for a company proposing to make video games, which were not yet a known product and were associated with the mobbed-up pinball industry.* Banks de-

---

*Three years later, in 1975, when Atari decided to build a home version of *Pong*, the venture capital industry had caught fire, and Bushnell was able to get $20 million in

murred as well when Bushnell ambled in for a loan. Only Wells Fargo came through, providing a credit line of $50,000, which was far less than Bushnell had requested.

With the money, Bushnell was able to open up a production facility in an abandoned roller-skating rink a few blocks from Atari's Santa Clara office. The *Pong* games were put together not on an assembly line but in the middle of the floor, with young workers ambling up to stick in the various components. Workers were dragooned from unemployment centers nearby. After weeding out the hires that were heroin addicts or stole the television monitors, the operation scaled up rapidly. At first they were making ten units a day, but within two months they could make almost a hundred. The economics were improved as well; the cost of each game was held to just over $300, but the sales price was raised to $1,200.

The atmosphere was what you might expect from the fun-loving Bushnell and Alcorn, both still in their twenties, and it took to the next level the casual style of Silicon Valley startups. Every Friday there would be a beer bash and pot-smoking party, sometimes capped by skinny-dipping, especially if that week's numbers had been made. "We found out our employees would respond to having a party for hitting quotas as much as having a bonus," Bushnell said.

Bushnell bought himself a nice house in the hills of nearby Los Gatos, where he sometimes held board meetings or staff parties in his hot tub. When he built a new engineering facility, he decreed that it should have its own hot tub. "It was a recruiting tool," he insisted. "We found out that our lifestyle and the parties were hugely good for attracting workers. If we were trying to hire somebody, we'd invite him to one of our parties."[26]

In addition to being a recruiting tool, the culture at Atari was a natural outgrowth of Bushnell's personality. But it was not simply self-indulgent. It was based on a philosophy that drew from the hippie movement and would help define Silicon Valley. At its core were certain principles: authority should be questioned, hierarchies should

---

funding from Don Valentine, who had just founded Sequoia Capital. Atari and Sequoia helped to launch each other.

be circumvented, nonconformity should be admired, and creativity should be nurtured. Unlike at East Coast corporations, there were no fixed working hours and no dress code, either for the office or the hot tub. "At that time in IBM you had to wear a white shirt, dark pants and a black tie with your badge stapled to your shoulder or something," said Steve Bristow, an engineer. "At Atari the work people did counted more than how they looked."[27]

The success of *Pong* prompted a lawsuit from Magnavox, which marketed the Odyssey home-television game that Bushnell had played at a trade show. The Magnavox game had been devised by an outside engineer named Ralph Baer. He could not claim to have invented the concept; its roots went back at least to 1958, when William Higinbotham at the Brookhaven National Lab rigged up an oscilloscope on an analog computer to knock a blip back and forth in what he called Tennis for Two. Baer, however, was one of those innovators, like Edison, who believed that filing for patents was a key element of the invention process. He had more than seventy of them, including for various aspects of his games. Instead of fighting the lawsuit, Bushnell came up with a clever deal that was a win for both companies. He paid a rather low flat fee, $700,000, for perpetual rights to make the game on the condition that Magnavox enforce its patents and demand a percentage royalty from the other companies, including his former partners Bally Midway and Nutting Associates, that wanted to make similar games. That helped put Atari at a competitive advantage.

Innovation requires having at least three things: a great idea, the engineering talent to execute it, and the business savvy (plus deal-making moxie) to turn it into a successful product. Nolan Bushnell scored a trifecta when he was twenty-nine, which is why he, rather than Bill Pitts, Hugh Tuck, Bill Nutting, or Ralph Baer, goes down in history as the innovator who launched the video game industry. "I am proud of the way we were able to engineer *Pong*, but I'm even more proud of the way I figured out and financially engineered the business," he said. "Engineering the game was easy. Growing the company without money was hard."[28]

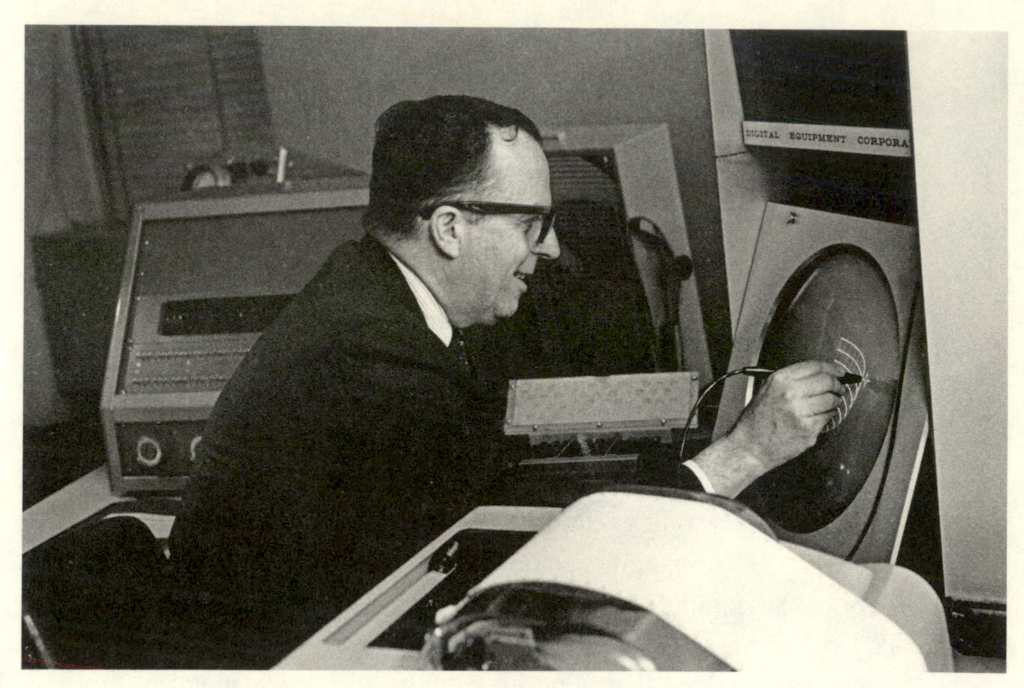

J. C. R. Licklider (1915–90).

Bob Taylor (1932–).

Larry Roberts (1937–).

CHAPTER SEVEN

THE INTERNET

VANNEVAR BUSH'S TRIANGLE

Innovations often bear the imprint of the organizations that created them. For the Internet, this was especially interesting, for it was built by a partnership among three groups: the military, universities, and private corporations. What made the process even more fascinating was that this was not merely a loose-knit consortium with each group pursuing its own aims. Instead, during and after World War II, the three groups had been fused together into an iron triangle: the military-industrial-academic complex.

The person most responsible for forging this assemblage was Vannevar Bush, the MIT professor who in 1931 built the Differential Analyzer, the early analog computer described in chapter 2.[1] Bush was well suited to this task because he was a star in all three camps: dean of the MIT School of Engineering, a founder of the electronics company Raytheon, and America's top military science administrator during World War II. "No American has had greater influence in the growth of science and technology than Vannevar Bush," MIT's president Jerome Wiesner later proclaimed, adding that his "most significant innovation was the plan by which, instead of building large government laboratories, contracts were made with universities and industrial laboratories."[2]

Bush was born near Boston in 1890, the son of a Universalist minister who had begun his career as a cook on a mackerel smack. Both of Bush's grandfathers were whaling captains, which instilled in him a salty and forthright manner that helped make him a decisive manager and charismatic administrator. Like many successful technology leaders, he was an expert in both engineering products and making crisp decisions. "All of my recent ancestors were sea captains, and they have a way of running things without any doubt," he once said. "That left me with some inclination to run a show once I was in it."[3]

Also like many good technology leaders, he grew up loving both the humanities and the sciences. He could quote Kipling and Omar Khayyam "by the yard," played the flute, loved symphonies, and read philosophy for pleasure. His family, too, had a basement workshop, where he built little boats and mechanical toys. As *Time* later reported in its inimitable old style, "Lean, sharp, salty, Van Bush is a Yankee whose love of science began, like that of many American boys, in a passion for tinkering with gadgets."[4]

He went to Tufts, where in his spare time he built a surveying machine that used two bicycle wheels and a pendulum to trace the perimeter of an area and calculate its size, thus being an analog device for doing integral calculus. He got a patent on it, which became the first of forty-nine that he would accumulate. While at Tufts, he and his roommates consulted with a series of small companies and then, after graduating, founded Raytheon, which grew into a sprawling defense contractor and electronics firm.

Bush earned a PhD in electrical engineering jointly from MIT and Harvard, then became a professor and dean of engineering at MIT, where he built his Differential Analyzer. His passion was elevating the role of science and engineering in society at a time, the mid-1930s, when not much exciting seemed to be happening in either field. Televisions were not yet a consumer product, and the most notable new inventions put into the time capsule at the New York 1939 World's Fair were a Mickey Mouse watch and a Gillette Safety Razor. The advent of World War II would change that, producing an explosion of new technologies, with Vannevar Bush leading the way.

Worried that America's military was lagging in technology, he mobilized Harvard president James Bryant Conant and other scientific leaders to convince President Franklin Roosevelt to form the National Defense Research Committee and then the military's Office of Scientific Research and Development, both of which he headed. With an ever-present pipe in his mouth and a pencil in his hand, he oversaw the Manhattan Project to build the atom bomb as well as the projects to develop radar and air-defense systems. *Time* dubbed him "General of Physics" on its cover in 1944. "If we had been on our toes in war technology ten years ago," the magazine quoted him as saying as he banged his fist on his desk, "we would probably not have had this damn war."[5]

With his no-nonsense style tempered by a personal warmth, he was a tough but endearing leader. Once a group of military scientists, frustrated by some bureaucratic problem, walked into his office to resign. Bush couldn't figure out what the snafu was. "So I just told them," he recalled, "'One does not resign in time of war. You chaps get the hell out of here and get back to work, and I'll look into it.'"[6] They obeyed. As MIT's Wiesner later observed, "He was a man of strong opinions, which he expressed and applied with vigor, yet he stood in awe of the mysteries of nature, had a warm tolerance for human frailty, and was open-minded to change."[7]

When the war ended, Bush produced a report in July 1945 at Roosevelt's behest (which ended up being delivered to President Harry Truman) that advocated government funding of basic research in partnership with universities and industry. Bush chose an evocative and quintessentially American title, "Science, the Endless Frontier." His introduction deserves to be reread whenever politicians threaten to defund the research needed for future innovation. "Basic research leads to new knowledge," Bush wrote. "It provides scientific capital. It creates the fund from which the practical applications of knowledge must be drawn."[8]

Bush's description of how basic research provides the seed corn for practical inventions became known as the "linear model of innovation." Although subsequent waves of science historians sought to debunk the linear model for ignoring the complex interplay between theoretical research and practical applications, it had a popular appeal as well as an underlying truth. The war, Bush wrote, had made it "clear beyond all doubt" that basic science—discovering the fundamentals of nuclear physics, lasers, computer science, radar—"is absolutely essential to national security." It was also, he added, crucial for America's economic security. "New products and new processes do not appear full-grown. They are founded on new principles and new conceptions, which in turn are painstakingly developed by research in the purest realms of science. A nation which depends upon others for its new basic scientific knowledge will be slow in its industrial progress and weak in its competitive position in world trade." By the end of his report, Bush had reached poetic heights in extolling the practical payoffs of basic scientific research: "Advances in science when put to practical use mean more jobs, higher wages, shorter hours, more abundant crops, more leisure for recreation, for study, for learning how to live without the deadening drudgery which has been the burden of the common man for past ages."[9]

Based on this report, Congress established the National Science Foundation. At first Truman vetoed the bill because it mandated that the director be appointed by an independent board rather than the president. But Bush turned Truman around by explaining that this would buffer him from those seeking political favors. "Van, you should be a politician," Truman told him. "You have some of the instincts." Bush replied, "Mr. President, what the hell do you think I've been doing around this town for five or six years?"[10]

The creation of a triangular relationship among government, industry, and academia was, in its own way, one of the significant innovations that helped produce the technological revolution of the late twentieth century. The Defense Department and National Science Foundation soon became the prime funders of much of America's basic research, spending as much as private industry during the 1950s through the 1980s.* The return on that investment was huge, leading

---

*By 2010, federal spending on research had dropped to half of what was spent by private industry.

not only to the Internet but to many of the pillars of America's postwar innovation and economic boom.[11]

A few corporate research centers, most notably Bell Labs, existed before the war. But after Bush's clarion call produced government encouragement and contracts, hybrid research centers began to proliferate. Among the most notable were the RAND Corporation, originally formed to provide research and development (hence the name) to the Air Force; Stanford Research Institute and its offshoot, the Augmentation Research Center; and Xerox PARC. All would play a role in the development of the Internet.

Two of the most important of these institutes sprang up around Cambridge, Massachusetts, just after the war: Lincoln Laboratory, a military-funded research center affiliated with MIT, and Bolt, Beranek and Newman, a research and development company founded and populated by MIT (and a few Harvard) engineers. Closely associated with both of them was an MIT professor with a Missouri drawl and an easygoing talent for teambuilding. He would become the single most important person in creating the Internet.

J. C. R. LICKLIDER

In searching for fathers of the Internet, the best person to start with is a laconic yet oddly charming psychologist and technologist, with an open-faced grin and show-me attitude, named Joseph Carl Robnett Licklider, born in 1915 and known to everyone as "Lick." He pioneered the two most important concepts underlying the Internet: decentralized networks that would enable the distribution of information to and from anywhere, and interfaces that would facilitate human-machine interaction in real time. Plus, he was the founding director of the military office that funded the ARPANET, and he returned for a second stint a decade later when protocols were created to weave it into what became the Internet. Said one of his partners and protégés, Bob Taylor, "He was really the father of it all."[12]

Licklider's father was a poor Missouri farm boy who became a successful insurance salesman in St. Louis and then, when the Depression wiped him out, a Baptist minister in a tiny rural town. As a doted-upon only child, Lick turned his bedroom into a model plane production facility and rebuilt clunker cars with his mother standing by his side handing him tools. Nevertheless, he felt trapped growing up in an isolated rural area filled with barbed-wire fences.

He escaped first to Washington University in St. Louis and then, after getting a doctorate in psychoacoustics (how we perceive sounds), joined Harvard's psychoacoustics lab. Increasingly interested in the relationship between psychology and technology, how human brains and machines interacted, he moved to MIT to start a psychology section based in the Electrical Engineering Department.

At MIT Licklider joined the eclectic circle of engineers, psychologists, and humanists gathered around Professor Norbert Wiener, a theorist who studied how humans and machines worked together and coined the term *cybernetics*, which described how any system, from a brain to an artillery aiming mechanism, learned through communications, control, and feedback loops. "There was tremendous intellectual ferment in Cambridge after World War II," Licklider recalled. "Wiener ran a weekly circle of forty or fifty people who got together. They would gather together and talk for a couple of hours. I was a faithful adherent to that."[13]

Unlike some of his MIT colleagues, Wiener believed that the most promising path for computer science was to devise machines that would work well with human minds rather than try to replace them. "Many people suppose that computing machines are replacements for intelligence and have cut down the need for original thought," Wiener wrote. "This is not the case."[14] The more powerful the computer, the greater the premium that will be placed on connecting it with imaginative, creative, high-level human thinking. Licklider became an adherent of this approach, which he later called "man-computer symbiosis."

Licklider had a mischievous but friendly sense of humor. He loved watching the Three Stooges and was childishly fond of sight gags. Sometimes, when a colleague was about to give a slide presentation, Licklider would slip a photo of a beautiful woman into the projector's carousel. At work he energized himself with a steady supply of Cokes and candies from the vending machines, and he gave out Hershey bars to his kids and students whenever they delighted him. He was also devoted to his graduate students, whom he would invite to dinners at his home in the Boston suburb of Arlington. "To him, collaboration was what it was all about," his son Tracy said. "He wandered around setting up islands of people and encouraging them to be inquisitive and solve problems." That was one reason he became interested in networks. "He knew that getting good answers involved distant collaboration. He loved spotting talented people and tying them together in a team."[15]

His embrace, however, did not extend to people who were pretentious or pompous (with the exception of Wiener). When he thought a speaker was spouting nonsense, he would stand up and ask what seemed to be innocent but were in fact devilish questions. After a few moments, the speaker would realize he had been deflated and Licklider would sit down. "He didn't like poseurs or pretenders," Tracy recalled. "He was never mean, but he slyly pricked people's pretensions."

One of Licklider's passions was art. Whenever he traveled he would spend hours at museums, sometimes dragging along his two reluctant children. "He became a nut about it, couldn't get enough of it," said Tracy. Sometimes he would spend five hours or more in a museum marveling at each brushstroke, analyzing how each picture came together, and attempting to fathom what it taught about creativity. He had an instinct for spotting talent in all fields, arts as well as sciences, but he felt that it was most easy to discern in its purest forms, such as the brushstroke of a painter or the melodic refrain of a composer. He said he looked for the same creative strokes in the designs of computer or network engineers. "He became a really skilled scout of creativity. He often discussed what made people creative. He felt it was easier to see in an artist, so he tried even harder to spot it in engineering, where you can't see the brushstrokes quite as readily."[16]

Most important, Licklider was kind. When he worked at the Pentagon later in his career, according to his biographer Mitchell Waldrop, he noticed the cleaning woman admiring the art prints on his wall late one evening. She told him, "You know, Dr. Licklider, I always leave your room until last because I like to have time by myself, with nothing pressing, to look at the pictures." He asked which print she liked most, and she pointed to a Cézanne. He was thrilled, since it was his favorite, and he promptly gave it to her.[17]

Licklider felt that his love of art made him more intuitive. He could process a wide array of information and sniff out patterns. Another attribute, which would serve him well when he helped put together the team that laid the foundations for the Internet, was that he loved to share ideas without craving credit for them. His ego was so tamed that he seemed to enjoy giving away rather than claiming credit for ideas that were developed in conversation. "For all his considerable influence on computing, Lick retained his modesty," said Bob Taylor. "His favorite kind of joke was one at his own expense."[18]

TIME-SHARING AND MAN-COMPUTER SYMBIOSIS

At MIT Licklider collaborated with the artificial intelligence pioneer John McCarthy, in whose lab the hackers of the Tech Model Railroad Club had invented *Spacewar*. With McCarthy in the lead, they helped to develop, during the 1950s, systems for computer time-sharing.

Up until then, when you wanted a computer to perform a task, you had to submit a stack of punch cards or a tape to the computer's operators, as if handing an offering to the priests who shielded an oracle. This was known as "batch processing," and it was annoying. It could take hours or even days to get results back; any little mistake might mean having to resubmit your cards for another run; and you might not be able to touch or even see the computer itself.

Time-sharing was different. It allowed a whole lot of terminals to be hooked up to the same mainframe, so that many users could type in commands directly and get a response almost instantly. Like a grandmaster playing dozens of games of chess simultaneously, the mainframe's core memory would keep track of all the users, and its operating system would be capable of multitasking and running many programs. This provided users with an enchanting experience: you could have a hands-on and real-time interaction with a computer, like a conversation. "We had a kind of little religion growing here about how this was going to be totally different from batch processing," said Licklider.[19]

It was a key step toward a direct human-computer partnership or symbiosis. "The invention of *interactive* computing through time-sharing was even more important than the invention of computing itself," according to Bob Taylor. "Batch processing was like exchanging letters with someone, while interactive computing was like talking to them."[20]

The importance of interactive computing became apparent at Lincoln Laboratory, the military-funded research center that Licklider helped to build at MIT in 1951. There he forged a team, half of them psychologists and half engineers, to look at ways that humans could interact more intuitively with computers and information could be presented with a friendlier interface.

One mission at Lincoln Laboratory was developing computers for an air defense system that would provide early warning of an enemy attack and coordinate a response. It was known as SAGE, for Semi-Automatic Ground Environment, and it cost more money and employed more people than the Manhattan Project that built the atom bomb. For it to work, the SAGE system needed to enable its users to have instant interactions with its computers. When an enemy missile or bomber was on the way, there would be no time for batch processing of the calculations.

The SAGE system included twenty-three tracking centers across the United States, which were connected by long-distance phone lines. It was able to disseminate information on up to four hundred fast-moving planes at once. This required powerful interactive computers, networks that could transmit vast amounts of information, and displays that could present this information in an easy-to-understand graphical fashion.

Because of his background in psychology, Licklider was called upon to help design the human-machine interfaces (what users saw on the screen). He formulated a set of theories about ways to cultivate a symbiosis, an intimate partnership, that would allow humans and machines to work cooperatively to solve problems. Particularly important was figuring out ways to visually convey changing situations. "We wanted methods of saving the air situation for successive seconds, and plotting tracks, not blips, and coloring the tracks so that we could see which was the recent information and tell which way the thing was going," he explained.[21] America's fate might depend on the ability of a console jockey to assess data correctly and respond instantly.

The interactive computers, intuitive interfaces, and high-speed networks showed how people and machines could work together in a collaborative partnership, and Licklider imagined ways this could happen for more than air defense systems. He began to talk about what he called a "truly SAGE system" that would network not only air defense centers but also "thinking centers" incorporating vast libraries of knowledge, which people could interact with on friendly display consoles—in other words, the digital world we have today.

These ideas formed the basis for one of the most influential papers in the history of postwar technology, titled "Man-Computer Symbiosis," which Licklider published in 1960. "The hope is that, in not too many years, human brains and computing machines will be coupled together very tightly," he wrote, "and that the resulting partnership will think as no human brain has ever thought and process data in a way not approached by the information-handling machines we know today." This sentence bears rereading, because it became one of the seminal concepts of the digital age.[22]

Licklider sided with Norbert Wiener, whose theory of cybernetics was based on humans and machines working closely together, rather than with their MIT colleagues Marvin Minsky and John McCarthy, whose quest for artificial intelligence involved creating machines that could learn on their own and replicate human cognition. As Licklider explained, the sensible goal was to create an environment in which humans and machines "cooperate in making decisions." In other words, they would augment each other. "Men will set the goals, formulate the hypotheses, determine the criteria, and perform the evaluations. Computing machines will do the routinizable work that must be done to prepare the way for insights and decisions in technical and scientific thinking."

THE INTERGALACTIC COMPUTER NETWORK

As he combined his interests in psychology and engineering, Licklider became even more focused on computers. That led him in 1957 to sign on with the fledgling Cambridge-based firm of Bolt, Beranek and Newman (BBN), a commercial-academic research company where many of his friends worked. As at Bell Labs when the transistor was being invented, BBN gathered a galvanizing mix of talent that included theorists, engineers, technicians, computer scientists, psychologists, and the occasional Army colonel.[23]

One of Licklider's assignments at BBN was to lead a team tasked with figuring out how computers could transform libraries. He dictated his final report, "Libraries of the Future," in the course of five hours while sitting by the pool during a conference in Las Vegas.[24] It explored the potential "of devices and techniques for on-line man-computer interaction," a concept that foreshadowed the Internet. He envisioned the accumulation of a huge database of information that was curated and weeded so that "it doesn't get too diffuse, overwhelming, or unreliable."

In a colorful part of the paper, he presented a fictional scenario in which he posed questions to the machine. He imagined the machine's activity: "Over the week-end it retrieved over 10,000 documents, scanned them all for sections rich in relevant material, analyzed all the rich sections into statements in a high-order predicate calculus, and entered the statements into the data base." Licklider realized that the approach he described would eventually be superseded. "Certainly, a more sophisticated approach will be feasible before 1994," he wrote, looking three decades into the future.[25] He was remarkably prescient. In 1994 the first text-crawling search engines, WebCrawler and Lycos, were developed for the Internet, and they were quickly followed by Excite, Infoseek, AltaVista, and Google.

Licklider also predicted something that was counterintuitive but has turned out to be pleasantly true: that digital information would not completely replace print. "As a medium for the display of information, the printed page is superb," he wrote. "It affords enough resolution to meet the eye's demand. It presents enough information to occupy the reader for a convenient quantum of time. It offers great flexibility of font and format. It lets the reader control the mode and rate of inspection. It is small, light, movable, cuttable, clippable, pastable, replicable, disposable, and inexpensive."[26]

In October 1962, while he was still working on his "Libraries of the Future" project, Licklider was recruited to Washington to head up a new office dealing with information processing at the Defense Department's Advanced Research Projects Agency, then known as ARPA.* Embedded in the Pentagon, it was empowered to fund basic research at universities and corporate institutes, thus becoming one of many ways the government implemented Vannevar Bush's vision. It also had a more proximate cause. On October 4, 1957, the Russians launched *Sputnik*, the first man-made satellite. The connection that Bush had made between science and defense was now twinkling in the sky every night. When Americans squinted to see it, they could also see that Bush was right: the nation that funded the best science would produce the best rockets and satellites. A ripple of healthy public panic ensued.

President Eisenhower liked scientists. Their culture and their mode of thinking, their ability to be nonideological and rational, appealed to him. "Love of liberty means the guarding of every resource that makes freedom possible—from the sanctity of our families and the wealth of our soil to the genius of our scientists," he had proclaimed in his first inaugural address. He threw White House dinners for scientists, the way that the Kennedys would do for artists, and gathered many around him in advisory roles.

Sputnik gave Eisenhower the opportunity to formalize his embrace. Less than two weeks after it was launched, he gathered fifteen top science advisors who had worked with the Office of Defense Mobilization and asked them, his aide Sherman Adams recalled, "to tell him where scientific research belonged in the structure of the federal

*The government has repeatedly changed whether there should be a "D" for "Defense" in the acronym. The agency was created in 1958 as ARPA. It was renamed DARPA in 1972, then reverted to ARPA in 1993, and then became DARPA again in 1996.

government."[27] He then met for breakfast with James Killian, the president of MIT, and appointed him to be his full-time science advisor.[28] Together with the defense secretary, Killian worked out a plan, announced in January 1958, to put the Advanced Research Projects Agency in the Pentagon. As the historian Fred Turner wrote, "ARPA marked an extension of the defense-oriented military-university collaborations that began in World War II."[29]

The office within ARPA that Licklider was recruited to lead was called Command and Control Research. Its mission was to study how interactive computers could help facilitate the flow of information. There was another job opening to lead a group studying psychological factors in military decision making. Licklider argued that these two topics should be put together. "I started to wax eloquent on my view that the problems of command and control were essentially problems of man-computer interaction," he later said.[30] He agreed to take both jobs and renamed his combined group ARPA's Information Processing Techniques Office (IPTO).

Licklider had a lot of exciting ideas and passions, most notably ways to encourage time-sharing, real-time interactivity, and interfaces that would nurture man-machine symbiosis. All of these tied together into a simple concept: a network. With his wry sense of humor, he began referring to his vision with the "intentionally grandiloquent" phrase "the Intergalactic Computer Network."[31] In an April 1963 memo addressed to "members and affiliates" of that dream network, Licklider described its goals: "Consider the situation in which several different centers are netted together . . . Is it not desirable, or even necessary for all the centers to agree upon some language or, at least, upon some conventions for asking such questions as 'What language do you speak?'"[32]

BOB TAYLOR AND LARRY ROBERTS

Unlike many other partners who advanced the digital age, Bob Taylor and Larry Roberts were never friends, either before or after their time together at IPTO. Indeed, in later years they would bitterly disparage each other's contributions. "Larry claims that he laid out the network himself, which is totally false," Taylor complained in 2014. "Don't trust what he says. I feel sorry for him."[33] For his part, Roberts claims that Taylor is bitter because he did not get enough credit: "I don't know what to give him credit for other than hiring me. That's the only important thing Bob did."[34]

But during the four years they worked together at ARPA in the 1960s, Taylor and Roberts complemented each other well. Taylor was not a brilliant scientist; he didn't even have a doctorate. But he had an affable, persuasive personality and was a magnet for talent. Roberts, by contrast, was an intense engineer with an abrupt manner, bordering on curt, who used to measure on a stopwatch the time it took to walk alternative routes between offices in the sprawling Pentagon. He didn't charm his colleagues, but he often awed them. And his brusque direct manner made him a competent if not beloved manager. Taylor cajoled people, while Roberts impressed them with his intellect.

Bob Taylor was born in 1932 in a home for unwed mothers in Dallas, put on a train to an orphanage in San Antonio, and adopted when he was twenty-eight days old by an itinerant Methodist minister and his wife. The family uprooted every couple of years to pulpits in such towns as Uvalde, Ozona, Victoria, San Antonio, and Mercedes.[35] His upbringing, he said, left two imprints on his personality. As with Steve Jobs, who was also adopted, Taylor's parents repeatedly emphasized that he had been "chosen, specially picked out." He joked, "All the other parents had to take what they got, but I was chosen. That probably gave me an undeserved sense of confidence." He also had to learn repeatedly, with each family move, to forge new relationships, learn new lingo, and secure his place in a small-town social order. "You've got to make a new set of friends and interact with a new set of prejudices every time."[36]

Taylor studied experimental psychology at Southern Methodist University, served in the Navy, and got a bachelor's and a master's degree from the University of Texas. While doing a paper on psychoacoustics, he had to submit his data on punch cards for batch processing at the university computing system. "I had to carry around stacks of cards that took days to get processed, and then they would say I had some comma wrong on card 653 or something and needed to have it all redone," he said. "It made me angry." He realized that there could be a better way when he read Licklider's paper on interactive machines and man-computer symbiosis, which elicited a Eureka moment. "Yes, that's how it should be!" he recalled saying to himself.[37]

After teaching at a prep school and working at a defense contractor in Florida, Taylor got a job at NASA headquarters in Washington, DC, overseeing research on flight-simulation displays. Licklider was by then running the Information Processing Techniques Office at ARPA, where he began a regular series of meetings with other government researchers doing similar work. When Taylor showed up in late 1962, Licklider surprised him by knowing of the psychoacoustics paper he had written at the University of Texas. (Taylor's advisor was a friend of Licklider.) "I was terribly flattered," Taylor recalled, "so I became an admirer and really good friend of Lick's from then on."

Taylor and Licklider sometimes traveled to conferences together, further sealing their friendship. On a trip to Greece in 1963, Licklider took Taylor to one of Athens's art museums and demonstrated his technique for studying brushstrokes by squinting at a painting. At a taverna late that evening, Taylor got himself invited to sit in with the band and taught them to play Hank Williams songs.[38]

Unlike some engineers, both Licklider and Taylor understood human factors; they had studied psychology, could relate to people, and took joy in appreciating art and music. Although Taylor could be blustery and Licklider tended to be gentle, they both loved working with other people, befriending them, and nurturing their talents. This love of human interaction and appreciation for how it worked made them well suited to designing the interfaces between humans and machines.

When Licklider stepped down from IPTO, his deputy, Ivan Sutherland, took over temporarily, and at Licklider's urging Taylor moved over from NASA to become Sutherland's deputy. Taylor was among the few who realized that information technology could be more exciting than the space program. After Sutherland resigned in

1966 to become a tenured professor at Harvard, Taylor was not everyone's first choice to replace him, since he did not have a PhD and wasn't a computer scientist, but he eventually got the job.

Three things at IPTO struck Taylor. First, every one of the universities and research centers that had a contract with ARPA wanted the latest computers with the most capabilities. That was wasteful and duplicative. There might be a computer that did graphics in Salt Lake City and another that mined data at Stanford, but a researcher who needed to perform both tasks either had to go back and forth by plane or ask IPTO to fund another computer. Why couldn't they be connected by a network that allowed them to time-share each other's computer? Second, on his travels to talk to young researchers, Taylor discovered that those in one place were intensely interested in learning about the research happening at other places. He realized that it would make sense to connect them electronically so they could share more easily. Third, Taylor was struck by the fact that there were three terminals in his Pentagon office, each with its own passwords and commands, connected to different computer centers ARPA was funding. "Well this is silly," he thought. "I should be able to access any of these systems from a single terminal." His need for three terminals, he said, "led to an epiphany."[39] All three of these problems could be solved by building a data network to connect research centers, that is, if he could implement Licklider's dream of an Intergalactic Computer Network.

He walked across to the E-ring of the Pentagon to see his boss, ARPA director Charles Herzfeld. With his Texas twang, Taylor knew how to charm Herzfeld, an intellectual Viennese refugee. He brought no presentation or memos, instead just launching into an ebullient pitch. A network funded and imposed by ARPA could permit research centers to share computing resources, collaborate on projects, and allow Taylor to jettison two of his office terminals.

"Great idea," Herzfeld said. "Get it going. How much money do you need?"

Taylor allowed that it might take a million dollars just to get the project organized.

"You've got it," Herzfeld said.

As he headed back to his office, Taylor looked at his watch. "Jesus Christ," he murmured to himself. "That only took twenty minutes."[40]

It was a story Taylor told often in interviews and oral histories. Herzfeld liked the tale, but he later felt compelled to confess that it was a little misleading. "He leaves out the fact that I'd been studying the problem with him and with Licklider for three years," said Herzfeld. "It was not hard to get the million dollars because I was kind of waiting for him to ask for it."[41] Taylor conceded that was the case, and he added his own lagniappe: "What really pleased me was that Charlie took the money out of funds that were supposed to go to developing a missile defense system, which I thought was the stupidest and most dangerous idea."[42]

Taylor now needed someone to run the project, which is how Larry Roberts came into the picture. He was an obvious choice.

Roberts seemed born and bred to help build the Internet. Both of his parents had doctorates in chemistry, and as a kid growing up near Yale he had built a television, Tesla coil,* ham radio, and telephone system from scratch. He went to MIT, where he got his bachelor's, master's, and doctoral degrees in engineering. Impressed by Licklider's papers on man-computer symbiosis, he went to work with him at Lincoln Laboratory and became his protégé in the fields of time-sharing, networks, and interfaces. One of his experiments at Lincoln Laboratory involved connecting two distant computers; it had been funded by Bob Taylor at ARPA. "Licklider inspired me with his vision of linking computers into a network," Roberts recalled, "and I decided that would be my work."

But Roberts kept turning down Taylor's offer to come to Washington to be his deputy. He liked his job at Lincoln Laboratory and didn't especially respect Taylor. There was also something Taylor didn't know: a year earlier, Roberts had been offered Taylor's job. "When Ivan was leaving, he asked me to come to IPTO as the next director,

---

*A high-frequency transformer that can take ordinary voltage, like the 120 volts in a U.S. outlet, and step it up to superhigh voltages, often discharging energy in cool-looking electrical arcs.

but it was a management job, and I preferred research," he said. Having declined the top post, Roberts was not about to be Taylor's deputy. "Forget it," he told Taylor. "I'm busy. I'm having fun with this wonderful research."[43]

There was another reason Roberts resisted, which Taylor could sense. "Larry was from MIT with a doctorate, and I was from Texas with just a master's," Taylor later said. "So I suspect he didn't want to work for me."[44]

Taylor, however, was a clever and stubborn Texan. In the fall of 1966, he asked Herzfeld, "Charlie, doesn't ARPA fund 51 percent of Lincoln Laboratory?" Herzfeld confirmed that. "Well, you know this networking project that I want to do, I'm having a hard time getting the program manager that I want, and he works at Lincoln Laboratory." Perhaps Herzfeld could call the head of the lab, Taylor suggested, and say that it would be in its interest to convince Roberts to accept the job. It was a Texas way of doing business, as the president at the time, Lyndon Johnson, would have appreciated. The lab's chief was no dummy. "It would probably be a nice thing for all of us if you'd consider this," he pointed out to Roberts after getting Herzfeld's call.

So in December 1966, Larry Roberts went to work at ARPA. "I blackmailed Larry Roberts into becoming famous," Taylor later said.[45]

When Roberts first moved to Washington, around Christmas, he and his wife stayed for a few weeks with Taylor while looking for a home. Even though they were not destined to be personal pals, the relationship between the two men was cordial and professional, at least during their years at ARPA.[46]

Roberts was not as genial as Licklider, nor as extroverted as Taylor, nor as congregational as Bob Noyce. "Larry's a cold fish," according to Taylor.[47] Instead he had a trait that was just as useful in promoting collaborative creativity and managing a team: he was decisive. More important, his decisiveness was based not on emotion or personal favoritism but rather on a rational and precise analysis of options. His colleagues respected his decisions, even if they disagreed with them, because he was clear, crisp, and fair. It was one of the advantages of having a true product engineer in charge. Uncomfortable at being Taylor's deputy, Roberts was able to work out an arrangement with

ARPA's top boss, Charlie Herzfeld, to be designated the agency's chief scientist instead. "I managed contracts during the day and did my networking research at night," he recalled.[48]

Taylor, on the other hand, was jocular and gregarious, sometimes to a fault. "I'm an outgoing person," he observed. Each year he would convene a conference of the ARPA-funded researchers and another for their best graduate students, usually in fun places like Park City, Utah, and New Orleans. He made each researcher give a presentation, and then everyone could pile on with questions and suggestions. In that way he got to know the rising stars around the country, making him a magnet for talent that would later serve him well when he went to work at Xerox PARC. It also helped him accomplish one of the most important tasks in building a network: getting everyone to buy into the idea.

ARPANET

Taylor knew that he needed to sell the time-sharing network idea to the people it was intended to help, namely the researchers who were getting ARPA funding. So he invited them to a meeting at the University of Michigan in April 1967, where he had Roberts present the plan. The computer sites would be connected, Roberts explained, by leased phone lines. He described two possible architectures: a hub system with a central computer in a place like Omaha that would route information, or a weblike system that looked like a highway map with lines crisscrossing as they were spun from place to place. Roberts and Taylor had begun to favor the decentralized approach; it would be safer. The information could be passed along from node to node until it reached its destination.

Many of the participants were reluctant to join the network. "The universities in general did not want to share their computers with anybody," Roberts said. "They wanted to buy their own machines and hide in the corner."[49] Nor did they want the valuable processing time of their computers to be nibbled away by having to handle the traffic routing that would come with being on the network. The first to dissent were Marvin Minsky of the MIT Artificial Intelligence Lab and his former colleague John McCarthy, who had moved to Stanford. Their computers, they said, were already being used to the max. Why would they want to allow others to tap into them? In addition they would have the burden of routing network traffic from computers they didn't know and whose language they didn't speak. "Both complained they would lose computing power and said they didn't want to participate," Taylor recalled. "I told them they had to, because it would let me cut my funding of computers by a factor of three."[50]

Taylor was persuasive and Roberts persistent, and they pointed out to the participants that they were all being funded by ARPA. "We are going to build a network and you are going to participate in it," Roberts declared flatly. "And you are going to connect it to your machines."[51] They would get no more funding to buy computers until they were hooked into the network.

Ideas are often sparked by the exchanges at meetings, and one popped up at the end of the Michigan session that helped to defuse opposition to the network. It came from Wes Clark, who had conceived a personal computer at Lincoln Laboratory dubbed the LINC. He was more interested in developing computers designed for individual use than he was in promoting time-sharing of large computers, so he hadn't been paying much attention. But as the meeting was ending he realized why it was hard getting the research centers to accept the network idea. "Just before we broke up, I do remember suddenly realizing what the meta-problem was," he said. "I passed Larry a note saying that I thought I saw how to solve the problem."[52] On the ride to the airport, in a rental car that Taylor was driving, Clark explained his idea to Roberts, along with two other colleagues. ARPA should not force the research computers at each site to handle the routing of data, Clark argued. Instead ARPA should design and give each site a standardized minicomputer that would do the routing. The big research computer at each site would then have only the simple task of establishing a connection with its ARPA-supplied routing minicomputer. This had three advantages: it would take most of the burden off the host site's mainframe, give ARPA the power to standardize the network, and allow the routing of data to be completely distributed rather than controlled by a few big hubs.

Taylor embraced the idea right away. Roberts asked a few questions and then agreed. The network would be managed by the standardized minicomputers that Clark had suggested, which became known as Interface Message Processors, or IMPs. Later they would simply be called "routers."

When they got to the airport, Taylor asked who should build these IMPs. Clark said it was obvious: the task should be assigned to Bolt, Beranek and Newman, the Cambridge firm where Licklider had worked. But also in the car was Al Blue, who was in charge of compliance issues at ARPA. He reminded the group that the project would have to be sent out for bids in accordance with federal contracting standards.[53]

At a follow-up conference in Gatlinburg, Tennessee, in October 1967, Roberts presented the revised plan for the network. He also gave it a name, ARPA Net, which later morphed into ARPANET. But one issue remained unresolved: Would communication between two places on the network require a dedicated line between them, as a phone call did? Or was there some practical way to allow multiple data streams to share lines simultaneously, sort of like a time-sharing system for phone lines? Potential specifications for such a data network had been proposed earlier that month by a committee at the Pentagon.

That is when a young engineer from England, Roger Scantlebury, got up to present a paper describing the research of his boss, Donald Davies of Britain's National Physical Laboratory. It provided an answer: a method of breaking messages into small units that Davies had dubbed "packets." Scantlebury added that the idea had been developed independently by a researcher named Paul Baran at RAND. After the talk, Larry Roberts and others gathered around Scantlebury to learn more, then moved on to the bar to discuss it late into the night.

PACKET SWITCHING: PAUL BARAN, DONALD DAVIES, AND LEONARD KLEINROCK

There are many ways of sending data through a network. The simplest, known as *circuit switching*, is the way a phone system does it: a set of switches creates a dedicated circuit for signals to go back and forth for the duration of the conversation, and the connection remains open, even during long pauses. Another method is *message switching* or, as the telegraph operators called it, *store-and-forward switching*. In this system, an entire message is given an address header, sent into the network, and then passed along from node to node as it wends its way to its destination.

An even more efficient method is *packet switching*, a special type of store-and-forward switching in which the messages are broken into bite-size units of the exact same size, called packets, which are given address headers describing where they should go. These packets are then sent hopping through the network to their destination by being passed along from node to node, using whatever links are most available at that instant. If certain links start getting clogged with too much data, some of the packets will be routed to alternative paths. When all the packets get to their destination node, they are reassembled based on the instructions in the headers. "It's like breaking a long letter into dozens of postcards, each numbered and addressed to the same place," explained Vint Cerf, one of the Internet's pioneers. "Each may take different routes to get to the destination, and then they're reassembled."[54]

As Scantlebury explained in Gatlinburg, the person who first fully conceived a packet-switched network was an engineer named Paul Baran (pronounced BEAR-en). His family had immigrated from Poland when he was two and settled in Philadelphia, where his father opened a small grocery store. After graduating from Drexel in 1949, Baran joined Presper Eckert and John Mauchly in their new computer company, where he tested components for UNIVAC. He moved to Los Angeles, took night classes at UCLA, and eventually got a job at the RAND Corporation.

When the Russians tested a hydrogen bomb in 1955, Baran found his life mission: to help prevent a nuclear holocaust. One day at RAND he was looking at the weekly list sent by the Air Force of topics it needed researched, and he seized on one that related to building a military communications system that would survive an enemy attack. He knew that such a system could help prevent a nuclear exchange, because

Donald Davies (1924–2000).

Paul Baran (1926–2011).

Leonard Kleinrock (1934–).

Vint Cerf (1943–) and Bob Kahn (1938–).

if one side feared that its communications system could be knocked out it would be more likely to launch a preemptive first strike when tensions mounted. With survivable communications systems, nations would not feel the need to adopt a hair-trigger posture.

Baran came up with two key ideas, which he began publishing in 1960. His first was that the network should not be centralized; there should be no main hub that controlled all the switching and routing. Nor should it even be merely decentralized, with the control in many regional hubs, like AT&T's phone system or the route map of a major airline. If the enemy took out a few such hubs, the system could be incapacitated. Instead control should be completely *distributed.* In other words, each and every node should have equal power to switch and route the flow of data. This would become the defining trait of the Internet, the ingrained attribute that would allow it to empower individuals and make it resistant to centralized control.

He drew a network that looked like a fishnet. All of the nodes would have the power to route traffic, and they were each connected to a few other nodes. If any one of the nodes was destroyed, then the traffic would just be routed along other paths. "There is no central control," Baran explained. "A simple local routing policy is performed at each node." He figured out that even if each node had only three or four links, the system would have almost unlimited resilience and survivability. "Just a redundancy level of maybe three or four would permit almost as robust a network as the theoretical limit."[55]

"Having figured out how to get robustness, I then had to tackle the problem of getting signals through this fishnet type of network," Baran recounted.[56] This led to his second idea, which was to break up the data into standard-size little blocks. A message would be broken into many of these blocks, each of which would scurry along different paths through the network's nodes and be reassembled when they got to their destination. "A universally standardized message block would be composed of perhaps 1024 bits," he wrote. "Most of the message block would be reserved for whatever type data is to be transmitted, while the remainder would contain housekeeping information such as error detection and routing data."

Baran then collided with one of the realities of innovation, which was that entrenched bureaucracies are resistant to change. RAND recommended his packet-switched network idea to the Air Force, which, after a thorough review, decided to build one. But then the Department of Defense decreed that any such undertaking should be handled by the Defense Communications Agency so that it could be used by all of the service branches. Baran realized that the Agency would never have the desire or the ability to get it done.

So he tried to convince AT&T to supplement its circuit-switched voice network with a packet-switched data network. "They fought it tooth and nail," he recalled. "They tried all sorts of things to stop it." They would not even let RAND use the maps of its circuits, so Baran had to use a leaked set. He made several trips to AT&T headquarters in lower Manhattan. On one of them, a senior executive who was an old-fashioned analog engineer looked stunned when Baran explained that his system would mean that data could go back and forth without a dedicated circuit remaining open the whole time. "He looked at his colleagues in the room while his eyeballs rolled up sending a signal of his utter disbelief," according to Baran. After a pause, the executive said, "Son, here's how a telephone works," and proceeded with a patronizing and simplistic description.

When Baran continued to push his seemingly preposterous notion that messages could be chopped up and skedaddle through the net as tiny packets, AT&T invited him and other outsiders to a series of seminars explaining how its system really worked. "It took ninety-four separate speakers to describe the entire system," Baran marveled. When it was over, the AT&T executives asked Baran, "Now do you see why packet switching wouldn't work?" To their great disappointment, Baran simply replied, "No." Once again, AT&T was stymied by the innovator's dilemma. It balked at considering a whole new type of data network because it was so invested in traditional circuits.[57]

Baran's work eventually culminated in eleven volumes of detailed engineering analysis, *On Distributed Communications*, completed in 1964. He insisted that it not be classified as secret because he realized such a system worked best if the Russians had one as well. Although Bob Taylor read some of it, no one else at ARPA did, so Baran's idea had little impact until it was brought to the attention of Larry Roberts at the 1967 Gatlinburg conference. When he returned to Washington, Roberts unearthed Baran's reports, dusted them off, and began to read.

Roberts also got hold of the papers written by Donald Davies's group in England, which Scantlebury had summarized in Gatlinburg. Davies was the son of a Welsh coal mine clerk who died a few months after his son was born, in 1924. Young Davies was raised in Portsmouth by his mother, who worked for Britain's General Post Office, which ran the nation's telephone system. He spent his childhood playing with telephone circuits, then earned degrees in math and physics at Imperial College in London. During the war he worked at Birmingham University creating alloys for nuclear weapons tubes as an assistant to Klaus Fuchs, who turned out to be a Soviet spy. He went on to work with Alan Turing building the Automatic Computing Engine, a stored-program computer, at the National Physical Laboratory.

Davies developed two interests: computer time-sharing, which he had learned about during a 1965 visit to MIT, and the use of phone lines for data communications. Combining these ideas in his head, he hit upon the goal of finding a method similar to time-sharing for maximizing the use of communications lines. This led him to the same concepts that Baran had developed about the efficiency of bite-size message units. He also came up with a good old English word for them: *packets*. In trying to convince the General Post Office to adopt the system, Davies ran into the same problem that Baran had when knocking on the door of AT&T. But they both found a fan in Washington. Larry Roberts not only embraced their ideas; he also adopted the word *packet*.[58]

A third and somewhat more controversial contributor in this mix was Leonard Kleinrock, a joyful, affable, and occasionally self-promoting expert on the flow of data in networks, who became close friends with Larry Roberts when they shared an office as doctoral students at MIT. Kleinrock grew up in New York City in a family of poor immigrants. His interest in electronics was sparked when, at the age of six, he was reading a Superman comic and saw instructions for building a crystal radio with no battery. He pieced together a toilet paper roll, one of his father's razor blades, some wire, and graphite from a pencil, and then convinced his mother to take him on the subway to lower Manhattan to buy a variable capacitor at an electronics store. The contraption worked, and a lifelong fascination with electronics blossomed. "I still am awed by it," he recalled of the radio. "It still seems magical." He began scoring radio tube manuals from surplus stores and scavenging discarded radios from Dumpsters, picking apart their components like a vulture so he could build his own radios.[59]

Unable to afford college, even at tuition-free City College of New York, he worked days at an electronics firm and took night courses. The instructors at night were more practical than those during the day; instead of being taught the theory of a transistor, Kleinrock remembered his teacher telling him how heat-sensitive they were and how to adjust for the expected temperature when designing a circuit. "You'd never learn such practical things in the day session," he recalled. "The instructors just wouldn't know that."[60]

After graduating, he won a fellowship to do his doctorate at MIT. There he studied queuing theory, which looks at such questions as what an average wait time in a line might be depending on a variety of factors, and in his dissertation he formulated some of the underlying math that analyzed how messages would flow and bottlenecks arise in switched data networks. In addition to sharing an office with Roberts, Kleinrock was a classmate of Ivan Sutherland and went to lectures by Claude Shannon and Norbert Wiener. "It was a real hotbed of intellectual brilliance," he recalled of MIT at the time.[61]

Late one night at the MIT computer lab, a tired Kleinrock was running one of the machines, a huge experimental computer known as the TX-2, and heard an unfamiliar "psssssss" sound. "I began to get very worried," he recalled. "There was an empty slot where a piece of the machine had been removed to be repaired, and my eyes raised up and I looked at that slot and looking back were two eyes!" It was Larry Roberts, playing a prank on him.[62]

The effervescent Kleinrock and the tightly controlled Roberts remained pals, despite (or maybe because of) their difference in personalities. They enjoyed going to Las Vegas casinos together to try to outsmart the house. Roberts came up with a card-counting scheme for blackjack, based on tracking both high and low cards, and he taught it to Kleinrock. "We got kicked out once, playing with my wife at the Hilton, when the casino managers were watching us through the ceiling and became suspicious when I bought insurance on a hand when you normally wouldn't unless you knew there weren't many high cards left," Roberts recalled. Another ploy involved trying to calculate the trajectory of the ball at the roulette table using a counter made from transistors and an oscillator. It would measure the velocity of the ball and predict which side of the wheel it would end up on, allowing them to bet with more favorable odds. To gather the necessary data, Roberts had his hand wrapped in gauze to hide a recorder. The croupier, figuring something was afoot, looked at them and asked, "Would you like me to break your other arm?" He and Kleinrock decided not, and left.[63]

In his MIT dissertation proposal, written in 1961, Kleinrock proposed exploring the mathematical basis for predicting traffic jams in a weblike network. In this and related papers, he described a store-and-forward network—"communication nets in which there is storage at each of the nodes"—but not a purely packet-switched network, in which the messages would be broken up into very small units of the exact same size. He addressed the issue of "the average delay experienced by a message as it passes through the net" and analyzed how imposing a priority structure that included breaking messages into pieces would help solve the problem. He did not, however, use the term *packet* nor introduce a concept that closely resembled one.[64]

Kleinrock was a gregarious and eager colleague, but he was never known for emulating Licklider in being reticent about claiming credit. He would later alienate many of the other developers of the Internet by asserting that, in his PhD thesis and his paper proposing it (both written after Baran began formulating packet switching at RAND), he had "developed the basic principles of packet switching" and "the mathematical theory of packet networks, the technology underpinning the Internet."[65] Beginning in the mid-1990s, he began an energetic campaign to be recognized "as the Father of Modern Data

Networking."[66] He claimed in a 1996 interview, "My dissertation laid out the basic principles for packet switching."[67]

This led to an outcry among many of the other Internet pioneers, who publicly attacked Kleinrock and said that his brief mention of breaking messages into smaller pieces did not come close to being a proposal for packet switching. "Kleinrock is a prevaricator," said Bob Taylor. "His claim to have anything to do with the invention of packet switching is typical incorrigible self-promotion, which he has been guilty of from day one."[68] (Countered Kleinrock, "Taylor is disgruntled because he never got the recognition he thought he deserved."[69])

Donald Davies, the British researcher who coined the term *packet*, was a gentle and reticent researcher who never boasted of his accomplishments. People called him humble to a fault. But as he was dying, he wrote a paper to be published posthumously that attacked Kleinrock in surprisingly strong terms. "The work of Kleinrock before and up to 1964 gives him no claim to have originated packet switching," Davies wrote after an exhaustive analysis. "The passage in his book on time-sharing queue discipline, if pursued to a conclusion, might have led him to packet switching, but it did not. . . . I can find no evidence that he understood the principles of packet switching."[70] Alex McKenzie, an engineer who managed BBN's network control center, would later be even more blunt: "Kleinrock claims to have introduced the idea of packetization. This is utter nonsense; there is NOTHING in the entire 1964 book that suggests, analyzes, or alludes to the idea of packetization." He called Kleinrock's claims "ludicrous."[71]

The backlash against Kleinrock was so bitter that it became the subject of a 2001 *New York Times* article by Katie Hafner. In it she described how the usual collegial attitude of the Internet pioneers had been shattered by Kleinrock's claim of priority for the concept of packet switching. Paul Baran, who did deserve to be known as the father of packet switching, came forward to say that "the Internet is really the work of a thousand people," and he pointedly declared that most people involved did not assert claims of credit. "It's just this one little case that seems to be an aberration," he added, referring disparagingly to Kleinrock.[72]

Interestingly, until the mid-1990s Kleinrock had credited others with coming up with the idea of packet switching. In a paper published in November 1978, he cited Baran and Davies as pioneers of the concept: "In the early 1960's, Paul Baran had described some of the properties of data networks in a series of RAND Corporation papers. . . . In 1968 Donald Davies at the National Physical Laboratories in England was beginning to write about packet-switched networks."[73] Likewise, in a 1979 paper describing the development of distributed networks, Kleinrock neither mentioned nor cited his own work from the early 1960s. As late as 1990 he was still declaring that Baran was the first to conceive of packet switching: "I would credit him [Baran] with the first ideas."[74] However, when Kleinrock's 1979 paper was reprinted in 2002, he wrote a new introduction that claimed, "I developed the underlying principles of packet switching, having published the first paper on the subject in 1961."[75]

In fairness to Kleinrock, whether or not he had claimed that his work in the early 1960s devised packet switching, he would have been (and still should be) accorded great respect as an Internet pioneer. He was indisputably an important early theorist of data flow in networks and also a valued leader in building the ARPANET. He was one of the first to calculate the effect of breaking up messages as they were passed from node to node. In addition, Roberts found his theoretical work valuable and enlisted him to be part of the implementation team for the ARPANET. Innovation is driven by people who have both good theories and the opportunity to be part of a group that can implement them.

The Kleinrock controversy is interesting because it shows that most of the Internet's creators preferred—to use the metaphor of the Internet itself—a system of fully distributed credit. They instinctively isolated and routed around any node that tried to claim more significance than the others. The Internet was born of an ethos of creative collaboration and distributed decision making, and its founders liked to protect that heritage. It became ingrained in their personalities—and in the DNA of the Internet itself.

WAS IT NUKE-RELATED?

One of the commonly accepted narratives of the Internet is that it was built to survive a nuclear attack. This enrages many of its architects, including Bob Taylor and Larry Roberts, who insistently and repeatedly debunked this origin myth. However, like many of the innovations of the digital age, there were multiple causes and origins. Different players have different perspectives. Some who were higher in the chain of command than Taylor and Roberts, and who have more knowledge of why funding decisions were actually made, have begun to debunk the debunking. Let's try to peel away the layers.

There is no doubt that when Paul Baran proposed a packet-switched network in his RAND reports, nuclear survivability was one of his rationales. "It was necessary to have a strategic system that could withstand a first attack and then be able to return the favor in kind," he explained. "The problem was that we didn't have a survivable communications system, and so Soviet missiles aimed at U.S. missiles would take out the entire telephone-communication system."[76] That led to an unstable hair-trigger situation; a nation was more likely to launch a preemptive strike if it feared that its communications and ability to respond would not survive an attack. "The origin of packet switching is very much Cold War," he said. "I got very interested in the subject of how the hell you build a reliable command and control system."[77] So in 1960 Baran set about devising "a communication network which will allow several hundred major communications stations to talk with one another after an enemy attack."[78]

That may have been Baran's goal, but remember that he never convinced the Air Force to build such a system. Instead his concepts were adopted by Roberts and Taylor, who insisted that they were merely seeking to create a resource-sharing network for ARPA researchers, not one that would survive an attack. "People have been taking what Paul Baran wrote about a secure nuclear defense network and applying it to the ARPANET," said Roberts. "Of course, they had nothing to do with each other. What I told Congress was that this was for the future of science in the world—the civilian world as well as the military—and the military would benefit just as much as the rest of the world. But it clearly wasn't for military purposes. And I didn't mention nuclear war."[79] At one point *Time* magazine reported that the Internet had been built to assure communications after a nuclear attack, and Taylor wrote a letter to the editors correcting them. *Time* didn't print it. "They sent me back a letter insisting that their sources were correct," he recalled.[80]

Time's sources were higher in the chain of command than Taylor. Those who worked at ARPA's Information Processing Techniques Office, which was responsible for the network project, may have sincerely believed that their project had nothing to do with nuclear survivability, but some of the higher-ups at ARPA believed that was, in fact, one of its critical missions. And that is how they convinced Congress to keep funding it.

Stephen Lukasik was the deputy director of ARPA from 1967 to 1970 and then director until 1975. In June 1968 he was able to get the formal authorization and appropriation for Roberts to proceed with building the network. That was just a few months after the Tet Offensive and the My Lai Massacre in Vietnam. Antiwar protests were at their height, and students had rioted at top universities. Defense Department money was not flowing freely to costly programs designed merely to allow collaboration among academic researchers. Senator Mike Mansfield and others had begun demanding that only projects directly relevant to a military mission get funding. "So in this environment," Lukasik said, "I would have been hard pressed to plow a lot of money into the network just to improve the productivity of the researchers. That rationale would just not have been strong enough. What was strong enough was this idea that packet switching would be more survivable, more robust under damage to a network. . . . In a strategic situation—meaning a nuclear attack—the president could still communicate to the missile fields. So I can assure you, to the extent that I was signing the checks, which I was from 1967 on, I was signing them because *that* was the need I was convinced of."[81]

In 2011 Lukasik was amused and somewhat annoyed by what had become the conventional dogma, that the ARPANET had *not* been built for strategic military reasons. So he wrote a piece entitled "Why the Arpanet Was Built," which he circulated to colleagues. "ARPA's existence and its sole purpose was to respond to new national security concerns," he explained. "In the instant case it was the command and control of military forces, especially those deriving from the existence of nuclear weapons and deterring their use."[82]

This directly contradicted the statements of one of his predecessors as ARPA director, Charles Herzfeld, the Viennese refugee who approved Bob Taylor's proposal of a time-sharing research network in 1965. "The ARPANET was not started to create a Command and Control System that would survive a nuclear attack, as many now claim," Herzfeld insisted many years later. "To build such a system was, clearly, a major military need, but it was not ARPA's mission to do this."[83]

Two semiofficial histories authorized by ARPA come down on opposite sides. "It was from the RAND study that the false rumor started claiming that the ARPANET was somehow related to building a network resistant to nuclear war," said the history written by the Internet Society. "This was never true of the ARPANET, only the unrelated RAND study."[84] On the other hand, the "Final Report" by the National Science Foundation in 1995 declared, "An outgrowth of the Department of Defense's Advanced Research Projects Agency, the ARPANET's packet-switching scheme was meant to provide reliable communications in the face of nuclear attack."[85]

So which view is correct? In this case, both are. For the academics and researchers who were actually building the network, it had only a peaceful purpose. For some of those who were overseeing and funding the project, especially in the Pentagon and Congress, it also had a military rationale. Stephen Crocker was a graduate student in the late 1960s who became integrally involved in coordinating how the ARPANET would be designed. He never considered nuclear survivability to be part of his mission. Yet when Lukasik sent around his 2011 paper, Crocker read it, smiled, and revised his thinking. "I was on top and you were on the bottom, so you really had no idea of what was going on and why we were doing it," Lukasik told him. To which Crocker replied, with a dab of humor masking a dollop of wisdom, "I was on the bottom and you were on the top, so you had no idea of what was going on or what we were doing."[86]

As Crocker finally realized, "You can't get all the guys involved to agree on why it was built." Leonard Kleinrock, who had been his supervisor at UCLA, came to the same conclusion: "We will never know if nuclear survivability was the motivation. It was an unanswerable question. For me, there was no notion of a military rationale. But if you go up the chain of command, I am sure that some were saying that surviving a nuclear attack was a reason."[87]

The ARPANET ended up representing an interesting conjunction of military and academic interests. It was funded by the Defense Department, which tended to want hierarchal command systems with centralized controls. But the Pentagon had delegated the design of the network to a bunch of academics, some of whom were avoiding being drafted and most of whom had a distrust of centralized authority. Because they opted for a structure of limitless nodes, each with its own router, rather than one based on a few centralized hubs, the network would be hard to control. "My bias was always to build decentralization into the net," Taylor said. "That way it would be hard for one group to gain control. I didn't trust large central organizations. It was just in my nature to distrust them."[88] By picking people like Taylor to build its network, the Pentagon was spawning one that it would not be able to fully control.

There was yet another layer of irony. The decentralized and distributed architecture meant that the network would be more reliable. It could even withstand a nuclear attack. Building a resilient and attack-proof military command-and-control system was not what motivated the ARPA researchers. It wasn't even in the back of their minds. But that was one reason they ended up getting a steady stream of Pentagon and congressional funding for the project.

Even after the ARPANET morphed into the Internet in the early 1980s, it would continue to serve both a military and a civilian purpose. Vint Cerf, a gentle and reflective thinker who helped create the Internet, recalled, "I wanted to demonstrate that our technology could survive a nuclear attack." So in 1982 he ran a series of tests that replicated a nuclear attack artificially. "There were a number of such simulations or demonstrations like that, some of which were extremely ambitious. They involved the Strategic Air Command. At one point we put airborne packet radios in the field while using the airborne systems to sew together fragments of Internet that had been segregated by a simulated nuclear attack." Radia Perlman, one of the foremost women network engineers, developed at MIT protocols that would assure network robustness in the face of malicious attacks, and she helped Cerf come up with ways to partition and reconstruct the ARPANET when necessary to make it more survivable.[89]

This interplay of military and academic motives became ingrained in the Internet. "The design of both the ARPANET and the Internet favored military values, such as survivability, flexibility, and high performance, over commercial goals, such as low cost, simplicity, or consumer appeal," the technology historian Janet Abbate noted. "At the same time, the group that designed and built ARPA's networks was dominated by academic scientists, who incorporated their own values of collegiality, decentralization of authority, and open exchange of information into the system."[90] These academic researchers of the late 1960s, many of whom associated with the antiwar counterculture, created a system that resisted centralized command. It would route around any damage from a nuclear attack but also around any attempt to impose control.

ONE GIANT LEAP: THE ARPANET HAS LANDED, OCTOBER 1969

In the summer of 1968, when much of the world, from Prague to Chicago, was being rocked by political unrest, Larry Roberts sent out a solicitation for bids to companies that might want to build the minicomputers that would be sent to each research center to serve as the routers, or Interface Message Processors, of the proposed ARPANET. His plan incorporated the packet-switching concept of Paul Baran and Donald Davies, the suggestion for standardized IMPs proposed by Wes Clark, the theoretical insights of J. C. R. Licklider, Les Earnest, and Leonard Kleinrock, and the contributions of many other inventors.

Of the 140 companies that received the request, only a dozen decided to submit bids. IBM, for example, didn't. It doubted that the

IMPs could be built at a reasonable price. Roberts convened a committee meeting in Monterey, California, to assess the bids that had been submitted, and Al Blue, the compliance officer, took pictures of each with measuring sticks showing how thick they were.

Raytheon, the large Boston-area defense contractor that had been cofounded by Vannevar Bush, emerged as the frontrunner, and even entered into price negotiations with Roberts. But Bob Taylor stepped in and expressed the view, already being pushed by Wes Clark, that the contract should go to BBN, which was not burdened with a multilayer corporate bureaucracy. "I said that the corporate culture between Raytheon and the research universities would be bad, like oil and water," Taylor recalled.[91] As Clark put it, "Bob overruled the committee." Roberts went along. "Raytheon had a good proposal that competed equally with BBN, and the only distinguishing thing in the long run for my final decision was that BBN had a tighter team organized in a way that I thought would be more effective," he recalled.[92]

In contrast to the bureaucracy-laden Raytheon, BBN had a nimble band of brilliant engineers, led by two refugees from MIT, Frank Heart and Robert Kahn.[93] They helped to improve Roberts's proposal by specifying that when a packet was passed from one IMP to the next, the sending IMP would keep it stored until it got an acknowledgment from the receiving IMP, and it would resend the message if the acknowledgment didn't come promptly. That became a key to the net's reliability. At each step, the design was being improved by collective creativity.

Just before Christmas, Roberts surprised many by announcing the selection of BBN rather than Raytheon. Senator Ted Kennedy sent the usual telegram that goes to a constituent who lands a big federal project. In it, he congratulated BBN for being chosen to build the *Interfaith* Message Processor, which in some ways was an apt description of the ecumenical role of the Interface Message Processors.[94]

Roberts selected four research centers to be the first ARPANET nodes: UCLA, where Len Kleinrock worked; Stanford Research Institute (SRI), with the visionary Douglas Engelbart; the University of Utah, with Ivan Sutherland; and the University of California at

Santa Barbara. They were given the task of figuring out how their big "host" computers would connect to the standardized IMPs that would be shipped to them. Like typical senior professors, the researchers at these centers enlisted a motley crew of graduate students to do the work.

The members of this young work team gathered in Santa Barbara to figure out how to proceed, and they discovered a verity that would remain true even in the age of digital social networks: it was useful—and fun—to get together in person, interfacing in the literal sense of that word. "There was a kind of cocktail-party phenomenon where you find you have a lot of rapport with each other," recalled Stephen Crocker, a graduate student on the UCLA team who had driven up with his best friend and colleague, Vint Cerf. So they decided to meet regularly, rotating among their sites.

The polite and deferential Crocker, with his big face and bigger smile, had just the right personality to be the coordinator of what became one of the digital age's archetypical collaborative processes. Unlike Kleinrock, Crocker rarely used the pronoun *I*; he was more interested in distributing credit than claiming it. His sensitivity toward others gave him an intuitive feel for how to coordinate a group without trying to centralize control or authority, which was well suited to the network model they were trying to invent.

Months passed, and the graduate students kept meeting and sharing ideas while they waited for some Powerful Official to descend upon them and give them marching orders. They assumed that at some point the authorities from the East Coast would appear with the rules and regulations and protocols engraved on tablets to be obeyed by the mere managers of the host computer sites. "We were nothing more than a self-appointed bunch of graduate students, and I was convinced that a corps of authority figures or grownups from Washington or Cambridge would descend at any moment and tell us what the rules were," Crocker recalled. But this was a new age. The network was supposed to be distributed, and so was the authority over it. Its invention and rules would be user-generated. The process would be open. Though it was funded partly to facilitate military command and control, it would do so by being resistant to centralized command

and control. The colonels had ceded authority to the hackers and academics.

So after an especially fun gathering in Utah in early April 1969, this gaggle of graduate students, having named itself the Network Working Group, decided that it would be useful to write down some of what they had conjured up.[95] And Crocker, who with his polite lack of pretense could charm a herd of hackers into consensus, was tapped for the task. He was anxious to find an approach that did not seem presumptuous. "I realized that the mere act of writing down what we were talking about could be seen as a presumption of authority and someone was going to come and yell at us—presumably some adult out of the east." His desire to be deferential kept him up at nights, literally. "I was living with my girlfriend and her baby from a previous relationship at her parents' house. The only place to work at night without disturbing people was the bathroom, and I would stand there naked and scribble down notes."[96]

Crocker realized that he needed an unassertive name for the list of suggestions and practices. "To emphasize the informal nature, I hit upon this silly little idea of calling every one of them a 'Request for Comments'—no matter whether it really was a request." It was the perfect phrase to encourage Internet-era collaboration—friendly, not bossy, inclusive, and collegial. "It probably helped that in those days we avoided patents and other restrictions; without any financial incentive to control the protocols, it was much easier to reach agreement," Crocker wrote forty years later.[97]

The first RFC went out on April 7, 1969, mailed in old-fashioned envelopes through the postal system. (There was no such thing as email, since they hadn't invented the network yet.) In a warm and casual tone, devoid of any officiousness, Crocker set forth the task of figuring out how the host computer at each institution should connect to the new network. "During the summer of 1968, representatives from the initial four sites met several times to discuss the host software," he wrote. "I present here some of the tentative agreements reached and some of the open questions encountered. Very little of what is here is firm and reactions are expected."[98] The people who received RFC 1 felt that they were being included in a fun process rather than being dictated to by a bunch of protocol czars. It was a *network* they were talking about, so it made sense to try to loop everyone in.

The RFC process pioneered open-source development of software, protocols, and content. "That culture of open processes was essential in enabling the Internet to grow and evolve as spectacularly as it has," Crocker said later.[99] Even more broadly, it became the standard for collaboration in the digital age. Thirty years after RFC 1, Vint Cerf wrote a philosophical RFC called "The Great Conversation" that began, "A long time ago, in a network far, far away . . ." After describing the informal way RFCs had begun, Cerf continued, "Hiding in the history of the RFCs is the history of human institutions for achieving cooperative work."[100] It was a grand statement, and it would have seemed overblown except that it was true.

The RFCs produced a set of host-to-IMP standards by the end of August 1969, just when the first IMP was shipped to Kleinrock's lab. When it arrived at the UCLA loading dock, a dozen people were there to greet it: Crocker, Kleinrock, a few other team members, and Cerf and his wife, Sigrid, who had brought champagne. They were surprised to see that the IMP was the size of a refrigerator and was clad, as per the specifications of the military machine that it was, in battleship-gray steel. It was wheeled into the computer room, plugged in, and started right up. BBN had done great work, delivering on time and on budget.

One machine does not a network make. It was not until a month later, when a second IMP was delivered to SRI on the edge of the Stanford campus, that the ARPANET could truly get up and running. On October 29 the connection was ready to be made. The event was appropriately casual. It had none of the drama of the "one small step for man, one giant leap for mankind" that had occurred on the moon a few weeks earlier, with a half billion people watching on television. Instead it was an undergraduate named Charley Kline, under the eye of Crocker and Cerf, who put on a telephone headset to coordinate with a researcher at SRI while typing in a login sequence that he hoped would allow his terminal at UCLA to connect through the network to the computer 354 miles away in Palo Alto. He typed in "L." The guy at SRI told him that it had been received. Then he typed in "O." That, too, was confirmed. When he typed in "G," the system hit a memory snag because of an auto-complete feature and crashed. Nevertheless, the first message had been sent across the ARPANET, and if it wasn't as eloquent as "The Eagle has landed" or "What has God wrought," it was suitable in its understated way: "Lo." As in "Lo and behold." In his logbook, Kline recorded, in a memorably minimalist notation, "22:30. Talked to SRI Host to Host. CSK."[101]

It was thus that in the second half of 1969—amid the static of Woodstock, Chappaquiddick, Vietnam War protests, Charles Manson, the Chicago Eight trial, and Altamont—the culmination was reached for three historic enterprises, each in the making for almost a decade. NASA was able to send a man to the moon. Engineers in Silicon Valley were able to devise a way to put a programmable computer on a chip called a microprocessor. And ARPA created a network that could connect distant computers. Only the first of these (perhaps the least historically significant of them?) made headlines.

THE INTERNET

The ARPANET was not yet the Internet. It was just one network. Within a few years, there were other packet-switched networks that were similar but not interconnected. For example, engineers at Xerox's Palo Alto Research Center (PARC) wanted a local-area network to connect the office workstations they were designing in the early 1970s, and a recent Harvard PhD there named Bob Metcalfe created a way to use coaxial cable (the type that plugs into cable TV boxes) to create a high-bandwidth system that he named "Ethernet." It was modeled on a wireless network developed in Hawaii known as ALOHAnet, which sent packet data through UHF and satellite signals. In addition, there was a packet radio network in San Francisco, known as PRNET, and also a satellite version called SATNET. Despite their similarities, these packet-switched networks were not compatible or interoperable.

In early 1973 Robert Kahn set out to remedy that. There should be a way, he decided, to allow all these networks to interconnect, and he was in a position to make that happen. He had left BBN, where he had helped develop the IMPs, to become a project manager at ARPA's Information Processing Techniques Office. Having worked on the ARPANET and then PRNET, he made it his mission to create a method to connect them and other packet networks, a system that he and his colleagues began calling an "internetwork." After a while, that word got shortened a bit, to "internet."

To be his partner in this endeavor, Kahn tapped Vint Cerf, who had been Steve Crocker's sidekick on the group writing Requests for Comments and figuring out the protocols of the ARPANET. Cerf was raised in Los Angeles, where his father worked for a company that made engines for the Apollo space program. Like Gordon Moore, he grew up playing with a chemistry set in the days when they were delightfully dangerous. "We had things like powdered magnesium, powdered aluminum, and sulfur and glycerin and potassium permanganate," he recalled. "When you pour them together, they burst into flame." In fifth grade he was bored with math, so his teacher gave him a seventh-grade algebra book. "I spent the whole summer working every single problem in the book," he said. "I liked the word problems the best because they were like little mystery stories. You had to figure out who 'x' was, and I was always curious to find out what's 'x' going to turn out to be." He also became deeply immersed in science fiction, especially the stories of Robert Heinlein, and began his lifelong practice of rereading J. R. R. Tolkien's *Lord of the Rings* trilogy almost every year.[102]

Because he had been born prematurely, Cerf was hearing impaired, and he began using a hearing aid at age thirteen. Around that time he also started wearing a coat and tie to school and carrying a briefcase. "I didn't want to fit in with everybody else," he said. "I wanted to look different, be noticed. That was a very effective way to do it, and it was better than wearing a nose ring, which I figured my Dad would not have put up with in the 1950's."[103]

In high school he became best friends with Crocker, and they spent weekends together doing science projects and playing 3-D chess. After graduating from Stanford and working for IBM for two years, he became a doctoral student at UCLA, where he worked in Kleinrock's group. There he met Bob Kahn, and they remained close after Kahn went to work at BBN and then ARPA.

When Kahn embarked on his internetwork endeavor in the spring of 1973, he visited Cerf and described all the packet-switched networks that had sprung up in addition to the ARPANET. "How are we going to hook these different kinds of packet-nets to each other?" Kahn asked. Cerf seized on the challenge, and the two of them launched into a three-month burst of collaboration that would lead to the creation of the Internet. "He and I clicked instantly on this," Kahn later said. "Vint is the kind of guy who likes to roll up his sleeves and say let's get on with it. I thought that was a breath of fresh air."[104]

They began by organizing a meeting at Stanford in June 1973 to gather ideas. As a result of this collaborative approach, Cerf later said, the solution "turned out to be the open protocol that everybody had a finger in at one time or another."[105] But most of the work was done as a duet by Kahn and Cerf, who holed up for intense sessions at Rickeys Hyatt House in Palo Alto or at a hotel next to Dulles Airport. "Vint liked to get up and draw these spider drawings," Kahn recalled. "Often times we would have a conversation back and forth and he would say, 'Let me draw a picture of that.'"[106]

One day in October 1973, Cerf made a simple sketch in a San Francisco hotel lobby that codified their approach. It showed various networks like the ARPANET and PRNET, each with lots of host computers connected to them, and a set of "gateway" computers that would pass packets between each of the networks. Finally, they spent an entire weekend together at the ARPA office near the Pentagon, where they stayed up almost two nights straight and then ended up at a nearby Marriott for a triumphal breakfast.

They rejected the idea that the networks could each keep their own different protocols, although that would have been easier to sell. They wanted a common protocol. That would allow the new internetwork to scale up explosively, since any computer or network using the new protocol could link in without requiring a translation system. The traffic between ARPANET and any other network should be seamless. So they came up with the idea of having every computer adopt the same method and template for addressing its packets. It was as if every postcard mailed in the world had to have a four-line address specifying street number and city and country using the Roman alphabet.

The result was an Internet Protocol (IP) that specified how to put the packet's destination in its header and helped determine how it would travel through networks to get there. Layered above it was a higher-level Transmission Control Protocol (TCP) that instructed how to put the packets back together in the right order, checked to see if any of them was missing, and requested retransmission of any information that had been lost. These became known as TCP/IP. Kahn and Cerf published them as a paper called "A Protocol for Packet Network Interconnection." The Internet was born.

On the twentieth anniversary of the ARPANET in 1989, Kleinrock and Cerf and many of the other pioneers gathered at UCLA, where the network's first node had been installed. There were poems and songs and doggerel written to celebrate the occasion. Cerf performed a parody of Shakespeare, titled "Rosencrantz and Ethernet," that raised to a Hamlet-like question the choice between packet switching and dedicated circuits:

All the world's a net! And all the data in it merely packets
come to store-and-forward in the queues a while and then are
heard no more. 'Tis a network waiting to be switched!

To switch or not to switch? That is the question:
Whether 'tis wiser in the net to suffer
The store and forward of stochastic networks,
Or to raise up circuits against a sea of packets,
And by dedication serve them?[107]

A generation later, in 2014, Cerf was working at Google in Washington, DC, still enjoying himself and marveling at the wonders they had wrought by creating the Internet. Wearing Google Glass, he noted that every year brings something new. "Social networks—I joined Facebook as an experiment—business apps, mobile, new things keep piling onto the Internet," he said. "It has scaled up a million times over. Not many things can do that without breaking. And yet those old protocols we created are doing just fine."[108]

NETWORKED CREATIVITY

So who does deserve the most credit for inventing the Internet? (Hold the inevitable Al Gore jokes. We will get to his role—yes, he did have one—in chapter 10.) As with the question of who invented the computer, the answer is that it was a case of collaborative creativity. As Paul Baran later explained to the technology writers Katie Hafner and Matthew Lyon, using a beautiful image that applies to all innovation:

> The process of technological development is like building a cathedral. Over the course of several hundred years new people come along and each lays down a block on top of the old foundations, each saying, "I built a cathedral." Next month another block is placed atop the previous one. Then comes along an historian who asks, "Well, who built the cathedral?" Peter added some stones here, and Paul added a few more. If you are not careful, you can con yourself into believing that you did the most important part. But the reality is that each contribution has to follow onto previous work. Everything is tied to everything else.[109]

The Internet was built partly by the government and partly by private firms, but mostly it was the creation of a loosely knit cohort of academics and hackers who worked as peers and freely shared their creative ideas. The result of such peer sharing was a network that facilitated peer sharing. This was not mere happenstance. The Internet was built with the belief that power should be distributed rather than centralized and that any authoritarian diktats should be circumvented. As Dave Clark, one of the early participants in the Internet Engineering Task Force, put it, "We reject kings, presidents, and voting. We believe in rough consensus and running code."[110] The result was a networked commons, a place where innovations could be crowdsourced and open-source.

Innovation is not a loner's endeavor, and the Internet was a prime example. "With computer networks, the loneliness of research is supplanted by the richness of shared research," proclaimed the first issue of *ARPANET News*, the new network's official newsletter.

The network pioneers J. C. R. Licklider and Bob Taylor realized that the Internet, because of how it was built, had an inherent tendency to encourage peer-to-peer connections and the formation of online communities. This opened up beautiful possibilities. "Life will be happier for the on-line individual because the people with whom one interacts most strongly will be selected more by commonality of interests and goals than by accidents of proximity," they wrote in a visionary 1968 paper titled "The Computer as a Communication Device." Their optimism verged on utopianism. "There will be plenty of opportunity for everyone (who can afford a console) to find his calling, for the whole world of information, with all its fields and disciplines, will be open to him."[111]

But it didn't happen right away. After the Internet was created in the mid-1970s, there were a few more innovations necessary before it could become a transformative tool. It was still a gated community, open primarily to researchers at military and academic institutions. It wasn't until the early 1980s that civilian counterparts to ARPANET were fully opened, and it would take yet another decade before most ordinary home users could get in.

There was, in addition, one other major limiting factor: the only people who could use the Internet were those who had hands-on access to computers, which were still big, intimidating, costly, and not something you could run down to Radio Shack and buy. The digital age could not become truly transformational until computers became truly personal.

Ken Kesey (1935–2001) holding a flute on the bus.

Stewart Brand (1938–).

The first issue, fall 1968.

CHAPTER EIGHT

THE PERSONAL COMPUTER

"AS WE MAY THINK"

The idea of a *personal* computer, one that ordinary individuals could get their hands on and take home, was envisioned in 1945 by Vannevar Bush. After building his big analog computer at MIT and helping to create the military-industrial-academic triangle, he wrote an essay for the July 1945 issue of the *Atlantic* titled "As We May Think."*[1] In it he conjured up the possibility of a personal machine, which he dubbed a *memex*, that would store and retrieve a person's words, pictures, and other information: "Consider a future device for individual use, which is a sort of mechanized private file and library. . . . A memex is a device in which an individual stores all his books, records, and communications, and which is mechanized so that it may be consulted with exceeding speed and flexibility. It is an enlarged intimate supplement to his memory." The word *intimate* was important. Bush and his followers focused on ways to make close, personal connections between man and machine.

---

*It appeared the same month that he presented to President Truman his other groundbreaking essay, "Science, the Endless Frontier," which proposed the creation of a research collaboration among government, industry, and universities. See chapter 7.

Bush imagined that the device would have a "direct entry" mechanism, such as a keyboard, so you could put information and your records into its memory. He even predicted hypertext links, file sharing, and ways to collaborate on projects. "Wholly new forms of encyclopedias will appear, ready made with a mesh of associative trails running through them, ready to be dropped into the memex and there amplified," he wrote, anticipating Wikipedia by a half century.

As it turned out, computers did not emerge the way that Bush envisioned, at least not initially. Instead of becoming *personal* tools and memory banks for individuals to use, they became hulking industrial and military colossi that researchers could time-share but the average person could not touch. By the early 1970s innovative companies such as DEC were making minicomputers the size of a small refrigerator, but they dismissed the idea that there would be a market for desktop models that could be owned and operated by ordinary folks. "I can't see any reason that anyone would want a computer of his own," DEC president Ken Olsen declared at a May 1974 meeting where his operations committee was debating whether to create a smaller version of its PDP-8 for personal consumers.[2] As a result, the personal computer revolution, when it erupted in the mid-1970s, was led by scruffy entrepreneurs in strip malls and garages who started companies with names like Altair and Apple.

THE CULTURAL BREW

The personal computer was made possible by a number of technological advances, most notably the microprocessor, a circuit etched on a tiny chip that integrated all of the functions of a computer's central processing unit. But social forces also help drive and shape innovations, which then bear the imprint of the cultural milieu into which they were born. Rarely has there been a more potent cultural amalgam than the one that bubbled up in the San Francisco Bay Area beginning in the 1960s, and it turned out to be ripe for producing homebrewed computers.

What were the tribes that formed that cultural mix?[3] It began with the pocket protector–wearing engineers who migrated to the area with the growth of defense contractors, such as Westinghouse and Lockheed. Next there arose an entrepreneurial startup culture, exemplified by Intel and Atari, where creativity was encouraged and stultifying bureaucracies disdained. The hackers who moved west from MIT brought their craving for hands-on computers that they could touch and play with. There was also a subculture populated by wireheads, phreakers, and hard-core hobbyists who got their kicks hacking into the Bell System's phone lines or the time-shared computers of big corporations. And emanating from San Francisco and Berkeley were idealists and community organizers who sought ways, in the words of one of them, Liza Loop, "to co-opt technological advances for progressive purposes and thereby triumph over the bureaucratic mindset."[4]

Added to this mix were three countercultural strands. There were the hippies, born out of the Bay Area's beat generation, whose merry rebelliousness was fueled by psychedelics and rock music. There were the New Left activists, who spawned the Free Speech Movement at Berkeley and the antiwar protests on campuses around the world. And interwoven with them were the Whole Earth communalists, who believed in controlling their own tools, sharing resources, and resisting the conformity and centralized authority imposed by power elites.

As different as some of these tribes were from each other, their worlds intermingled and they shared many values. They aspired to a do-it-yourself creativity that was nurtured by building Heathkit radios as kids, reading the *Whole Earth Catalog* in college, and fantasizing about someday joining a commune. Ingrained in them was the very American belief, so misunderstood by Tocqueville, that rugged individualism and the desire to form associations were totally compatible, even complementary, especially when it involved creating things collaboratively. The maker culture in America, ever since the days of community barn raisers and quilting bees, often involved do-it-ourselves rather than do-it-yourself. In addition, many of these Bay Area tribes of the late 1960s shared a resistance to power elites and a desire to control their own access to information. Technology should be made open and friendly and convivial rather than daunting and mysterious and Orwellian. As Lee Felsenstein, one of the avatars of many of these cultural strands, put it, "We wanted there to be personal computers so that we could free ourselves from the constraints of institutions, whether government or corporate."[5]

Ken Kesey was a muse of the hippie strand of this cultural tapestry. After graduating from the University of Oregon, he went to the Bay Area in 1958 as a graduate student in Stanford's creative writing program. While there, he worked the overnight shift at a mental hospital and signed up to be a guinea pig in a CIA-funded series of experiments, Project MKUltra, testing the effects of the psychedelic drug LSD. Kesey ended up liking the drug, very much. The combustible combination of creative writing, dropping acid for pay, and working as an orderly in an asylum led to his first novel, *One Flew Over the Cuckoo's Nest.*

While others were starting electronics companies in the neighborhood around Stanford, Kesey used the proceeds from his book, combined with some acid he had been able to liberate from the CIA experiments, to form a commune of early hippies called the Merry Pranksters. In 1964 he and his posse embarked on a psychedelic cross-country odyssey in an old International Harvester school bus dubbed *Furthur* (spelling later corrected) painted in Day-Glo colors.

Upon his return, Kesey began hosting a series of Acid Tests at his home, and at the end of 1965 he decided, since he was an entrepreneur as well as a hippie, to take them public. One of the earliest took place that December at Big Ng's, a music club in San Jose. Kesey enlisted a bar band that he liked, led by Jerry Garcia, which had just changed its name from the Warlocks to the Grateful Dead.[6] Flower power was born.

Concurrently there arose a companion cultural phenomenon, the peace movement, that shared this rebellious spirit. The confluence of hippie and antiwar sensibilities led to memorable period pieces, amusing in retrospect but considered deep at the time, such as psychedelic posters exhorting "Make love not war" and tie-dyed T-shirts featuring peace symbols.

The hippie and antiwar movements were both wary of computers, at least initially. The hulking mainframes with whirring tapes and blinking lights were seen as depersonalizing and Orwellian, tools of Corporate America, the Pentagon, and the Power Structure. In *The Myth of the Machine*, the sociologist Lewis Mumford warned that the rise of computers could mean that "man will become a passive, purposeless, machine-conditioned animal."[7] At peace protests and hippie communes, from Sproul Plaza at Berkeley to Haight-Ashbury in San Francisco, the injunction printed on punch cards, "Do not fold, spindle or mutilate," became an ironic catchphrase.

But by the early 1970s, when the possibility of *personal* computers arose, attitudes began to change. "Computing went from being dismissed as a tool of bureaucratic control to being embraced as a symbol of individual expression and liberation," John Markoff wrote in his history of the period, *What the Dormouse Said*.[8] In *The Greening of America*, which served as a manifesto for the new era, a Yale professor, Charles Reich, denounced the old corporate and social hierarchies and called for new structures that encouraged collaboration and personal empowerment. Instead of deploring computers as tools of the old power structure, he argued that they could aid the shift in social consciousness if they were made more personal: "The machine, having been built, may now be turned to human ends, in order that man once more can become a creative force, renewing and creating his own life."[9]

A technotribalism began to emerge. Tech gurus such as Norbert Wiener, Buckminster Fuller, and Marshall McLuhan became required reading in communes and dorms. By the 1980s the LSD evangelist Timothy Leary would update his famous mantra "Turn on, tune in, drop out" to proclaim instead "Turn on, boot up, jack in."[10] Richard Brautigan was the poet-in-residence in 1967 at Caltech, and that year he captured the new ethos in a poem, "All Watched Over by Machines of Loving Grace."[11] It began:

I like to think (and
the sooner the better!)
of a cybernetic meadow
where mammals and computers

live together in mutually
programming harmony
like pure water
touching clear sky.

STEWART BRAND

The person who best embodied and most exuberantly encouraged this connection between techies and hippies was a lanky enthusiast with a toothy smile named Stewart Brand, who popped up like a gangly sprite at the intersection of a variety of fun cultural movements over the course of many decades. "The counterculture's scorn for centralized authority provided the philosophical foundations of the entire personal-computer revolution," he wrote in a 1995 *Time* essay titled "We Owe It All to the Hippies."

> Hippie communalism and libertarian politics formed the roots of the modern cyberrevolution. . . . Most of our generation scorned computers as the embodiment of centralized control. But a tiny contingent—later called "hackers"—embraced computers and set about transforming them into tools of liberation. That turned out to be the true royal road to the future . . . youthful computer programmers who deliberately led the rest of civilization away from centralized mainframe computers.[12]

Brand was born in 1938 in Rockford, Illinois, where his father was a partner in an ad agency and, like so many fathers of digital entrepreneurs, a ham radio operator. After graduating as a biology major from Stanford, where he was in Army ROTC, Brand served two years as an infantry officer, including airborne training and a stint as an Army photographer. He then began a joyful life meandering among different communities at that exciting juncture where performance art and technology intermingle.[13]

Not surprisingly, life on that techno/creative edge led Brand to become one of the early experimenters with LSD. After being introduced to the drug in a pseudoclinical setting near Stanford in 1962, he became a regular at Kesey's Merry Prankster gatherings. He was a photographer, technician, and producer at a multimedia art c lective called USCO, which produced events that involved acid roc music, technological wizardry, strobe lights, projected images, and performances that enlisted audience participation. Occasionally they featured talks by Marshall McLuhan, Dick Alpert, and other new age prophets. A promotional piece on the group noted that it "unites the cults of mysticism and technology as a basis for introspection and communication," a phrase that served as a suitable credo for techno-spiritualists. Technology was a tool for expression that could expand the boundaries of creativity and, like drugs and rock, be rebellious.

For Brand, the 1960s protest slogan "Power to the people" began to ring hollow when used by New Left political activists, but computers offered a true opportunity for individual empowerment. "Power to the people was a romantic lie," he later said. "Computers did more than politics did to change society."[14] He visited the Stanford Artificial Intelligence Lab and wrote an article for *Rolling Stone* in 1972 calling it "the most bzz-bzz-busy scene I've been around since Merry Prankster Acid Tests." This counterculture and cyberculture combination, he realized, was a recipe for a digital revolution. "The freaks who design computer science" would wrest power away from the "rich and powerful institutions," he wrote. "Ready or not, computers are coming to the people. That's good news, maybe the best since psychedelics." This utopian vision, he added, was "in line with the romantic fantasies of the forefathers of the science, such as Norbert Wiener, J. C. R. Licklider, John von Neumann, and Vannevar Bush."[15]

All of these experiences led Brand to become the impresario and techie for one of the seminal events of the 1960s counterculture, the January 1966 Trips Festival at Longshoreman's Hall in San Francisco. After the joys of the Acid Tests, which had been held weekly throughout December, Brand proposed to Kesey that they throw a blowout version that would last for three days. The extravaganza opened with Brand's own troupe, America Needs Indians, performing a "sensorium" that included a high-tech light show, slide projectors, music, and Native American dancers. It was followed by what the program described as "revelations, audioprojections, the endless osion, the congress of wonders, liquid projections, and the jazz ce." And that was just the opening night. The next night was kicked ff by Kesey, who had been busted for drugs a few days earlier on Brand's North Beach roof but was out on bail and orchestrating the event from a command scaffold. Featured were the Merry Pranksters and their Psychedelic Symphony, Big Brother and the Holding Company, the Grateful Dead, and members of the Hells Angels motorcycle gang. The writer Tom Wolfe tried to recapture the technodelic essence in his seminal work of New Journalism, *The Electric Kool-Aid Acid Test*:

> Lights and movies sweeping around the hall; five movie projectors going and God knows how many light machines, interferrometrics, the intergalactic science-fiction seas all over the walls, loudspeakers studding the hall all the way around like flaming chandeliers, strobes exploding, black lights with Day-Glo objects under them and Day-Glo paint to play with, street lights at every entrance flashing red and yellow, and a troop of weird girls in leotards, leaping around the edges blowing dog whistles.

The final night celebrated technology even more enthusiastically. "Since the common element of all shows is ELECTRICITY, this evening will be programmed live from stimuli provided by a PINBALL MACHINE," the program exulted. "The audience is invited to wear ECSTATIC DRESS & bring their own GADGETS (a.c. outlets will be provided)."[16]

Yes, the Trip Festival's conjunction of drugs, rock, and technology—*acid and a.c. outlets!*—was jarring. But it turned out to be, significantly, a quintessential display of the fusion that shaped the personal computer era: technology, counterculture, entrepreneurship, gadgets, music, art, and engineering. From Stewart Brand to Steve Jobs, those ingredients fashioned a wave of Bay Area innovators who were comfortable at the interface of Silicon Valley and Haight-Ashbury. "The Trips Festival marked Stewart Brand's emergence as a countercultural entrepreneur—but in a deeply technocratic mold," wrote the cultural historian Fred Turner.[17]

A month after the Trips Festival, in February 1966, Brand was sitting on his gravelly rooftop in San Francisco's North Beach enjoying the effects of 100 micrograms of LSD. Staring at the skyline, he ruminated on something that Buckminster Fuller had said: our perception that the world is flat and stretches indefinitely, rather than round and small, is because we have never seen it from outer space. Abetted by the acid, he began to grok the smallness of the earth and the importance of other people appreciating that as well. "It had to be broadcast, this fundamental point of leverage on the world's ills," he recalled. "A photograph would do it—a color photograph from space of the earth. There it would be for all to see, the earth complete, tiny, adrift, and no one would ever perceive things the same way."[18] It would, he believed, promote big-picture thinking, empathy for all the earth's inhabitants, and a sense of connectedness.

He resolved to convince NASA to take such a picture. So, with the offbeat wisdom that comes from acid, he decided to produce hundreds of buttons so that people in the pre-Twitter age could spread the word. "Why haven't we seen a photograph of the whole Earth yet?" they read. His plan was goofy-simple: "I prepared a Day-Glo sandwich board with a little sales shelf on the front, decked myself out in a white jump suit, boots and costume top hat with crystal heart and flower, and went to make my debut at the Sather Gate of the University of California in Berkeley, selling my buttons for twenty-five cents." University officials did him the favor of throwing him off campus, which prompted a story in the *San Francisco Chronicle*, thus helping publicize his one-man crusade. He took it on the road to other colleges across the country, ending at Harvard and MIT. "Who the hell's that?" asked an MIT dean as he watched Brand give an impromptu lecture while selling his buttons. "That's my brother," said Peter Brand, an MIT instructor.[19]

In November 1967 NASA complied. Its ATS-3 satellite took a picture of Earth from twenty-one thousand miles up, which served as the cover image and title inspiration for Brand's next venture, the *Whole Earth Catalog*. As its name implied, it was (or at least dressed itself in the guise of) a catalogue, one that cleverly blurred the distinction between consumerism and communalism. Its subtitle was "Access to Tools," and it combined the sensibilities of the back-to-the-land counterculture with the goal of technological empowerment. Brand wrote on the first page of the first edition, "A realm of intimate, personal power is developing—power of the individual to conduct his own education, find his own inspiration, shape his own environment, and share his adventure with whoever is interested. Tools that aid this process are sought and promoted by the *Whole Earth Catalog*." Buckminster Fuller followed with a poem that began, "I see God in the instruments and mechanisms that work reliably." The first edition featured such items as Norbert Wiener's book *Cybernetics* and a programmable HP calculator, along with buckskin jackets and beads. The underlying premise was that a love of the earth and a love of technology could coexist, that hippies should make common cause with engineers, and that the future should be a festival where a.c. outlets would be provided.[20]

Brand's approach was not New Left political. Nor was it even antimaterialist, given his celebration of games and gadgets you could buy. But he did pull together, better than anyone, many of the cultural strands of that period, from acid-dropping hippies to engineers to communal idealists who sought to resist the centralized control of technology. "Brand did the marketing work for the concept of the personal computer through the *Whole Earth Catalog*," said his friend Lee Felsenstein.[21]

DOUGLAS ENGELBART

Shortly after the first edition of the *Whole Earth Catalog* came out, Brand helped to produce a happening that was an odd echo of his techno-choreography of the January 1966 Trips Festival. Dubbed "the Mother of All Demos," the December 1968 extravaganza became the seminal event of the personal computer culture, just as the Trips Festival had been for the hippie culture. It happened because, like a magnet, Brand naturally attracted and attached himself to interesting people. This time it was an engineer named Douglas Engelbart, who had taken on as his life's passion inventing ways that computers could augment human intelligence.

Engelbart's father, an electrical engineer, had a shop in Portland, Oregon, where he sold and repaired radios; his grandfather, who operated hydropower dams in the Pacific Northwest, liked to take the family inside the goliath plants to see how the turbines and generators worked. So it was natural that Engelbart developed a passion for electronics. In high school he heard that the Navy had a program, cloaked in secrecy, to train technicians in a mysterious new technology called radar, and he studied hard to make sure he could get in it, which he did.[22]

His great awakening came while serving in the Navy. He was loaded onto a ship that set sail from just south of the Bay Bridge in San Francisco, and as they were waving good-bye, an announcement came on the public address system that the Japanese had surrendered and World War II was over. "We all shouted," Engelbart recounted, "'Turn around! Let us go back and celebrate!'" But the ship kept sailing, "right out into the fog, into the seasickness," on to Leyte Gulf in the Philippines.[23] On Leyte Island, Engelbart secluded himself whenever possible in a Red Cross library in a thatched hut on stilts, and there he became enthralled by a heavily illustrated *Life* magazine reprint of Vannevar Bush's *Atlantic* article "As We May Think," the one that envisioned the memex personal information system.[24] "The whole concept of helping people work and think that way just excited me," he recalled.[25]

After his Navy service, he got an engineering degree from Oregon State and then worked at the forerunner to NASA at the Ames Research Center in Silicon Valley. Painfully shy, he joined an intermediate Greek folk-dancing class at the Palo Alto Community Center in order to meet a woman he could marry, which he did. On the day after his engagement, as he was driving to work, he felt a frightening, life-altering apprehension: "By the time I got to work, I had this realization that I didn't have any more goals."[26]

For the next two months, he assiduously tended to the task of finding for himself a worthy life goal. "I looked at all the crusades people could join, to find out how I could retrain myself." What struck him was that any effort to improve the world was complex. He thought about people who tried to fight malaria or increase food production in poor areas and discovered that led to a complex array of other issues, such as overpopulation and soil erosion. To succeed at any ambitious project, you had to assess all of the intricate ramifications of an action, weigh probabilities, share information, organize people, and more. "Then one day, it just dawned on me—BOOM—that *complexity* was the fundamental thing," he recalled. "And it just went *click*. If in some way, you could contribute significantly to the way humans could handle complexity and urgency, that would be universally helpful."[27] Such an endeavor would address not just one of the world's problems; it would give people the tools to take on *any* problem.

The best way to help people handle complexity was along the lines that Bush had proposed, Engelbart decided. As he tried to imagine conveying information on graphic screens in real time, his radar training came in handy. "It was within an hour that I had the image of sitting at a big screen with all kinds of symbols," he recalled, "and you could be operating all kinds of things to drive the computer."[28] That day he set out on a mission to find ways to allow people to visually portray the thinking they were doing and link them to other people so they could collaborate—in other words, networked interactive computers with graphic displays.

This was in 1950, five years before Bill Gates and Steve Jobs were born. Even the very first commercial computers, such as UNIVAC, were not yet publicly available. But Engelbart bought into Bush's vision that someday people would have their own terminals, which they could use to manipulate, store, and share information. This expansive conception needed a suitably grand name, and Engelbart came up with one: *augmented intelligence*. In order to serve as the pathfinder for this mission, he enrolled at Berkeley to study computer science, earning his doctorate in 1955.

Engelbart was one of those people who could project intensity by speaking in an eerily calm monotone. "When he smiles, his face is wistful and boyish, but once the energy of his forward motion is halted and he stops to ponder, his pale blue eyes seem to express sadness or loneliness," a close friend said. "His voice, as he greets you, is low and soft, as though muted from having traveled a long distance. There is something diffident yet warm about the man, something gentle yet stubborn."[29]

To put it more bluntly, Engelbart sometimes gave the impression that he had not been born on this planet, which made it difficult for him to get funding for his project. He finally was hired in 1957 to work on magnetic storage systems at the Stanford Research Institute, an independent nonprofit set up by the university in 1946. A hot topic at SRI was artificial intelligence, especially the quest to create a system that mimicked the neural networks of the human brain.

But the pursuit of artificial intelligence didn't excite Engelbart, who never lost sight of his mission to augment human intelligence by creating machines like Bush's memex that could work closely with people and help them organize information. This goal, he later said, was born out of his respect for the "ingenious invention" that was the human mind. Instead of trying to replicate that on a machine, Engelbart focused on how "the computer could interact with the different capabilities that we've already got."[30]

For years he worked on draft after draft of a paper describing his vision, until it grew to forty-five thousand words, the length of a small book. He published it as a manifesto in October 1962 titled "Augmenting Human Intellect." He began by explaining that he was not seeking to replace human thought with artificial intelligence. Instead he argued that the intuitive talents of the human mind should be combined with the processing abilities of machines to produce "an integrated domain where hunches, cut-and-try, intangibles, and the human 'feel for a situation' usefully co-exist with powerful concepts, streamlined terminology and notation, sophisticated methods, and high-powered electronic aids." In painstaking detail, he gave many examples of how this human-computer symbiosis would work, including an architect using a computer to design a building and a professional putting together an illustrated report.[31]

As he was working on the paper, Engelbart wrote a fan letter to Vannevar Bush, and he devoted an entire section of his paper to describing the memex machine.[32] Seventeen years after Bush had written "As We May Think," there was still a radical feel to his concept that humans and computers should interact in real time through simple interfaces that included graphical screens, pointers, and input devices. Engelbart emphasized that his system wouldn't be just for math: "Every person who does his thinking with symbolized concepts (whether in the form of the English language, pictographs, formal logic, or mathematics) should be able to benefit significantly." Ada Lovelace would have been thrilled.

Engelbart's treatise appeared the same month that Licklider, who had explored the same concepts two years earlier in his "Man-Computer Symbiosis" paper, took over ARPA's Information Processing Techniques Office. Part of Licklider's new job was to give out federal grants to promising projects. Engelbart got in line. "I was standing at the door with this 1962 report and a proposal," he recalled. "I thought, 'Oh boy, with all the things he's saying he wants to do, how can he refuse me?'"[33] He couldn't, so Engelbart got an ARPA grant. Bob Taylor, who was then still at NASA, also gave Engelbart some funding. Thus it was that he was able to create his own Augmentation Research Center at SRI. It became another example of how government funding of speculative research eventually paid off hundreds of times over in practical applications.

THE MOUSE AND NLS

The NASA grant from Taylor was supposed to be applied to a stand-alone project, and Engelbart decided to use it to find an easy way for humans to interact with machines.[34] "Let's go after some screen-select devices," he suggested to his colleague Bill English.[35] His goal was to find the simplest way for a user to point to and select something on a screen. Dozens of options for moving an on-screen cursor were being tried by researchers, including light pens, joysticks, trackballs, trackpads, tablets with styli, and even one that users were supposed to control with their knees. Engelbart and English tested each. "We timed how long it took each user to move the cursor to the object," Engelbart said.[36] Light pens seemed the simplest, for example, but they required a user to pick them up and put them down each time, which was tiresome.

They made a chart of all the advantages and drawbacks of each device, which helped Engelbart imagine devices that hadn't yet been conceived. "Just as the periodic table's rules have led to the discovery of certain previously unknown elements, this grid ultimately defined the desirable characteristics of a device that didn't yet exist," he said. One day in 1961 he was at a conference and began to daydream. He recalled a mechanical device that had fascinated him in high school, a planimeter, that could calculate the area of a space by being rolled around its perimeter. It used two perpendicular wheels, one horizontal and the other vertical, to tote up the distance it was rolled in each direction. "Just thinking about those two wheels, soon the rest of it was very simple, so I went and made a sketch," he recalled.[37] In his pocket notebook he showed how the device could roll around a desktop and its two wheels would register higher or lower voltages as they turned in each direction. That voltage could be transmitted through a cord to the computer screen to move a cursor up and down and back and forth.

The result, at once both simple and profound, was a classic physical expression of the augmentation ideal and the hands-on imperative. It made use of the human talent of mind-hand-eye coordination (something robots are not good at) to provide a natural interface with a computer. Instead of acting independently, humans and machines would act in harmony.

Engelbart gave his sketch to Bill English, who carved a piece of mahogany to make the first model. When they tried it on their focus group, it tested better than any other device. At first, the cord was in front, but they quickly realized it worked better coming out of the back end, like a tail. They dubbed the device a "mouse."

Most true geniuses (Kepler, Newton, Einstein, and even Steve Jobs, to name a few) have an instinct for simplicity. Engelbart didn't. Desiring to cram a lot of functionality into any system he built, he wanted the mouse to have many buttons, perhaps up to ten. But to his disappointment, the testing determined that the optimum number of buttons the mouse should have was three. As it turned out, even that was at least one button too many, or perhaps, as the simplicity-freak Jobs would later insist, two buttons too many.

Over the next six years, culminating in 1968, Engelbart went on to devise a full-fledged augmentation system that he called "oNLine System," or NLS. In addition to the mouse, it included many other advances that led to the personal computer revolution: on-screen graphics, multiple windows on a screen, digital publishing, blog-like journals, wiki-like collaborations, document sharing, email, instant messaging, hypertext linking, Skype-like videoconferencing, and the formatting of documents. One of his technocharged protégés, Alan Kay, who would later advance each of these ideas at Xerox PARC, said of Engelbart, "I don't know what Silicon Valley will do when it runs out of Doug's ideas."[38]

THE MOTHER OF ALL DEMOS

Engelbart was more into Greek folk dances than Trips Festivals, but he had gotten to know Stewart Brand when they experimented with LSD at the same lab. Brand's succession of ventures, including the *Whole Earth Catalog*, were based just a few blocks from Engelbart's Augmentation Research Center. Thus it was natural that they team up for a demonstration in December 1968 of Engelbart's oNLine System. Thanks to Brand's instincts as an impresario, the demo, which later became known as the Mother of All Demos, became a multimedia extravaganza, like an Electric Kool-Aid Acid Test on silicon. The event turned out to be the ultimate melding of hippie and hacker culture, and it has remained unchallenged, even by Apple product launches, as the most dazzling and influential technology demonstration of the digital age.[39]

The year had been turbulent. In 1968 the Tet Offensive turned America against the Vietnam War, Robert Kennedy and Martin Luther King were assassinated, and Lyndon Johnson announced he would not seek reelection. Peace protests shut down major universities and disrupted the Democratic National Convention in Chicago. The Russians crushed Prague Spring, Richard Nixon was elected

Doug Engelbart (1925–2013).

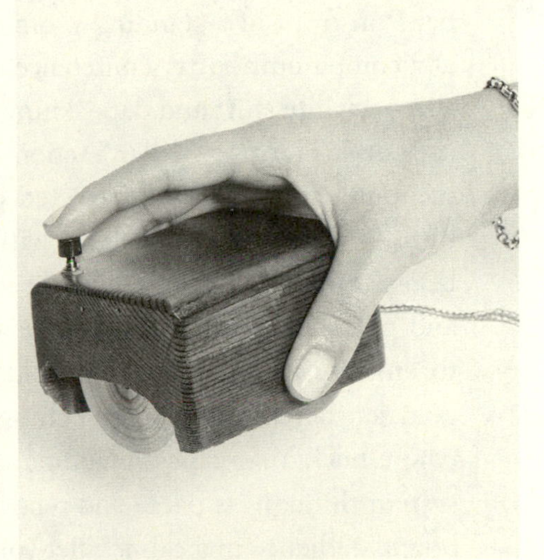

Englebart's first mouse.

Stewart Brand (*center*) assisting at the Mother of All Demos in 1968.

president, and Apollo 8 orbited the moon. Also that year, Intel was founded and Stewart Brand published the first *Whole Earth Catalog*.

Engelbart's ninety-minute demonstration occurred on December 9 in front of a standing-room-only crowd of close to a thousand at a computer industry conference in San Francisco. Wearing a short-sleeved white shirt and dark skinny tie, he sat on the right of the stage in a sleek Herman Miller "Action Office" console. The display of his computer terminal was projected onto a twenty-foot screen behind him. "I hope you'll go along with this rather unusual setting," he began. He wore a microphone headset that a fighter pilot might use, and he spoke in a monotone, like a computer-generated voice trying to emulate the narrator in an old movie newsreel. Howard Rheingold, a cyberculture guru and chronicler, later said that he looked like "the Chuck Yeager of the computer cosmos, calmly putting the new system through its paces and reporting back to his astonished earth-bound audience in a calm, quiet voice."[40]

"If in your office," Engelbart intoned, "you as an intellectual worker were supplied with a computer display backed up by a computer that was alive for you all day and was instantly responsive to every action you have, how much value could you derive from that?" He promised that the combination of technologies he was about to demonstrate would "all go very interesting," and then he muttered under his breath, "I think."

A camera mounted on his terminal provided a video stream of his face, while another camera overhead showed his hands controlling the mouse and keyboard. Bill English, the mouse crafter, sat in the back of the auditorium like a newsroom producer selecting which images were mixed, matched, and projected on the big screen.

Stewart Brand was thirty miles south at Engelbart's lab near Stanford, generating computer images and working cameras. Two leased microwave lines and a telephone hookup transmitted to the lab every mouse and keyboard click Engelbart made, then sent images and information back to the auditorium. The audience watched incredulously as Engelbart collaborated with distant colleagues to create a document; different people made edits, added graphics, changed the layout, built a map, and embedded audio and visual elements in real time. They even were able to create hypertext links together. In short, Engelbart showed, back in 1968, nearly everything that a networked personal computer does today. The demo gods were with him, and to his amazement there were no glitches. The crowd gave him a standing ovation. Some even rushed up to the stage as if he were a rock star, which in some ways he was.[41]

Down the hall from Engelbart, a competing session was being presented by Les Earnest, who had cofounded, with the MIT refugee John McCarthy, the Stanford Artificial Intelligence Lab. As reported by John Markoff in *What the Dormouse Said*, their session featured a film about a robot that acted as if it could hear and see things. The two demos presented a clear contrast between the goal of artificial intelligence and that of augmented intelligence. The latter mission had seemed rather quirky when Engelbart began working on it, but when he showed off all of its elements in his December 1968 demo—a personal computer that humans could easily interact with in real time, a network that allowed collaborative creativity—it overshadowed the robot. The headline of the story from the conference in the next day's San Francisco *Chronicle* was "Fantastic World of Tomorrow's Computer." It was about Engelbart's oNLine System, not about the robot.[42]

As if to seal the marriage of the counterculture and cyberculture, Brand brought Ken Kesey to Engelbart's lab to experience the oNLine System. Kesey, by then famous from Tom Wolfe's *The Electric Kool-Aid Acid Test*, got a full tour of how the system could cut, paste, retrieve, and collaboratively create books and other documents. He was impressed. "It's the next thing after acid," Kesey pronounced.[43]

ALAN KAY

Alan Kay struggled to make sure that he got to Engelbart's Mother of All Demos. He had a 102-degree fever and strep throat, but he was able to drag himself onto a plane from Utah, where he was a graduate student. "I was shivering and sick and could barely walk," he recalled, "but I was determined to get there."[44] He had already seen and embraced Engelbart's ideas, but the drama of the demonstration struck him like a clarion call. "To me he was Moses opening the Red Sea," Kay said. "He showed us a promised land that needed to be found, and the seas and rivers we needed to cross to get there."[45]

Like Moses, Engelbart would not actually make it to that promised land. Instead it would be Kay and a merry band of colleagues at the research center of a copier company who would be at the forefront of bringing the ideas of Licklider and Engelbart into the paradise of personal computing.

Kay had learned to love both the sciences and humanities as a child in central Massachusetts, where he was born in 1940. His father was a physiologist who designed artificial legs and arms. On long walks with him, Kay developed a love of science. But he also became passionate about music. His mother was an artist and musician, just as was her father, Clifton Johnson, a noted illustrator and author who played the pipe organ at their local church. "Since my father was a scientist and my mother was an artist, the atmosphere during my early years was full of many kinds of ideas and ways to express them. I did not distinguish between 'art' and 'science' and still don't."[46]

At seventeen he went away to music camp, where he played guitar and was a member of the jazz band. He also, like his grandfather, loved pipe organs, and he eventually helped a master builder construct one in the Spanish baroque style for a Lutheran seminary. He was a savvy and well-read student who often got in trouble in school, mainly for insubordination, a trait of many tech innovators. He was almost expelled, but he also starred on the national radio show *Quiz Kids*.

Kay enrolled in Bethany College in West Virginia to study math and biology, but he was kicked out during the spring of his first year after "excessive unexcused absences." For a while he hung out in Denver, where a friend had taken a job tending to United Airlines' reservation computer system. Kay was struck that the computers seemed to increase rather than reduce drudge work for humans.

Faced with being drafted, he enlisted in the Air Force, where his top scores on an aptitude test led him to be tapped for training as a computer programmer. He worked on the IBM 1401, the first widely marketed computer for small businesses. "This was back in the days when programming was a low-status profession and most of the programmers were women," he said. "They were really good. My boss was a woman."[47] When his service was over, he enrolled at the University of Colorado, where he indulged all of his passions: he studied biology, mathematics, music, and theater while programming supercomputers at the National Center for Atmospheric Research.

He wandered off to graduate school at the University of Utah, which he ended up considering "the best luck I ever had." The computer science pioneer David Evans was building there the best graphics program in the country. On the day that Kay arrived in the fall of 1966, Evans handed him a document from a stack on his desk and told him to read it. It was the MIT doctoral dissertation of Ivan Sutherland, who was then teaching at Harvard but would soon move to Utah. Written under the supervision of the information theorist Claude Shannon, the thesis was titled "Sketchpad: A Man-Machine Graphical Communications System."[48]

Sketchpad was a computer program that pioneered the use of a graphical user interface, one that displayed icons and graphics on the display screen, the way today's computers do. The graphics, which could be created and manipulated with a light pen, provided a charming new way for humans and computers to interact. "The Sketchpad system makes it possible for a man and a computer to converse rapidly through the medium of line drawings," Sutherland wrote. The realization that art and technology could combine to create a delightful computer interface appealed to Kay's childlike enthusiasm for making sure that the future would turn out to be fun. Sutherland's ideas, he said, were "a glimpse of heaven" and "imprinted" him with a passion for creating friendly personal computers.[49]

His first contact with Engelbart came in early 1967, a few months after being turned on to Sutherland's Sketchpad ideas. Engelbart was on a tour of universities, lecturing on the ideas that he would eventually put on display in his Mother of All Demos and lugging a Bell & Howell projector so that he could show a film of his oNLine System. "He would freeze frame and run it at different speeds forward and backward," Kay recalled. "He would say, 'Here's the cursor. Watch what it's going to do next!'"[50]

The field of computer graphics and natural user interfaces was on fire, and Kay soaked up ideas from many sources. He heard a lecture by MIT's Marvin Minsky on artificial intelligence and the dreadful way that schools were crushing the creativity of young students by not teaching them to deal imaginatively with complexity. "He put forth a terrific diatribe against traditional education methods," Kay remembered.[51] He then met Minsky's colleague Seymour Papert, who had created a programming language called LOGO that was simple enough for a school kid to use. Among its many tricks was allowing students to use simple commands to control a robotic turtle moving around the classroom. After hearing Papert, Kay began drawing sketches of what a kid-friendly personal computer might look like.

At a conference at the University of Illinois, Kay saw a rudimentary flat-screen display, made of thin glass with neon gas. Putting that together in his mind with Engelbart's demonstrations of the oNLine System, and roughly calculating the effect of Moore's Law, he realized that graphical displays with windows, icons, hypertext, and a mouse-controlled cursor could be incorporated in small computers within a decade. "I was almost frightened by the implications," he said, indulging his flair for dramatic storytelling. "It must have been the same kind of disorientation people had after reading Copernicus and first looked up from a different Earth to a different Heaven."

Kay saw the future with great clarity, and he became impatient to invent it. "There would be millions of personal machines and users, mostly outside of direct institutional control," he realized. This would require the creation of small personal computers with graphical displays easy enough for a kid to use and cheap enough for every individual to own. "It all came together to form an image of what a personal computer really should be."

In his doctoral thesis he described some of its traits, most notably that it should be simple ("It must be learnable in private") and friendly ("Kindness should be an integral part"). He was designing a computer as if he were a humanist as well as an engineer. He drew inspiration from an Italian printer in the early sixteenth century named Aldus Manutius, who realized that personal books would need to fit into saddlebags and thus produced ones of the size now common.

Likewise, Kay recognized that the ideal personal computer had to be no larger than a notebook. "It was easy to know what to do next," he recalled. "I built a cardboard model of it to see what it would look and feel like."[52]

Kay had been inspired by what Engelbart was trying to do at his Augmentation Research Center. But instead of taking a job there, he joined the Stanford Artificial Intelligence Laboratory (SAIL), run by Professor John McCarthy. It was not a good fit. Because McCarthy was focused on artificial intelligence, rather than on ways to augment human intelligence, he had little interest in *personal* computers. He believed instead in large computers that would be time-shared.

In an academic paper he delivered in 1970, just after Kay came to SAIL, McCarthy described his vision of time-sharing systems that used terminals with little processing power or memory of their own. "The terminal is to be connected by the telephone system to a time-shared computer which, in turn, has access to files containing all books, magazines, newspapers, catalogs, airline schedules," he wrote. "Through the terminal the user can get any information he wants, can buy and sell, could communicate with persons and institutions, and process information in other useful ways."[53]

McCarthy foresaw that this could lead to a proliferation of new information sources that would compete with traditional media, though he mistakenly thought that these would be supported by customer payments rather than advertising. "Since the cost of keeping a file of information in the computer and making it publicly available will be small, even a high school student could compete with the *New Yorker* if he could write well enough and if word of mouth and mention by reviewers brought him to public attention." He also predicted crowdsourced content: a user would be able to "tell the system whether last year's cure for baldness worked and get a summary of the opinions of those who bothered to record their opinions of the cure he contemplates trying now." McCarthy had a rosy view of what turned out to be a raucous blogosphere: "Public controversy can be carried out more expeditiously than at present. If I read something that seems controversial, I can ask the system if anyone has filed a reply. This, together with an author's ability to revise his original statement, will lead people to converge on considered positions more quickly."

McCarthy's vision was prescient, but it differed in one major way from Kay's vision, and from the networked world that we have today. It was not based on *personal* computers with their own memory and processing power. Instead McCarthy believed that people would have inexpensive, dumb terminals that would be connected to powerful distant computers. Even after hobbyist clubs began springing up to celebrate personal computers, McCarthy was pushing a plan for a "Home Terminal Club" that would lease to people for $75 a month simple Teletype-like terminals that would allow them to time-share on a distant powerful mainframe.[54]

Kay's contrasting vision was that powerful small computers, complete with their own memory and processing power, would become personal tools for individual creativity. He dreamed of kids wandering into the woods and using them under trees, just as they might use crayons and a pad of paper. So after two years of toiling among the time-sharing evangelists at SAIL, Kay accepted an offer in 1971 to join a corporate research center two miles away that was attracting young innovators who wanted to make computers that were personal, friendly, and geared to individuals. McCarthy would later dismiss these goals as "Xerox heresies,"[55] but they ended up setting the course for the era of personal computers.

XEROX PARC

In 1970 the Xerox Corporation followed in the footsteps of the Bell System by launching a lab dedicated to pure research. In order that it not be contaminated by the mind-set of the corporation's bureaucracy or the day-to-day demands of its business, it was located in the Stanford industrial park, some three thousand miles from the company's Rochester, New York, headquarters.[56]

Among those recruited to lead Xerox's Palo Alto Research Center, known as Xerox PARC, was Bob Taylor, who had recently left ARPA's Information Processing Techniques Office after helping to build the ARPANET. Through his visits to ARPA-funded research centers and the conferences he hosted for the brightest graduate students, he had developed a radar for talent. "Taylor had worked with and funded many of the leading computer science research groups during this period," recalled Chuck Thacker, who was one of Taylor's recruits. "As a result, he was in a unique position to attract a staff of the highest quality."[57]

Taylor had another leadership skill that he had refined at his meetings with ARPA researchers and graduate students: he was able to provoke "creative abrasion," in which a team of people can question each other, even try to eviscerate each other's ideas, but then are expected to articulate the other side of the dispute. Taylor did that at what he called "Dealer" meetings (evoking people trying to beat the dealer at blackjack), in which one person had to present an idea while others engaged in constructive (usually) criticism. Taylor was not a technology wizard himself, but he knew how to get a group of them to sharpen their sabers in friendly duels.[58] His flair for playing a master of ceremonies allowed him to prod, cajole, stroke, and jolly up temperamental geniuses and get them to collaborate. He was much better at tending to the egos of people who worked under him than catering to his bosses, but that was part of his charm—especially if you weren't one of his bosses.

Among Taylor's first recruits was Alan Kay, whom he knew from ARPA conferences. "I met Alan when he was a doctoral student at Utah, and I liked him a lot," Taylor said.[59] He didn't, however, hire Kay for his own lab at PARC but instead recommended him to another group there. It was Taylor's way of seeding the whole place with people who impressed him.

When he went to PARC for his formal interview, Kay was asked what he hoped his great achievement there would be. "A personal computer," he answered. Asked what that was, he picked up a notebook-size portfolio, flipped open its cover, and said, "This will be a flat-panel display. There'll be a keyboard here on the bottom, and enough power to store your mail, files, music, artwork, and books. All in a package about this size and weighing a couple of pounds. That's what I'm talking about." His interviewer scratched his head and muttered to himself, "Yeah, right." But Kay got the job.

With his twinkling eyes and lively mustache, Kay came to be seen as a disruptor, which he was. He took impish pleasure in pushing the executives of a copier company to create a small and friendly computer for kids. Xerox's corporate planning director, Don Pendery, a dour New Englander, embodied what the Harvard professor Clay Christensen has labeled the innovator's dilemma: he saw the future filled with shadowy creatures that threatened to gnaw away at Xerox's copier business. He kept asking Kay and others for an assessment of "trends" that foretold what the future might hold for the company. During one maddening session, Kay, whose thoughts often seemed tailored to go directly from his tongue to wikiquotes, shot back a line that was to become PARC's creed: "The best way to predict the future is to invent it."[60]

For his 1972 *Rolling Stone* piece on the emerging tech culture in Silicon Valley, Stewart Brand visited Xerox PARC, causing agita back east at corporate headquarters when the article appeared. With literary gusto, he described how PARC's research had moved "away from hugeness and centrality, toward the small and the personal, toward putting maximum computer power in the hands of every individual who wants it." Among the people he interviewed was Kay, who said, "The people here are used to dealing lightning with both hands." Because of people like Kay, PARC had a playful sensibility that was derivative of the MIT Tech Model Railroad Club. "It's a place where you can still be an artisan," he told Brand.[61]

Kay realized that he needed a catchy name for the little personal computer he wanted to build, so he began calling it the Dynabook. He also came up with a cute name for its operating system software: Smalltalk. The name was meant to be unintimidating to users and not raise expectations among hard-core engineers. "I figured that Smalltalk was so innocuous a label that if it ever did anything nice, people would be pleasantly surprised," Kay noted.

He was determined that his proposed Dynabook would cost less than $500 "so that we could give it away in schools." It also had to be small and personal, so that "a kid could take it wherever he goes to hide," with a programming language that was user-friendly. "Simple things should be simple, complex things should be possible," he declared.[62]

Kay wrote a description of the Dynabook, titled "A Personal Computer for Children of All Ages," that was partly a product proposal but mostly a manifesto. He began by quoting Ada Lovelace's seminal insight about how computers could be used for creative tasks: "The Analytical Engine weaves algebraical patterns just as the Jacquard loom weaves flowers and leaves." In describing how children (of all ages) would use a Dynabook, Kay showed he was in the camp of those who saw personal computers primarily as tools for individual creativity rather than as networked terminals for collaboration. "Although it can be used to communicate with others through the 'knowledge utilities' of the future such as a school 'library,'" he wrote, "we think that a large fraction of its use will involve reflexive communication of the owner with himself through this personal medium, much as paper and notebooks are currently used."

The Dynabook, Kay continued, should be no larger than a notebook and weigh no more than four pounds. "The owner will be able to maintain and edit his own files of text and programs when and where he chooses. Need we add that it be usable in the woods?" In other words, it was not just a dumb terminal designed to be networked into a time-shared mainframe. However, he did envision a day when personal computers and digital networks would come together. "A combination of this 'carry anywhere' device and a global information utility such as the ARPA network or two-way cable TV will bring the libraries and schools (not to mention stores and billboards) to the home."[63] It was an enticing vision of the future, but one that would take another two decades to invent.

To advance his crusade for the Dynabook, Kay gathered around him a small team and crafted a mission that was romantic, aspirational, and vague. "I only hired people that got stars in their eyes when they heard about the notebook computer idea," Kay recalled. "A lot of daytime was spent outside of PARC, playing tennis, bike riding, drinking beer, eating Chinese food, and constantly talking about the Dynabook and its potential to amplify human reach and bring new ways of thinking to a faltering civilization that desperately needed it."[64]

In order to take the first step toward realizing the Dynabook, Kay proposed an "interim" machine. It would be about the size of a carry-on suitcase and would have a small graphical display screen. In May 1972 he made his pitch to Xerox PARC's hardware bosses to build thirty so that they could be tested in classrooms to see if students could do simple programming tasks on them. "The uses for a personal gadget as an editor, reader, take-home context, and intelligent terminal are fairly obvious," he told the engineers and managers sitting in beanbag chairs. "Now let's build thirty of these things so we can get on with it."

It was a romantic pitch confidently delivered, as tended to be the case with Kay, but it did not dazzle Jerry Elkind, the manager of PARC's computer lab. "Jerry Elkind and Alan Kay were like creatures from different planets, one an austere by-the-numbers engineer and the other a brash philosophical freebooter," according to Michael Hiltzik, who wrote a history of Xerox PARC. Elkind did not get stars in his eyes when imagining children programming toy turtles on Xerox machines. "Let me play devil's advocate," he responded. The other engineers perked up, sensing that a merciless evisceration was in the offing. PARC's mandate was to create the office of the future, Elkind noted, so why should it be in the business of child's play? The corporate environment lent itself to the time-sharing of corporate-run computers, so shouldn't PARC continue to pursue those opportunities? After a rapid-fire series of such questions, Kay felt like crawling away. When it was over, he cried. His request that a set of interim Dynabooks be built was denied.[65]

Bill English, who had worked with Engelbart and built the first mouse, was by then at PARC. After the meeting he pulled Kay aside, consoled him, and offered some advice. He needed to stop being a dreamy loner and instead should prepare a well-crafted proposal with a budget. "What's a budget?" Kay asked.[66]

Kay scaled back his dream and proposed an interim-interim plan. He would use $230,000 that he had in his budget to emulate the

Dynabook on a Nova, a footlocker-size minicomputer made by Data General. But the prospect didn't really thrill him.

That is when two stars from Bob Taylor's group at PARC, Butler Lampson and Chuck Thacker, popped into Kay's office with a different scheme.

"Do you have any money?" they asked.

"Yes, about $230K for Novas," Kay replied. "Why?"

"How would you like us to build your little machine for you?" they asked, referring to the interim Dynabook that Elkind had shot down.

"I'd like it fine," Kay allowed.[67]

Thacker wanted to build his own version of a personal computer, and he realized that Lampson and Kay also had the same general goal in mind. So the plot was to pool their resources and proceed without waiting for permission.

"What are you going to do about Jerry?" Kay asked about his nemesis Elkind.

"Jerry's out of the office for a few months on a corporate task force," said Lampson. "Maybe we can sneak it in before he gets back."[68]

Bob Taylor had helped hatch the plan because he wanted to push his team away from building time-sharing computers and devise instead "an interconnected collection of small display-based machines."[69] He was thrilled to get three of his favorite engineers—Lampson, Thacker, and Kay—collaborating on the project. The team had a push-pull dynamic: Lampson and Thacker knew what was possible, while Kay set his sights on the ultimate dream machine and challenged them to achieve the impossible.

The machine they designed was named the Xerox Alto (although Kay stubbornly continued to refer to it as "the interim Dynabook"). It had a bitmapped display, which meant that each pixel on the screen could be turned on or off to help render a graphic, a letter, a paintbrush swipe, or whatever. "We chose to provide a full bitmap, in which each screen pixel was represented by a bit of main storage," Thacker explained. That put a lot of demands on the memory, but the guiding principle was that Moore's Law would continue to rule and that memory would get cheaper in an exponential way. The user's

Alan Kay (1940–) at Xerox PARC in 1974.

Kay's 1972 sketch for a Dynabook.

Lee Felsenstein (1945–).

The first issue, October 1972.

interaction with the display was controlled by a keyboard and mouse, as Engelbart had designed. When it was completed in March 1973, it featured a graphic, painted by Kay, of Sesame Street's Cookie Monster holding the letter "C."

By keeping children (of all ages) in mind, Kay and his colleagues advanced Engelbart's concepts by showing that they could be implemented in a manner that was simple, friendly, and intuitive to use. Engelbart, however, did not buy into their vision. Instead he was dedicated to cramming as many functions as possible into his oNLine System, and thus he never had a desire to make a computer that was small and personal. "That's a totally different trip from where I'm going," he told colleagues. "If we cram ourselves in those little spaces, we'd have to give up a whole bunch."[70] That is why Engelbart, even though he was a prescient theorist, was not truly a successful innovator: he kept adding functions and instructions and buttons and complexities to his system. Kay made things easier, and in so doing showed why the ideal of simplicity—making products that humans find convivial and easy to use—was central to the innovations that made computers personal.

Xerox sent Alto systems to research centers around the country, spreading the innovations dreamed up by PARC engineers. There was even a precursor to the Internet Protocols, the PARC Universal Packet, that allowed different packet-switched networks to interconnect. "Most of the tech that makes the Internet possible was invented at Xerox PARC in the 1970s," Taylor later claimed.[71]

As things turned out, however, although Xerox PARC pointed the way to the land of personal computers—devices you could call your own—the Xerox Corporation did not lead the migration. It made two thousand Altos, mainly for use in Xerox offices or affiliated institutions, but it didn't market the Alto as a consumer product.* "The

---

*The Xerox Star workstation was not introduced until 1981, eight years after the Alto was invented, and even it was not initially marketed as a stand-alone computer but as part of an "integrated office system" that included a file server, a printer, and usually other networked workstations.

company wasn't equipped to handle an innovation," Kay recalled. "It would have meant completely new packaging, all new manuals, handling updates, training staff, localizing to different countries."[72]

Taylor recalled that he ran into a brick wall every time he tried to deal with the suits back east. As the head of a Xerox research facility in Webster, New York, explained to him, "The computer will never be as important to society as the copier."[73]

At a lavish Xerox corporate conference in Boca Raton, Florida (where Henry Kissinger was the paid keynote speaker), the Alto system was put on display. In the morning there was an onstage demo that echoed Engelbart's Mother of All Demos, and in the afternoon thirty Altos were set up in a showroom for everyone to use. The executives, all of whom were male, showed little interest, but their wives immediately started testing the mouse and typing away. "The men thought it was beneath them to know how to type," said Taylor, who had not been invited to the conference but showed up anyway. "It was something secretaries did. So they didn't take the Alto seriously, thinking that only women would like it. That was my revelation that Xerox would never get the personal computer."[74]

Instead, more entrepreneurial and nimble innovators would be the first to foray into the personal computer market. Some would eventually license or steal ideas from Xerox PARC. But at first the earliest personal computers were homebrewed concoctions that only a hobbyist could love.

THE COMMUNITY ORGANIZERS

Among the tribes of the Bay Area in the years leading up to the birth of the personal computer was a cohort of community organizers and peace activists who learned to love computers as tools for bringing power to the people. They embraced small-scale technologies, Buckminster Fuller's *Operating Manual for Spaceship Earth*, and many of the tools-for-living values of the Whole Earth crowd, without being enthralled by psychedelics or repeated exposures to the Grateful Dead.

Fred Moore was an example. The son of an Army colonel stationed at the Pentagon, he had gone west to study engineering at Berkeley in 1959. Even though the U.S. military buildup in Vietnam had not begun, Moore decided to become an antiwar protestor. He camped out on the steps of Sproul Plaza, soon to become the epicenter of student demonstrations, with a sign denouncing ROTC. His protest lasted only two days (his father came to take him home), but he reenrolled at Berkeley in 1962 and resumed his rebellious ways. He served two years in jail as a draft resister and then, in 1968, moved to Palo Alto, driving down in a Volkswagen van with a baby daughter whose mother had drifted away.[75]

Moore planned to become an antiwar organizer there, but he discovered the computers at the Stanford Medical Center and became hooked. Since nobody ever asked him to leave, he spent his days hacking around on the computers while his daughter wandered the halls or played in the Volkswagen. He acquired a faith in the power of computers to help people take control of their lives and form communities. If they could use computers as tools for personal empowerment and learning, he believed, ordinary folks could break free of the dominance of the military-industrial establishment. "Fred was a scrawny-bearded, intense-eyed radical pacifist," recalled Lee Felsenstein, who was part of the community organizing and computer scene in Palo Alto. "At the drop of a hat he would scurry off to pour blood on a submarine. You couldn't really shoo him away."[76]

Given his peacenik-tech passions, it is not surprising that Moore gravitated into the orbit of Stewart Brand and his Whole Earth crowd. Indeed, he ended up having a star turn at one of the weirdest events of the era: the 1971 demise party for the *Whole Earth Catalog*. Miraculously, the publication had ended its run with $20,000 in the bank, and Brand decided to rent the Palace of Fine Arts, an ersatz classical Greek structure in San Francisco's Marina District, to celebrate with a thousand kindred spirits who would decide how to give away the money. He brought a stack of hundred-dollar bills, harboring a fantasy that the rock- and drug-crazed crowd would come to a judicious consensus on what to do with it. "How can we ask anyone else in the world to arrive at agreements if we can't?" Brand asked the crowd.[77]

The debate lasted ten hours. Wearing a monk's black cassock with hood, Brand let each speaker hold the stack of money while addressing the crowd, and he wrote the suggestions on a blackboard. Paul Krassner, who had been a member of Ken Kesey's Merry Pranksters, gave an impassioned talk about the plight of the American Indians—"We ripped off the Indians when we came here!"—and said the money should be given to them. Brand's wife, Lois, who happened to be an Indian, came forward to declare that she and other Indians didn't want it. A person named Michael Kay said they should just give it out to themselves and started handing out the bills to the crowd; Brand retorted that it would be better to use it all together and asked people to pass the bills back up to him, which some of them did, prompting applause. Dozens of other suggestions, ranging from wild to wacky, were made. Flush it down the toilet! Buy more nitrous oxide for the party! Build a gigantic plastic phallic symbol to stick into the earth! At one point a member of the band Golden Toad shouted, "Focus your fucking energy! You've got nine million suggestions! Pick one! This could go on for the next fucking year. I came here to play music." This led to no decision but did prompt a musical interlude featuring a belly dancer who ended by falling to the floor and writhing.

At that point Fred Moore, with his scraggly beard and wavy hair, got up and gave his occupation as "human being." He denounced the crowd for caring about money, and to make his point he took the two dollar bills he had in his pocket and burned them. There was some debate about taking a vote, which Moore also denounced because it was a method for dividing rather than uniting people. By then it was 3 a.m., and the dazed and confused crowd had become even more so. Moore urged them to share their names so that they could stay together as a network. "A union of people here tonight is more important than letting a sum of money divide us," he declared.[78] Eventually he outlasted all but twenty or so diehards, and it was decided to give the money to him until a better idea came along.[79]

Since he didn't have a bank account, Moore buried the $14,905 that was left of the $20,000 in his backyard. Eventually, after much drama and unwelcome visits from supplicants, he distributed it as loans or grants to a handful of related organizations involved in providing computer access and education in the area. The recipients were part of the techno-hippie ecosystem that emerged in Palo Alto and Menlo Park around Brand and his *Whole Earth Catalog* crowd.

This included the catalogue's publisher, the Portola Institute, an alternative nonprofit that promoted "computer education for all grade levels." Its loose-knit learning program was run by Bob Albrecht, an engineer who had dropped out of corporate America to teach computer programming to kids and Greek folk dancing to Doug Engelbart and other adults. "While living in San Francisco at the top of the crookedest street, Lombard, I frequently ran computer programming, wine tasting, and Greek dancing parties," he recalled.[80] He and his friends opened a public-access computer center, featuring a PDP-8, and he took some of his best young students on field trips, most memorably to visit Engelbart at his augmentation lab. One of the early editions of the *Whole Earth Catalog* featured on its end page a picture of Albrecht, sporting a porcupine brush cut, teaching some kids to use a calculator.

Albrecht, who wrote self-teaching guides, including the popular *My Computer Likes Me (When I Speak BASIC)*, launched a publication called the *People's Computer Company*, which was not really a company but called itself one in honor of Janis Joplin's band, Big Brother and the Holding Company. The scraggly newsletter adopted as its motto "Computer power to the people." The first issue, in October 1972, had on its cover a drawing of a boat sailing into the sunset and the hand-scrawled declaration "Computers are mostly used against people instead of for people; used to control people instead of to free them; Time to change all that—we need a PEOPLE'S COMPUTER COMPANY."[81] Most issues featured lots of line drawings of dragons—"I loved dragons ever since I was thirteen," Albrecht recalled—and stories about computer education, BASIC programming, and various learning fairs and do-it-yourself technology festivals.[82] The newsletter helped to weave together electronic hobbyists, do-it-yourselfers, and community-learning organizers.

Another embodiment of this culture was Lee Felsenstein, an earnest antiwar protestor with an electrical engineering degree from Berkeley who became a featured character in Steven Levy's *Hackers*. Felsenstein was far from being a Merry Prankster. Even in the heady days of student unrest at Berkeley, he eschewed sex and drugs. He combined a political activist's instinct for community organizing with an electronic geek's disposition for building communications tools and networks. A faithful reader of the *Whole Earth Catalog*, he had an appreciation for the do-it-yourself strand in American community culture along with a faith that public access to communications tools could wrest power from governments and corporations.[83]

Felsenstein's community-organizing streak and love for electronics were instilled as a child in Philadelphia, where he was born in 1945. His father was a locomotive machinist who had become a sporadically employed commercial artist, and his mother was a photographer. Both were secret members of the Communist Party. "Their outlook was that what you were fed by the media was generally phony, which was one of my father's favorite words," Felsenstein recalled. Even after they left the Party, his parents remained left-wing organizers. As a kid, Felsenstein picketed visiting military leaders and helped organize demonstrations in front of a Woolworth's in support of the desegregation sit-ins in the South. "I always had a piece of paper to draw on when I was a kid, because my parents encouraged us to be creative and imaginative," he recalled. "And on the other side there was usually some mimeographed leaflet from an old block organization event."[84]

His technological interests were instilled partly by his mother, who repeatedly told of how her late father had created the small diesel engines used in trucks and trains. "I took the hint that she wanted me to be an inventor," he said. Once, when he was reprimanded by a teacher for daydreaming, he replied, "I'm not daydreaming, I'm inventing."[85]

In a household with a competitive older brother and an adopted sister, Felsenstein took refuge in going to the basement and playing with electronics. It instilled in him a sense that communications technology should enable individual empowerment: "The technology of electronics promised something I apparently wanted greatly—communication outside the hierarchical structure of the family."[86]

He took a correspondence course that came with booklets and test equipment, and he bought radio handbooks and ninety-nine-cent transistors so that he could learn how to turn schematic drawings into working circuits. One of the many hackers who grew up building Heathkits and other solder-it-yourself electronic projects, he later worried that subsequent generations were growing up with sealed devices that couldn't be explored.* "I learned electronics as a kid by messing around with old radios that were easy to tamper with because they were designed to be fixed."[87]

Felsenstein's political instincts and technological interests came together in a love for science fiction, particularly the writings of Robert Heinlein. Like generations of gamers and computer jockeys who helped to create the culture of the personal computer, he was inspired by the genre's most common motif, that of the hacker hero who uses tech wizardry to bring down evil authority.

He went to Berkeley in 1963 to study electrical engineering, just as the revolt against the Vietnam War was brewing. Among his first acts was to join a protest, along with the poet Allen Ginsberg, against the visit of a South Vietnamese dignitary. It ran late, and he had to get a cab to make it back in time for chemistry lab.

In order to pay his tuition, he entered a work-study program that got him a job with NASA at Edwards Air Force Base, but he was forced to quit when the authorities discovered that his parents had been communists. He called his father to ask if that was true. "I don't want to talk about it on the phone," his father replied.[88]

"Keep your nose clean, son, and you won't have any trouble getting your job back," Felsenstein was told by an Air Force officer. But it was not in his nature to keep his nose clean. The incident had inflamed his antiauthoritarian streak. He arrived back on campus in October 1964 just as the Free Speech Movement protests erupted, and, like a sci-fi hero, he decided to use his technology skills to engage in the fray. "We were looking for nonviolent weapons, and I suddenly

---

*In 2014 Felsenstein was working on a toy/kit for middle school students that would be like an electronic logic board Lego set that would help students visualize bits, electronic components, and logic functions such as *not*, *or*, and *and*.

realized that the greatest nonviolent weapon of all was information flow."[89]

At one point there was a rumor that the police had surrounded the campus, and someone shouted at Felsenstein, "Quick! Make us a police radio." It was not something he could do on the spot, but it resulted in another lesson: "I made up my mind, I had to be out front of everyone applying tech for societal benefit."[90]

His biggest insight was that creating new types of communications networks was the best way to wrest power from big institutions. That was the essence, he realized, of a *free speech* movement. "The Free Speech Movement was about bringing down the barriers to people-to-people communications and thus allowing the formation of connections and communities that were not handed down by powerful institutions," he later wrote. "It laid the ground for a true revolt against the corporations and governments that were dominating our lives."[91]

He began to think about what kind of information structures would facilitate this type of person-to-person communications. He first tried print, launching a newsletter for his student co-op, and then joined the underground weekly the *Berkeley Barb*. There he acquired the semi-ironic title of "military editor" after writing a story about a landing ship dock and using the initials "LSD" in a satirical fashion. He had hoped that "print could be the new community media," but he became disenchanted when he "saw it turn into a centralized structure that sold spectacle."[92] At one point he developed a bullhorn with a mesh network of input wires that allowed people in the crowd to talk back. "It had no center and thus no central authority," he said. "It was an Internet-like design, which was a way to distribute communications power to all the people."[93]

He realized that the future would be shaped by the distinction between broadcast media like television, which "transmitted identical information from a central point with minimal channels for return information," and nonbroadcast, "in which every participant is both a recipient and a generator of information." For him, networked computers would become the tool that would allow people to take control of their lives. "They would bring the locus of power down to the people," he later explained.[94]

In those pre-Internet days, before Craigslist and Facebook, there were community organizations known as Switchboards that served to make connections among people and link them to services they might be seeking. Most were low-tech, usually just a few people around a table with a couple of phones and a lot of cards and flyers tacked to the walls; they served as routers to create social networks. "It seemed that every subcommunity had one or more," Felsenstein recalled. "I visited them to see if there was any technology they could use to forward their efforts." At one point a friend accosted him on the street with some exciting news: one of these community groups had scored a mainframe computer by guilt-tripping some wealthy San Francisco liberals. That tip led him to a nonprofit called Resource One, which was reconfiguring the mainframe so that it could be time-shared by other Switchboards. "We had the idea that we were going to be the computer for the counterculture," he said.[95]

Around that time Felsenstein put a personal ad in the *Berkeley Barb* that read, "Renaissance Man, Engineer and Revolutionist, seeking conversation."[96] Through it he met one of the first female hackers and cyberpunks, Jude Milhon, who wrote under the name St. Jude. She, in turn, introduced him to her companion, Efrem Lipkin, a systems programmer. The Resource One computer had not been able to find any time-sharing clients, so at Lipkin's suggestion they embarked on a new effort, called Community Memory, to use the computer as a public electronic bulletin board. In August 1973 they set up a terminal, with a link via phone line to the mainframe, at Leopold's Records, a student-owned music store in Berkeley.[97]

Felsenstein had seized on a seminal idea: public access to computer networks would allow people to form communities of interest in a do-it-yourself way. The flyer-cum-manifesto advertising the project proclaimed that "non-hierarchical channels of communication—whether by computer and modem, pen and ink, telephone, or face-to-face—are the front line of reclaiming and revitalizing our communities."[98]

One smart decision Felsenstein and his friends made was to not have predefined keywords, such as *help wanted* or *cars* or *babysitting*, programmed into the system. Instead users could make up any keywords they wanted for their posting. This permitted the street to find its own uses for the system. The terminal became a bulletin board for posting poetry, organizing carpools, sharing restaurant ideas, and seeking compatible partners for chess, sex, studying, meditation, and just about anything else. With St. Jude leading the way, people created their own online persona and developed a literary flair not possible on cork-and-tack bulletin boards.[99] Community Memory became the forerunner to Internet bulletin board systems and online services such as The WELL. "We opened the door to cyberspace and found that it was hospitable territory," Felsenstein observed.[100]

Another insight, equally important for the digital age, came after a disagreement with his sometime friend Lipkin, who wanted to build a terminal that was iron-clad closed so that the people in the community couldn't break it. Felsenstein advocated the opposite approach. If the mission was to give computing power to the people, then the hands-on imperative needed to be honored. "Efrem said if people get their hands on it they will break it," Felsenstein recalled. "I took what became the Wikipedia philosophy, which was that allowing people to be hands-on would make them protective and fix it when broken." He believed computers should be playthings. "If you encourage people to tamper with the equipment, you will be able to grow a computer and a community in symbiosis."[101]

These instincts were crystallized into a philosophy when Felsenstein's father, just after the terminal had been set up in Leopold's, sent him a book called *Tools for Conviviality*, by Ivan Illich, an Austrian-born, American-raised philosopher and Catholic priest who criticized the domineering role of technocratic elites. Part of Illich's remedy was to create technology that would be intuitive, easy to learn, and "convivial." The goal, he wrote, should be to "give people tools that guarantee their right to work with high, independent efficiency."[102] Like Engelbart and Licklider, Illich spoke of the need for a "symbiosis" between the user and the tool.

Felsenstein embraced Illich's notion that computers should be built in a way that encouraged hands-on tinkering. "His writings encouraged me to be the pied piper leading people to equipment they could use." A dozen years later, when they finally met, Illich asked him, "If you want to connect people, why do you wish to interpose computers between them?" Felsenstein replied, "I want computers to be the tools that connect people and to be in harmony with them."[103]

Felsenstein wove together, in a very American way, the ideals of the maker culture—the fun and fulfillment that comes from an informal, peer-led, do-it-ourselves learning experience—with the hacker culture's enthusiasm for technological tools and the New Left's instinct for community organizing.* As he told a room full of earnest hobbyists at the Bay Area Maker Faire of 2013, after noting the odd but apt phenomenon of having a 1960s revolutionary as their keynote speaker, "The roots of the personal computer can be found in the Free Speech Movement that arose at Berkeley in 1964 and in the *Whole Earth Catalog*, which did the marketing for the do-it-yourself ideals behind the personal computer movement."[104]

In the fall of 1974, Felsenstein put together specifications for a "Tom Swift Terminal," which was, he said, "a Convivial Cybernetic Device" named after "the American folk hero most likely to be found tampering with the equipment."[105] It was a sturdy terminal designed to connect people to a mainframe computer or network. Felsenstein never got it fully deployed, but he mimeographed copies of the specs and handed them out to those who might embrace the notion. It helped nudge the Community Memory and *Whole Earth Catalog* crowd toward his creed that computers should be personal and convivial. That way they could become tools for ordinary people, not just the technological elite. In the poet Richard Brautigan's phrase, they should be "machines of loving grace," so Felsenstein named the consulting firm he formed Loving Grace Cybernetics.

---

*When *Wired* magazine featured maker culture in its April 2011 issue, it put a woman engineer on its cover for the first time, the MIT-trained do-it-yourself entrepreneur Limor Fried, whose moniker "ladyada" and company name Adafruit Industries were homages to Ada Lovelace.

Felsenstein was a natural-born organizer, so he decided to create a community of people who shared his philosophy. "My proposition, following Illich, was that a computer could only survive if it grew a computer club around itself," he explained. Along with Fred Moore and Bob Albrecht, he had become a regular at a potluck dinner hosted on Wednesday nights at the People's Computer Center. Another regular was Gordon French, a lanky engineer who loved to build his own computers. Among the topics they discussed was "What will personal computers really be like once they finally come into existence?" When the potluck dinners petered out early in 1975, Moore and French and Felsenstein decided to start a new club. Their first flyer proclaimed, "Are you building your own computer? Terminal? TV typewriter? I/O device? Or some other digital black-magic box? If so, you might like to come to a gathering of people with likeminded interests."[106]

The Homebrew Computer Club, as they dubbed it, ended up attracting a cross-section of enthusiasts from the many cultural tribes of the Bay Area digital world. "It had its psychedelic rangers (not many), its ham radio rule-followers, its white-shoe would-be industry potentates, its misfit second- and third-string techs and engineers, and its other offbeat folks—including a prim and proper lady who sat up front who had been, I was later told, President Eisenhower's personal pilot when she was a male," Felsenstein recalled. "They all wanted there to be personal computers, and they all wanted to throw off the constraints of institutions, be they government, IBM or their employers. People just wanted to get digital grit under their fingernails and play in the process."[107]

This first meeting of the Homebrew Computer Club was held on a rainy Wednesday, March 5, 1975, in Gordon French's Menlo Park garage. It occurred just when the first truly personal home computer became available, not from Silicon Valley but from a sagebrush-strewn strip mall in a silicon desert.

ED ROBERTS AND THE ALTAIR

There was one other character type that helped to create the personal computer: the serial entrepreneur. Eventually these overcaffeinated startup jockeys would come to dominate Silicon Valley, edging aside the hippies, Whole Earthers, community organizers, and hackers. But the first of this breed to be successful in creating a marketable personal computer was based far away from both Silicon Valley and the computer centers of the East Coast.

When the Intel 8080 microprocessor was about to come out in April 1974, Ed Roberts was able to score some handwritten data sheets describing it. A burly entrepreneur with an office in an Albuquerque, New Mexico, storefront, he came up with a perfectly simple idea of what he could make using this "computer on a chip": a computer.[108]

Roberts was not a computer scientist or even a hacker. He had no grand theories about augmenting intelligence or the symbiosis wrought by graphical user interfaces. He had never heard of Vannevar Bush or Doug Engelbart. He was instead a hobbyist. Indeed, he had a curiosity and passion that made him, in the words of one coworker, "the world's ultimate hobbyist."[109] Not the type who got all gooey talking about the maker culture but the type who catered to (and acted like an overgrown version of) the pimply-faced boys who loved to fly model airplanes and shoot off rockets in the backyard. Roberts helped usher in a period in which the world of personal computing was pushed forward not by whiz kids from Stanford and MIT but by Heathkit hobbyists who loved the sweet smell of smoldering solder.

Roberts was born in Miami in 1941, the son of a household appliance repairman. He joined the Air Force, which sent him to Oklahoma State to get an engineering degree and then assigned him to the laser division of a weapons lab in Albuquerque. There he began starting businesses, such as one that ran the animated characters in the Christmas display of a department store. In 1969 he and an Air Force mate named Forrest Mims launched a company aimed at the small but passionate market of model rocket enthusiasts. It produced do-it-yourself kits that allowed backyard space cadets to make miniature flashing lights and radio gadgets so that they could track their toy rockets.

Roberts had the buoyancy of a startup junkie. According to Mims, "He was utterly confident his entrepreneurial gifts would allow him to fulfill his ambitions of earning a million dollars, learning to fly, owning his own airplane, living on a farm, and completing medical school."[110] They named their company MITS, in order to evoke MIT, and then reverse-engineered it as an acronym for Micro Instrumentation and Telemetry Systems. Its $100-a-month office space, which had formerly been a snack bar, was wedged between a massage parlor and a Laundromat in a fraying strip mall. The old sign reading "The Enchanted Sandwich Shop" still, rather aptly, hung over the MITS door.

Following in the footsteps of Jack Kilby of Texas Instruments, Roberts next forayed into the electronic calculator business. Understanding the hobbyist mentality, he sold his calculators as unassembled do-it-yourself kits, even though assembled devices would not have cost much more. By then he'd had the good fortune of meeting Les Solomon, the technical editor of *Popular Electronics*, who had visited Albuquerque on a story-scouting tour. Solomon commissioned Roberts to write a piece, whose headline, "Electronic Desk Calculator You Can Build," appeared on the November 1971 cover. By 1973 MITS had 110 employees and $1 million in sales. But pocket calculator prices were collapsing, and there was no profit left to be made. "We went through a period where our cost to ship a calculator kit was $39, and you could buy one in a drugstore for $29," Roberts recalled.[111] By the end of 1974 MITS was more than $350,000 in debt.

Being a brash entrepreneur, Roberts responded to the crisis by deciding to launch a whole new business. He had always been fascinated by computers, and he assumed that other hobbyists felt the same. His goal, he enthused to a friend, was building a computer for the masses that would eliminate the Computer Priesthood once and for all. After studying the instruction set for the Intel 8080, Roberts concluded that MITS could make a do-it-yourself kit for a rudimentary computer that would be so cheap, under $400, that every enthusiast

Ed Roberts (1941–2010).

Altair on the cover, January 1975.

would buy it. "We thought he was off the deep end," a colleague later confessed.[112]

Intel was selling the 8080 for $360 at retail, but Roberts browbeat them down to $75 apiece on the condition that he would buy a thousand. He then got a bank loan based on his insistence that he would sell that many, though privately he feared that initial orders would be closer to the two hundred range. No matter. He had the entrepreneur's risk-loving outlook: either he would succeed and change history, or he would go bankrupt even faster than he already was.

The machine that Roberts and his motley crew built would not have impressed Engelbart, Kay, or the others in the labs around Stanford. It had only 256 bytes of memory and no keyboard or other input device. The only way to put data or instructions in was to toggle a row of switches. The wizards at Xerox PARC were building graphical interfaces to display information; the machine coming out of the old Enchanted Sandwich Shop could display binary-code answers only through a few lights on the front panel that flashed on and off. But even though it wasn't a technological triumph, it was what hobbyists had been yearning for. There was a pent-up demand for a computer that they could make and own, just like a ham radio.

Public awareness is an important component of innovation. A computer created in, say, a basement in Iowa that no one writes about becomes, for history, like a tree falling in Bishop Berkeley's uninhabited forest; it's not obvious that it makes a sound. The Mother of All Demos helped Engelbart's innovations catch on. That is why product launches are so important. The MITS machine might have languished with the unsold calculators in Albuquerque, if Roberts had not previously befriended Les Solomon of *Popular Electronics*, which was to the Heathkit set what *Rolling Stone* was for rock fans.

Solomon, a Brooklyn-born adventurer who as a young man had fought alongside Menachem Begin and the Zionists in Palestine, was eager to find a personal computer to feature on the cover of his magazine. A competitor had done a cover on a computer kit called the Mark-8, which was a barely workable box using the anemic Intel 8008. Solomon knew he had to top that story quickly. Roberts sent him the only workable prototype of his MITS machine via Railway Express Agency, which lost it. (The venerable shipping service went out of business a few months later.) So the January 1975 issue of *Popular Electronics* featured a fake version. As they were rushing the article into print, Roberts still hadn't picked a name for it. According to Solomon, his daughter, a *Star Trek* junkie, suggested it be named after the star that the spaceship *Enterprise* was visiting that night, Altair. And so the first real, working personal computer for home consumers was named the Altair 8800.[113]

"The era of the computer in every home—a favorite topic among science-fiction writers—has arrived!" the lede of the *Popular Electronics* story exclaimed.[114] For the first time, a workable and affordable computer was being marketed to the general public. "To my mind," Bill Gates would later declare, "the Altair is the first thing that deserves to be called a personal computer."[115]

The day that issue of *Popular Electronics* hit the newsstands, orders started pouring in. Roberts had to hire extra people in Albuquerque to answer the phones. In just one day they got four hundred orders, and within months five thousand kits had been sold (though not shipped, since MITS could not make them nearly that fast). People were sending checks to a company they had never heard of, in a town whose name they couldn't spell, in hopes of eventually getting a box of parts that they could solder together that would, if all went well, make some lights blink on and off based on information they had painstakingly entered using toggle switches. With the passion of hobbyists, they wanted a computer of their own—not a shared device or one that would network with other people but one that they could play with by themselves in their bedroom or basement.

As a result, electronics club hobbyists, in league with Whole Earth hippies and homebrew hackers, launched a new industry, personal computers, that would drive economic growth and transform how we live and work. In a power-to-the-people move, computers were wrested from the sole control of corporations and the military and placed into the hands of individuals, making them tools for personal enrichment, productivity, and creativity. "The dystopian society envisioned by George Orwell in the aftermath of World War II, at about the same time the transistor was invented, has completely failed to materialize," the historians Michael Riordan and Lillian Hoddeson wrote, "in large part because transistorized electronic devices have empowered creative individuals and nimble entrepreneurs far more than Big Brother."[116]

THE HOMEBREW DEBUT

At the first meeting of the Homebrew Computer Club in March 1975, the Altair was the centerpiece. MITS had sent it to the *People's Computer Company* for review, and it got passed around to Felsenstein, Lipkin, and others before being brought to the meeting. There it was exposed to a garage full of hobbyists, hippies, and hackers. Most of them were underwhelmed—"There was nothing to it but switches and lights," said Felsenstein—but they had an inkling that it heralded a new age. Thirty people gathered around and shared what they knew. "That may have been the moment at which the personal computer became a convivial technology," Felsenstein recalled.[117]

One hard-core hacker, Steve Dompier, told of going down to Albuquerque in person to pry loose a machine from MITS, which was having trouble fulfilling orders. By the time of the third Homebrew meeting, in April 1975, he had made an amusing discovery. He had written a program to sort numbers, and while he was running it, he was listening to a weather broadcast on a low-frequency transistor radio. The radio started going zip-*zzziiip*-ZZZIIIPP at different pitches, and Dompier said to himself, "Well what do you know! My first peripheral device!" So he experimented. "I tried some other programs to see what they sounded like, and after about eight hours of messing around I had a program that could produce musical tones and actually make music."[118] He charted the tones made by his different program loops, and eventually he was able to enter a program using the toggle switches that, when it ran, played the Beatles' "The Fool on the Hill" on his little radio.* The tones were not beautiful,

*To listen to Dompier's Altair play "Fool on the Hill," go to http://startup.nmnaturalhistory.org/gallery/story.php?ii=46.

but the Homebrew crowd reacted with a moment of awed silence, then cheers and a demand for an encore. Dompier then had his Altair produce a version of "Daisy Bell (Bicycle Built for Two)," which had been the first song ever played by a computer, at Bell Labs on an IBM 704 in 1961, and was reprised in 1968 by HAL when it was being dismantled in Stanley Kubrick's *2001: A Space Odyssey*. "Genetically inherited," was how Dompier described the song. The members of the Homebrew Club had found a computer they could take home and make do all sorts of beautiful things, including, as Ada Lovelace had predicted, rendering music.

Dompier published his musical program in the next issue of the *People's Computer Company*, which led to a historically noteworthy response from a mystified reader. "Steven Dompier has an article about the musical program that he wrote for the Altair in the *People's Computer Company* publication," Bill Gates, a Harvard student on leave writing software for MITS in Albuquerque, wrote in the Altair newsletter. "The article gives a listing of his program and the musical data for 'The Fool on the Hill' and 'Daisy.' He doesn't explain why it works and I don't see why. Does anyone know?"[119] The simple answer was that the computer, as it ran the programs, produced frequency interference that could be controlled by the timing loops and picked up as tone pulses by an AM radio.

By the time his query was published, Gates had been thrown into a more fundamental dispute with the Homebrew Computer Club. It became archetypal of the clash between the commercial ethic that believed in keeping information proprietary, represented by Gates, and the hacker ethic of sharing information freely, represented by the Homebrew crowd.

Paul Allen (1953–) and Bill Gates (1955–) in the Lakeside school's computer room.

Gates arrested for speeding, 1977.

The Microsoft team, with Gates at bottom left and Allen at bottom right, just before leaving Albuquerque in December 1978.

CHAPTER NINE

SOFTWARE

When Paul Allen wandered up to the cluttered news kiosk in the middle of Harvard Square and saw the January 1975 *Popular Electronics* cover with the Altair on it, he was both exhilarated and dismayed. Although thrilled that the era of the personal computer had arrived, he was afraid that he was going to miss the party. Slapping down seventy-five cents, he grabbed the issue and trotted through the slushy snow to the Harvard dorm room of Bill Gates, his high school buddy and fellow computer fanatic from Seattle, who had convinced him to drop out of college and move to Cambridge. "Hey, this thing is happening without us," Allen declared. Gates began to rock back and forth, as he often did during moments of intensity. When he finished the article, he realized that Allen was right. For the next eight weeks, the two of them embarked on a frenzy of code writing that would change the nature of the computer business.[1]

Unlike the computer pioneers before him, Gates, who was born in 1955, had not grown up caring much about the hardware. He had never gotten his thrills by building Heathkit radios or soldering circuit boards. A high school physics teacher, annoyed by the arrogance Gates sometimes displayed while jockeying at the school's time-sharing terminal, had once assigned him the project of assembling a Radio Shack electronics kit. When Gates finally turned it in, the teacher recalled, "solder was dripping all over the back" and it didn't work.[2]

For Gates, the magic of computers was not in their hardware circuits but in their software code. "We're not hardware gurus, Paul," he repeatedly pronounced whenever Allen proposed building a machine. "What we know is software." Even his slightly older friend Allen, who *had* built shortwave radios, knew that the future belonged to the coders. "Hardware," he admitted, "was not our area of expertise."[3]

What Gates and Allen set out to do on that December day in 1974 when they first saw the *Popular Electronics* cover was to create the software for personal computers. More than that, they wanted to shift the balance in the emerging industry so that the hardware would become an interchangeable commodity, while those who created the operating system and application software would capture most of the profits. "When Paul showed me that magazine, there was no such thing as a software industry," Gates recalled. "We had the insight that you could create one. And we did." Years later, reflecting on his innovations, he said, "That was the most important idea that I ever had."[4]

BILL GATES

The rocking motion that Gates exhibited when reading the *Popular Electronics* article had been a sign of his intensity since childhood. "As a baby, he used to rock back and forth in his cradle himself," recalled his father, a successful and gentle lawyer. His favorite toy was a springed hobbyhorse.[5]

Gates's mother, a respected civic leader from a prominent Seattle banking family, was known for her strong will, but she soon found that she was no match for her son. Often when she would summon him to dinner from his basement bedroom, which she had given up trying to make him clean, he wouldn't answer. "What are you doing?" she once demanded.

"I'm thinking," he shouted back.

"You're thinking?"

"Yes, Mom, I'm thinking," he replied. "Have you ever tried thinking?"

She sent him to a psychologist, who turned him on to books about Freud, which he devoured, but was unable to tame his attitude. After a year of sessions, he told Gates's mother, "You're going to lose. You had better just adjust to it because there's no use trying to beat him." His father recounted, "She came around to accepting that it was futile trying to compete with him."[6]

Despite such occasional rebellions, Gates enjoyed being part of a loving and close-knit family. His parents and his two sisters liked lively dinner table conversations, parlor games, puzzles, and cards. Because he was born William Gates III, his grandmother, an avid bridge player (and basketball star), dubbed him Trey, the card term for a 3, which became his childhood nickname. Along with family friends, they spent much of the summer and some weekends at a collection of cabins on the Hood Canal near Seattle, where the kids engaged in a "Cheerio Olympics" featuring a formal opening ceremony with torchlight parade followed by three-legged races, egg tosses, and similar games. "The play was quite serious," his father recalled. "Winning mattered."[7] It was there that Gates, at age eleven, negotiated his first formal contract; he drew up and signed a deal with one of his sisters giving him the nonexclusive but unlimited right to use her baseball glove for $5. "When Trey wants the mitt, he gets it" was one of the provisions.[8]

Gates tended to shy away from team sports, but he became a serious tennis player and water-skier. He also worked assiduously on perfecting fun tricks, such as being able to leap out of a trash can without touching the rim. His father had been an Eagle Scout (you could see in him throughout his life all twelve virtues of the Scout law), and young Bill in turn became an avid Scout, achieving Life Rank but falling three badges short of becoming an Eagle. At one jamboree he demonstrated how to use a computer, but that was before you could earn a badge for computing skill.[9]

Despite all these wholesome activities, Gates's extreme intellect, big glasses, skinny frame, squeaky voice, and wonkish style—shirt often buttoned to the neck—made him come across as seriously nerdy. "He was a nerd before the term was even invented," one teacher declared. His intellectual intensity was legendary. In fourth grade his science class was assigned a five-page paper, and he turned in thirty pages. That year he checked "scientist" when asked to select his future occupation. He also won a dinner atop Seattle's Space Needle by memorizing and reciting perfectly the Sermon on the Mount in a contest run by his family's pastor.[10]

In the fall of 1967, when Gates was just turning twelve but still looked about nine, his parents realized that he would be better off in a private school. "We became concerned about him when he was ready for junior high," said his father. "He was so small and shy, in need of protection, and his interests were so very different from the typical sixth grader's."[11] They chose Lakeside, which had an old brick campus that looked like a New England prep school and catered to the sons (and soon daughters) of Seattle's business and professional establishment.

A few months after he entered Lakeside, his life was transformed by the arrival of a computer terminal in a small downstairs room of the science and math building. It was not actually a true computer but instead a Teletype terminal that was connected by a phone line to a General Electric Mark II time-sharing computer system. The Lakeside Mothers Club, with $3,000 in proceeds from a rummage sale, had bought the right to use a block of time on the system at $4.80 per minute. It would turn out that they woefully underestimated how popular, and expensive, this new offering would be. When his seventh-grade math teacher showed him the machine, Gates was instantly hooked. "I knew more than he did for that first day," the teacher recalled, "but only that first day."[12]

Gates began going to the computer room whenever he could, every day, with a hard-core group of friends. "We were off in our own world," he remembered. The computer terminal became to him what a toy compass had been to the young Einstein: a mesmerizing object that animated his deepest and most passionate curiosities. In struggling to explain what he loved about the computer, Gates later said it was the simple beauty of its logical rigor, something that he had cultivated in his own thinking. "When you use a computer, you can't make fuzzy statements. You make only precise statements."[13]

The language the computer used was BASIC, Beginner's All-purpose Symbolic Instruction Code, which had been developed a few years earlier at Dartmouth to allow nonengineers to write programs. None of Lakeside's teachers knew BASIC, but Gates and his friends inhaled the forty-two-page manual and became wizards at it. Soon they were teaching themselves more sophisticated languages, such as Fortran and COBOL, but BASIC remained Gates's first love. While still in middle school, he produced programs that played tic-tac-toe and converted numbers from one mathematical base to another.

Paul Allen was two years ahead of Gates and physically far more mature (he could even grow sideburns) when they met in the Lakeside computer room. Tall and socially gregarious, he was not a typical wonk. He was immediately amused and charmed by Gates. "I saw a gangly, freckle-faced eighth-grader edging his way into the crowd around the Teletype, all arms and legs and nervous energy," Allen recalled. "His blond hair went all over the place." The two boys bonded and would often work late into the evening in the computer room. "He was really competitive," Allen said of Gates. "He wanted to show you how smart he was. And he was really, really persistent."[14]

One day Allen, who came from a more modest background (his father was a library administrator at the University of Washington), visited Gates at his home and was awed. "His parents subscribed to *Fortune* and Bill read it religiously." When Gates asked him what he thought it would be like to run a big company, Allen said he had no clue. "Maybe we'll have our own company someday," Gates declared.[15]

One trait that differentiated the two was focus. Allen's mind would flit among many ideas and passions, but Gates was a serial obsessor. "Where I was curious to study everything in sight, Bill would focus on one task at a time with total discipline," said Allen. "You could see it when he programmed—he'd sit with a marker clenched in his mouth, tapping his feet and rocking, impervious to distraction."[16]

On the surface, Gates could come across as both a nerd and a brat. He had a confrontational style, even with teachers, and when he was angry would throw a tantrum. He was a genius, knew it, and flaunted it. "That's stupid," he would tell classmates and teachers alike. Or he would escalate the insult to be "the stupidest thing I've ever heard" or "completely brain dead." At one point he laughed at a kid in class for being slow in figuring something out, causing a popular kid sitting in front of Gates to turn around and grab him by the buttoned-up collar and threaten to pummel him. The teacher had to step in.

But to those who knew him, Gates was more than merely nerdy or bratty. Intense and whip smart, he also had a sense of humor, loved adventures, took physical risks, and liked to organize activities. At sixteen he got a new red Mustang (he still had it more than forty years later, preserved in the garage of his mansion), and he took it on high-speed joy rides with his friends. He also brought his pals to his family compound on the Hood Canal, where he would kite-ski on a thousand-foot line behind a speedboat. He memorized James Thurber's classic story "The Night the Bed Fell" for a student performance, and he starred in a production of Peter Shaffer's *Black Comedy*. Around that time, he started informing people, in a matter-of-fact way, that he would make a million dollars before he turned thirty. He woefully underestimated himself; at age thirty he would be worth $350 million.

THE LAKESIDE PROGRAMMING GROUP

In the fall of 1968, when Gates was entering eighth grade, he and Allen formed the Lakeside Programming Group. Partly it was a geek's version of a gang. "At bottom, the Lakeside Programming Group was a boys' club, with lots of one-upmanship and testosterone in the air," said Allen. But it quickly morphed into a moneymaking business, and a competitive one at that. "I was the mover," declared Gates. "I was the guy who said, 'Let's call the real world and try to sell something to it.'"[17] As Allen later noted with a bit of an edge, "While we were all bent on showing our stuff, Bill was the most driven and competitive, hands down."[18]

The Lakeside Programming Group included two other denizens of the school's computer room. Ric Weiland, who was in Allen's tenth-grade class, was an altar boy at the local Lutheran church whose father was a Boeing engineer. Two years earlier, he had made his first computer in his basement. He looked much different from the other obsessives holed up in the computer room. Strikingly handsome, square jawed, tall, and muscular, he was coming to grips with the fact that he was gay, which was difficult to be open about in a conservative high school in the 1960s.

The other partner was Kent Evans, who was in Gates's eighth-grade class. The son of a Unitarian minister, he was gregarious and unfailingly friendly, with a lopsided but winning smile that came from being born with a cleft palate that had been surgically repaired. He was utterly fearless and uninhibited, whether it was cold-calling grown-up executives or climbing rocky cliffs. He had made up the name Lakeside Programming Group as a way to get free material from the companies that advertised in electronics magazines. He also loved business, and he and Gates read each issue of *Fortune* magazine together. He became Gates's best friend. "We were going to conquer the world," said Gates. "We used to talk on the phone forever. I still remember his phone number."[19]

The Lakeside Programming Group's first job came that fall of 1968. Some engineers from the University of Washington had formed a little time-sharing company, housed in an abandoned Buick dealership, called the Computer Center Corporation and nicknamed C-Cubed. They bought a DEC PDP-10—a versatile mainframe that was destined to become a workhorse of the burgeoning time-sharing industry and Gates's favorite machine—with the plan of selling time on it to customers, such as Boeing, who would hook in via Teletype and phone lines. One of the partners at C-Cubed was a Lakeside mother who came up with an offer for Gates's gang that was like asking a posse of third-graders to be tasters in a chocolate factory. The mission: to drive the new PDP-10 as hard as they could and as long as they wanted, programming and playing on it nights and weekends, to see what things they could do to make it crash. C-Cubed's deal with DEC was that it would not have to make lease payments for the machine until it was debugged and stable. DEC had not counted on its being tested by the pubescent hot-rodders of the Lakeside Programming Group.

There were two rules: whenever they crashed the machine, they had to describe what they had done, and they couldn't do the same trick again until they were told to. "They brought us in like we were monkeys to find bugs," Gates recalled. "So we would push the machine to the max in these totally brute-force ways." The PDP-10 had three magnetic tapes, and the Lakeside boys would get all of them spinning at once and then try to crash the system by launching a dozen or so programs to grab as much memory as they could. "It was goofy stuff," Gates said.[20] In return for performing their shakedown cruise, they got to use all the time they wanted to write programs of their own. They created a *Monopoly* game with random-number generators to roll the dice, and Gates indulged his fascination with Napoleon (also a math wizard) by concocting a complex war game. "You'd have these armies and you'd fight battles," Allen explained. "The program kept getting bigger and bigger, and finally when you stretched it all out, it was like fifty feet of Teletype paper."[21]

The boys would take the bus to C-Cubed and spend evenings and weekends hunkered down in the terminal room. "I became hard-core," Gates boasted. "It was day and night." They would program until they were starving, then walk across the street to a hippie hangout called Morningtown Pizza. Gates became obsessed. His room at home was strewn with clothes and Teletype printouts. His parents tried to impose a curfew, but it didn't work. "Trey got so into it," his father recalled, "that he would sneak out the basement door after we went to bed and spend most of the night there."[22]

The C-Cubed executive who became their mentor was none other than Steve "Slug" Russell, the creative and wry programmer who as a student at MIT had created *Spacewar*. The torch was being passed to a new generation of hackers. "Bill and Paul thought crashing the machine was so fun that I had to keep reminding them they weren't supposed to do it again until we told them so," Russell said.[23] "When I stuck my nose in on them, I'd get asked a question or five, and my natural inclination was to answer questions at considerable length."[24] What particularly amazed Russell was Gates's ability to associate different types of error with specific programmers back at DEC headquarters. A typical bug report from Gates read, "Well, Mr. Faboli's code at this line, he's made the same mistake of not checking the semaphore when he's changing the status. If we simply insert this line here, we can get rid of this problem."[25]

Gates and Allen came to appreciate the importance of the computer's operating system, which was akin to its nervous system. As Allen explained, "It does the logistical work that allows the central processing unit to compute: shifting from program to program; allocating storage to files; moving data to and from modems and disk drives and printers." The operating system software for the PDP-10 was called TOPS-10, and Russell allowed Gates and Allen to read, but not take home, the manuals. They would sometimes stay until dawn absorbing them.

In order to fully understand the operating system, Gates realized, they would have to get access to its source code, which programmers used to specify each action to be performed. But the source code was tightly held by the top engineers and was off-limits to the Lakeside boys. That made it all the more of a Holy Grail. One weekend they discovered that printouts of the programmers' work were discarded into a big Dumpster in the back of the building. So Allen clasped his hands to give Gates a boost—"He couldn't have weighed more than 110 pounds," Allen said—and he dove into the container to rummage among the coffee grinds and garbage to find the stacks of stained and crumpled fanfold printouts. "We took that precious hoard back to the terminal room and pored over it for hours," said Allen. "I had no Rosetta Stone to help me, and understood maybe one or two lines out of ten, but I was blown away by the source code's tightly written elegance."

That led Gates and Allen to want to drill down a level deeper. In order to grasp the architecture of the operating system they would have to master assembly code, the underlying commands—"Load B. Add C. Store in A."—that directly spoke to the machine's hardware. Allen recalled, "Seeing my interest, Steve Russell took me aside, handed me an assembler manual bound in glossy plastic, and told me, 'You need to read this.'"[26] He and Gates would read the manuals and sometimes still be confused. At that point Russell would hand them another one and say, "Now it's time to read this." After a while, they became masters of the complexities, and the simplicities, that can make an operating system so potent and graceful.

When the DEC software was finally determined to be stable, the

Lakeside boys lost their right to use the PDP-10 for free. "They basically said, 'Okay you monkeys, go home,'" Gates said.[27] The Lakeside Mothers Club came to the rescue, at least to some extent. It funded personal accounts for the boys, but there was a time and dollar limit. Gates and Allen knew they could never live within the limit, so they tried to beat the system by getting hold of an administrator's password, hacking into the internal accounting system file, and breaking the encryption code. That allowed them to tap into free accounts. But before they could wreak much havoc, they got caught: their math teacher found their roll of Teletype paper with all the account numbers and passwords. The matter went all the way up to the top echelons at C-Cubed and DEC, and a stern delegation visited the school for a meeting in the principal's office. Gates and Allen hung their heads, feigning deep contrition, but it didn't work. They were banned from using the system for the rest of the semester and the entire summer.

"I swore off computers for a while, and I tried to be normal," said Gates. "I decided to prove I could get all A's without ever taking a textbook home. Instead I read biographies of Napoleon and novels like *Catcher in the Rye.*"[28]

For almost a year the Lakeside Programming Group was on hiatus. Then, in the fall of 1970, the school started buying time on a PDP-10 from a company in Portland, Oregon, called Information Sciences, Inc. (ISI). It was expensive, $15 an hour. Gates and his friends quickly learned to hack in for free, but once again they got caught. So they took another approach: they sent ISI a letter offering their services in return for free time.

The executives at ISI were dubious, so the four boys went down to Portland carrying printouts and program codes to show how good they were. "We outlined our experience and submitted our résumés," Allen recalled. Gates, who had just turned sixteen, wrote his in pencil on lined notebook paper. They got an assignment to write a payroll program that would produce paychecks with correct deductions and taxes.[29]

That's when the first cracks appeared in the Gates-Allen relationship. The program had to be written not in BASIC (Gates's favorite language) but in COBOL, the more complex language that had been developed by Grace Hopper and others as a standard for businesses. Ric Weiland knew COBOL and wrote a program editor for the ISI system, which Allen quickly mastered. At that point the two older boys decided that they didn't need Gates or Kent Evans. "Paul and Rick decided there wasn't enough work to go around and said, we don't need you guys," Gates recalled. "They thought that they would do the work and get the computer time."[30]

Gates was frozen out for six weeks, during which time he read algebra books and avoided Allen and Weiland. "And then Paul and Rick realized, oh shit, this is a pain," said Gates. The program required not only coding skills but someone who could figure out social security deductions, federal taxes, and state unemployment insurance. "So then they say, 'hey, we're in trouble on this thing, can you come back and help?'" That's when Gates pulled a power play that would define his future relationship with Allen. As Gates described it, "That's when I say, 'Okay. But I'm going to be in charge. And I'll get used to being in charge, and it'll be hard to deal with me from now on unless I'm in charge. If you put me in charge, I'm in charge of this and anything else we do.'"[31]

And so he was, from then on. When he returned to the fold, Gates insisted on turning the Lakeside Programming Group into a legal partnership, using an agreement drawn up with his father's help. And though partnerships generally don't have presidents, Gates started calling himself that. He was sixteen. Then he divvied up the $18,000 worth of computer time that they were earning, screwing Allen in the process. "I gave 4/11ths to myself, 4/11ths to Kent, 2/11ths to Rick, and 1/11th to Paul," Gates recalled. "The guys thought it was really funny that I did 11ths. But Paul had been so lazy and had never done anything and I was just trying to decide, okay, there's a factor of two between what Paul did and what Rick did, and then there's more than a factor of two between what Rick did and what Kent and I did."[32]

At first Gates tried to give himself slightly more than Evans as well. "But Kent would never let me get away with that." Evans was as savvy about business as Gates was. When they finished the payroll program, Evans made a note in the meticulous journal that he kept: "Tuesday we go to Portland to deliver the program and as they have put it, 'hammer out an agreement for future work.' Everything so far has been done for its educational benefits and for large amounts of expensive computer time. Now we want to get some monetary benefits, too."[33] The negotiations were tense, and for a while ISI tried to hold back some of the computer time payment because they objected to the lack of documentation. But with the help of a letter written by Gates's father, the dispute was resolved and a new deal was negotiated.

In the fall of 1971, at the beginning of Gates's junior year, Lakeside merged with a girls' school. This created a class-scheduling nightmare, so the administrators asked Gates and Evans to write a program to solve it. Gates knew that a school schedule had scores of variables—required courses, teacher schedules, classroom space, honors classes, electives, staggered sections, double-period labs—that would make it extremely difficult, so he declined. Instead a teacher took on the challenge, while Gates and Evans taught his computer class for him. But that January, as he was still struggling to produce a workable program, the teacher was killed when a small plane he was riding in crashed. Gates and Evans agreed to take over the task. They spent hours in the computer room, often sleeping there overnight, trying to write a new program from scratch. In May they were still struggling, trying to finish so that it could be ready for the next school year.

That is when Evans, despite being exhausted, decided to go through with a mountain-climbing excursion he had signed up for. He was not an athlete. "It was really unusual that he signed up for this climbing course," Gates recalled. "I think he wanted to push himself." Evans's father, knowing how drained his son was, begged him to cancel: "The last conversation I had with him was trying to convince him not to go, but he had a commitment to finishing things." The class was learning how to belay on one of the more gentle slopes when Evans tripped. He tried to get up, then continued to roll more than two hundred yards across the snow and down a glacier, tucking in his arms to protect himself instead of splaying them out as he should have. His head smashed into several rocks, and he died aboard the helicopter that came to rescue him.

Lakeside's headmaster called the Gates home, and Bill was summoned into his parents' bedroom, where he was told the news.* The service was conducted by Lakeside's art teacher, Robert Fulghum, who was a Unitarian minister like Evans's father and who would later become a popular writer (*All I Really Need to Know I Learned in Kindergarten*). "I had never thought of people dying," Gates said. "At the service, I was supposed to speak, but I couldn't get up. For two weeks I couldn't do anything at all." He spent a lot of time afterward with Kent's parents. "Kent was the apple of their eye."[34]

Gates called Paul Allen, who had just finished his freshman year at Washington State, and asked him to come back to Seattle to help with the scheduling program. "I was going to do it with Kent," Gates told him. "I need help." He was in bad shape. "Bill stayed depressed for weeks," Allen recalled.[35] They brought cots to campus and, like old times, spent many nights in the computer room that summer of 1972, communing with a PDP-10. With his rigorous mind, Gates was able to take the problem posed by the Rubik's Cube of class-scheduling variables and break it into a series of small component problems that could be solved sequentially. He was also able to put himself into a history class with all the right girls and only one other boy ("a real wimp") and make sure that he and his senior class friends had Tuesday afternoons free. They had T-shirts made featuring a beer keg and the words "Tuesday Club" emblazoned on the front.[36]

That summer Gates and Allen became enchanted by Intel's new 8008 microprocessor, a powerful upgrade of its 4004 "computer on a chip." They were so excited by a story on it in *Electronics Magazine* that years later Gates would remember the page number it was on. If the chip really could act like a computer and be programmed, Allen asked Gates, why not write a programming language for it, specifically a version of BASIC? If they pulled off such a feat, Allen argued,

---

*After they became successful, Gates and Allen donated a new science building to Lakeside and named its auditorium after Kent Evans.

"ordinary people would be able to buy computers for their offices, even their homes." Gates dismissed the 8008 as not being up for such a task. "It would be dog-slow and pathetic," he replied. "And BASIC by itself would take up almost all the memory. There's just not enough horsepower." Allen realized that Gates was right, and they agreed to wait until, in accordance with Moore's Law, a microprocessor twice as powerful came out in a year or two. The parameters of their partnership were becoming clear. "I was the idea man, the one who'd conceive of things out of whole cloth," explained Allen. "Bill listened and challenged me, and then homed in on my best ideas to help make them a reality. Our collaboration had a natural tension, but mostly it worked productively and well."[37]

Gates had gotten a contract to analyze traffic patterns for a company that counted how many cars ran over rubber tubes laid across roads. He and Allen decided to create a special-purpose computer that would process the raw data. Showing his clunky taste, Gates chose the name Traf-O-Data for their new venture. They went to a nearby Hamilton Avnet electronics store and, with a great sense of the moment, shelled out $360 in cash to buy a single 8008 chip. Allen recalled the moment vividly: "The sales clerk handed us a small cardboard box, which we opened then and there for our first look at a microprocessor. Inside an aluminum foil wrapper, stuck into a small slab of nonconductive black rubber, was a thin rectangle about an inch long. For two guys who'd spent their formative years with massive mainframes, it was a moment of wonder." Gates told the clerk, "That's a lot of money for such a little thing," but he and Allen were suitably impressed, for they knew the little chip contained the brains of a whole computer. "These people thought it was the strangest thing ever to have these kids coming in and buying an 8008," Gates recalled. "And we were so worried as we unwrapped the foil that we would break the thing."[38]

In order to write a program that would work on the 8008, Allen devised a way to emulate the microprocessor on a mainframe computer. As he later explained, the emulation of the 8008 "reflected a truism in technology circles that harkened back to the theories of Alan Turing in the 1930s: any computer could be programmed to behave like any other computer." There was another lesson in this feat of alchemy, one that was at the core of what Gates and Allen contributed to the computer revolution: "Software trumped hardware," Allen later explained.[39]

Given their reverence for software over hardware, it is not surprising that Gates and Allen were able to write a good program for their proposed traffic tabulator but were never able to get the hardware components working properly, most notably the mechanism that was supposed to read the traffic tapes. One day, after they thought they had it running smoothly, an official with the Seattle engineering department came to Gates's family home to be given a sales demo. As they sat in the living room, the demo gods had their revenge and the tape reader kept failing. Gates ran to get his mother. "Tell him, Mom!" he implored. "Tell him it worked last night!"[40]

The final semester of Gates's senior year, in the spring of 1973, he and Allen were recruited by the Bonneville Power Administration, which was on a nationwide hunt for PDP-10 experts to help program its electrical grid management system. Gates and his parents talked to Lakeside's headmaster, who agreed that the job would be more educational than attending his last semester of school. Allen felt the same way about his semester at Washington State: "Here was a chance to work together again on a PDP-10, and for pay!" They piled into Gates's Mustang convertible, drove the 165 miles south from Seattle to the Bonneville command center in under two hours, and rented a cheap apartment together.

Their work was in an underground bunker on the Columbia River across from Portland. "They had this massive control room, which looks better than any TV show I've ever seen," Gates recalled. He and Allen would hunker down for coding sessions that lasted twelve hours or more. "When Bill felt himself flagging, he'd grab a jar of Tang, pour some powder on one hand, and lick it off for a pure sugar high," Allen recalled. "His palms had a chronic orange tinge that summer." Sometimes, after a two-day work binge, they would get "slept up," as Gates called it, by crashing for eighteen hours or so. "We had contests," Gates said, "to see who could stay in the building like three days straight, four days straight. Some of the more prudish people would say 'Go home and take a bath.' We were just hard-core, writing code."[41]

Occasionally Gates would take a break for some extreme waterskiing, including dry-dock starts from diving platforms, then go back to the bunker for more coding. He and Allen got along well, except when Allen's methodical chess-playing style would triumph over Gates's more reckless and aggressive approach. "When I beat him one day, he got so angry that he swept the pieces to the floor," said Allen. "After a few games like that, we stopped playing."[42]

Gates applied only to three colleges his senior year—Harvard, Yale, and Princeton—and he took different approaches to each. "I was born to apply for college," he boasted, fully aware of his ability to ace meritocratic processes. For Yale he cast himself as an aspiring political type and emphasized a monthlong summer internship he had done in Congress. For Princeton, he focused only on his desire to be a computer engineer. And for Harvard, he said his passion was math. He had also considered MIT, but at the last moment blew off the interview to play pinball. He was accepted to all three and chose Harvard.[43]

"You know, Bill," Allen warned him, "when you get to Harvard, there are going to be some people a lot better in math than you are."

"No way," Gates replied. "There's no way!"

"Wait and see," said Allen.[44]

GATES AT HARVARD

When Gates was asked to pick the types of roommates he preferred, he asked for an African American and an international student. He was assigned to Wigglesworth Hall, a freshman dorm in Harvard Yard, with Sam Znaimer, a science lover from a family of poor Jewish refugees in Montreal, and Jim Jenkins, a black student from Chattanooga. Znaimer, who had never known a privileged WASP before, found Gates very friendly and his study habits weirdly fascinating. "His habit was to do 36 hours or more at a stretch, collapse for ten hours, then go out, get a pizza, and go back at it," he said. "And if that meant he was starting again at three in the morning, so be it."[45] He marveled as Gates spent several nights filling out federal and state tax forms for Traf-O-Data's revenues. When working hard, Gates would rock back and forth. Then he would grab Znaimer for a frenzy of playing *Pong*, the Atari video game, in the dorm lounge, or *Spacewar* in Harvard's computer lab.

The computer lab was named after Howard Aiken, who had invented the Mark I and operated it during World War II with the help of Grace Hopper. It housed Gates's favorite machine: a PDP-10 from DEC, which had been destined for military use in Vietnam but was reassigned to assist military-funded research at Harvard. To avoid sparking an antiwar protest, it was smuggled into the Aiken Lab early one Sunday morning in 1969. It was funded by the Defense Department's Advanced Research Projects Agency (then known as DARPA), but that was kept quiet, so there was no written policy about who could use it. There was also a slew of PDP-1 computers on which to play *Spacewar*. For his freshman computer project, Gates linked the PDP-10 and a PDP-1 to create a video baseball game. "The logic was on the PDP-10, but I sent it down to the PDP-1 because I used the same display as *Spacewar*, a line-drawing display which you don't see anymore," he explained.[46]

Gates would stay up late writing the algorithms to direct the bounce of the ball and the angle of approach of the fielders. "The projects he worked on for the first year were not commercial," Znaimer said. "They were mostly done for the love of computing."[47] The professor who oversaw the lab, Thomas Cheatham, had mixed feelings: "He was a hell of a good programmer." He was, however, also a "pain in the ass" and "an obnoxious human being. . . . He'd put people down when it was not necessary, and just generally was not a pleasant fellow to have around."[48]

Allen's warning to Gates that he would not be the smartest kid in the class turned out to be true. There was a freshman who lived upstairs from him who was better at math, Andy Braiterman from Baltimore. They would wrestle with problem sets all night in Braiterman's room, eating pizza. "Bill was intense," Braiterman remembered, and also "a good arguer."[49] Gates was particularly forceful in arguing that soon everyone would have a home computer that could be used for calling up books and other information. The following year he and Braiterman roomed together.

Gates decided to major in applied math rather than pure math, and he was able to make a small mark on the field. In a class taught by the computer scientist Harry Lewis, he was introduced to a classical problem:

> The chef in our place is sloppy, and when he prepares a stack of pancakes they come out all different sizes. Therefore, when I deliver them to a customer, on the way to the table I rearrange them (so that the smallest winds up on top, and so on, down to the largest at the bottom) by grabbing several from the top and flipping them over, repeating this (varying the number I flip) as many times as necessary. If there are n pancakes, what is the maximum number of flips (as a function f(n) of n) that I will ever have to use to rearrange them?

The answer required coming up with a good algorithm, just as any computer program did. "I posed it in class and then I went on," Lewis recalled. "A day or two later, this smart sophomore comes into my office and explains that he's got a five-thirds N algorithm." In other words, Gates had figured out a way to do it with five-thirds flips per pancake in the stack. "It involved a complicated case analysis of what exactly the configuration of the top few pancakes might look like. It was quite clever." A teaching assistant in the class, Christos Papadimitriou, later published the solution in a scholarly paper coauthored with Gates.[50]

As Gates was preparing to begin his sophomore year in the summer of 1974, he convinced Allen to move to the Boston area and take a job with Honeywell that had originally been offered to Gates. Allen dropped out of Washington State, drove his Chrysler east, and urged Gates to drop out as well. We're going to miss the computer revolution, he argued. Over pizza they would fantasize about creating their own company. "If everything went right, how big do you think our company could be?" Allen asked at one point. Gates replied, "I think we could get it up to thirty-five programmers."[51] But Gates bowed to pressure from his parents to remain at Harvard, at least for the time being.

Like many innovators, Gates was rebellious just for the hell of it. He decided that he would not go to the lectures for any course in which he was enrolled, and he would audit lectures only of courses that he was not taking. He followed this rule carefully. "By my sophomore year, I was auditing classes that met at the same time as my actual classes just to make sure I'd never make a mistake," he recalled. "So I was this complete rejectionist."[52]

He also took up poker with a vengeance. His game of choice was Seven Card Stud, high low. A thousand dollars or more could be won or lost per night. Gates, whose IQ surpassed his EQ, was better at calculating the odds than in reading the thoughts of his fellow players. "Bill had a monomaniacal quality," Braiterman said. "He would focus on something and really stick with it." At one point he gave his checkbook to Allen in order to prevent himself from squandering more money, but he soon demanded it back. "He was getting some costly lessons in bluffing," said Allen. "He'd win three hundred dollars one night and lose six hundred the next. As Bill dropped thousands that fall, he kept telling me, 'I'm getting better.'"[53]

In a graduate-level economics class, he met a student who lived down the hall of his dorm. Steve Ballmer was very different from Gates on the surface. Big, boisterous, and gregarious, he was the type of campus activity junkie who liked to join or lead multiple organizations. He was in the Hasty Pudding Club, which wrote and produced musical theater shows, and served with a cheerleader's enthusiasm as manager of the football team. He was both the publisher of the *Advocate*, the campus literary magazine, and the advertising manager of the *Crimson*, the newspaper. He even joined one of the fraying men's clubs, and convinced his new best friend Gates to do so as well. "A bizarre experience," Gates called it. What bound them together was their shared superintensity. They would talk and argue and study together at high volume, each of them rocking back and forth. Then they would go to movies together. "We went and saw *Singin' in the*

Rain and *A Clockwork Orange*, which are only connected by the use of a common song," said Gates. "And then we got to be super-good friends."[54]

Gates's haphazard life at Harvard was suddenly upended in December 1974, halfway through his sophomore year, when Allen arrived at his Currier House room with the new issue of *Popular Electronics* featuring the Altair on the cover. Allen's rallying cry, "Hey, this thing is happening without us," jolted Gates into action.

BASIC FOR THE ALTAIR

Gates and Allen set out to write software that would make it possible for hobbyists to create their own programs on the Altair. Specifically, they decided to write an interpreter for the programming language BASIC that would run on the Altair's Intel 8080 microprocessor. It would become the first commercial native high-level programming language for a microprocessor. And it would launch the personal computer software industry.

Using some old stationery with the Traf-O-Data letterhead, they wrote a letter to MITS, the fledgling Albuquerque company that made the Altair, claiming that they had created a BASIC interpreter that could run on the 8080. "We are interested in selling copies of this software to hobbyists through you."[55] That wasn't exactly true. They had not yet written any software. But they knew they could scramble into action if MITS expressed interest.

When they did not hear back, they decided to phone. Gates suggested that Allen place the call, because he was older. "No, you should do it; you're better at this kind of thing," Allen argued. They came up with a compromise: Gates would call, disguising his squeaky voice, but he would use the name Paul Allen, because they knew it would be Allen who would fly out to Albuquerque if they got lucky. "I had my beard going and at least looked like an adult, while Bill still could pass for a high school sophomore," recalled Allen.[56]

When the gruff-sounding Ed Roberts answered the phone, Gates put on a deep voice and said, "This is Paul Allen in Boston. We've got a BASIC for the Altair that's just about finished, and we'd like to come out and show it to you." Roberts replied that he had gotten many such calls. The first person to walk through his door in Albuquerque with a working BASIC would get the contract. Gates turned to Allen and exulted, "God, we gotta get going on this!"

Because they did not have an Altair to work on, Allen had to emulate one on Harvard's PDP-10, a reprise of the tactic he had used to build their Traf-O-Data machine. So they bought a manual for the 8080 microprocessor, and within weeks Allen had the emulator and other development tools ready.

Meanwhile Gates was furiously writing the BASIC interpreter code on yellow legal pads. By the time Allen had finished the emulator, Gates had outlined the structure and much of the code. "I can still see him alternately pacing and rocking for long periods before jotting on a yellow legal pad, his fingers stained from a rainbow of felt-tip pens," Allen recalled. "Once my emulator was in place and he was able to use the PDP-10, Bill moved to a terminal and peered at his legal pad as he rocked. Then he'd type a flurry of code with those strange hand positions of his, and repeat. He could go like that for hours at a stretch."[57]

One night they were having dinner at Currier House, Gates's dorm, sitting at the table with the other math wonks, and they began complaining about facing the tedious task of writing floating-point math routines, which would give the program the ability to deal with both very small and very large numbers and decimal points in scientific notation.* A curly-haired kid from Milwaukee named Monte Davidoff piped up, "I've written those kinds of routines."[58] This was one of the benefits of being a geek at Harvard. Gates and Allen began peppering him with questions about his capacity to handle floating-point code. Satisfied that he knew what he was talking about, they brought him to Gates's room and negotiated a fee of $400 for his work. He became the third member of the team, and would eventually earn a lot more.

Gates ignored the exam cramming he was supposed to be doing

---

*Steve Wozniak's unwillingness to tackle this tedious task when he wrote BASIC for the Apple II would later force Apple to have to license BASIC from Allen and Gates.

and even stopped playing poker. For eight weeks he and Allen and Davidoff holed up day and night at Harvard's Aiken Lab making history on the PDP-10 that the Defense Department was funding. Occasionally they would break for dinner at Harvard House of Pizza or Aku Aku, an ersatz Polynesian restaurant. In the wee hours of the morning, Gates would sometimes fall asleep at the terminal. "He'd be in the middle of a line of code when he'd gradually tilt forward until his nose touched the keyboard," Allen said. "After dozing an hour or two, he'd open his eyes, squint at the screen, blink twice, and resume precisely where he'd left off—a prodigious feat of concentration."

They would scribble away at their notepads, competing at times to see who could execute a subroutine in the fewest lines. "I can do it in nine," one would shout. Another would shoot back, "Well, I can do it in five!" Allen noted, "We knew that each byte saved would leave that much more room for users to add to their applications." The goal was to get the program into less than the 4K of memory that an enhanced Altair would have, so there would still be room left for the consumer to use. (A 16GB smartphone has four million times that amount of memory.) At night they would fan out the printouts on the floor and search for ways to make it more elegant, compact, and efficient.[59]

By late February 1975, after eight weeks of intense coding, they got it down, brilliantly, into 3.2K. "It wasn't a question of whether I could write the program, but rather a question of whether I could squeeze it into under 4k and make it super fast," said Gates. "It was the coolest program I ever wrote."[60] Gates checked it for errors one last time, then commanded the Aiken Lab's PDP-10 to spew out a punch tape of it so Allen could take it to Albuquerque.

On the flight down, Allen remembered he hadn't written a loader, the sequence of commands that would instruct the Altair how to put the BASIC interpreter into its memory. As the plane was preparing to land, he grabbed a pad and wrote twenty-one lines in the machine language used by the Intel microprocessor, each line a three-digit number in base-8. He was sweating by the time he left the terminal, wearing a tan Ultrasuede polyester suit and looking for Ed Roberts. Eventually he spotted a jowly three-hundred-pound man in jeans and a string tie in a pickup truck. "I'd expected a high-powered executive from some cutting-edge entrepreneurial firm, like the ones clustered along Route 128, the high-tech beltway around Boston," Allen recalled.

The MITS world headquarters was likewise not quite what Allen expected. It was in a low-rent strip mall, and the only Altair with enough memory to run BASIC was still being tested. So they put off until the next morning trying out the program and headed off "to a three-dollar buffet at a Mexican place called Pancho's, where you got what you paid for," Allen said. Roberts drove him to the local Sheraton, where the desk clerk told him that his room would be $50. That was $10 more than Allen had brought with him, so after an awkward stare Roberts had to pay for the room. "I guess I wasn't what he'd been expecting, either," said Allen.[61]

The next morning, Allen returned to MITS for the big test. It took almost ten minutes to load in the code for the BASIC interpreter that he and Gates had written. Roberts and his colleagues exchanged amused glances, already suspecting that the show would be a fiasco. But then the Teletype clacked to life. "MEMORY SIZE?" it asked. "Hey, it typed something!" shouted one of the MITS team. Allen was happily flabbergasted. He typed in the answer: 7168. The Altair responded: "OK." Allen typed, "PRINT 2+2." It was the simplest of all commands, but it would test not only Gates's coding but also Davidoff's floating-point math routines. The Altair responded: "4."

Up until then, Roberts had been watching quietly. He had taken his failing company further into debt on the wild surmise that he could create a computer that a home hobbyist could use and afford. Now he was watching as history was being made. For the first time, a software program had run on a home computer. "Oh my God," he shouted. "It printed '4'!"[62]

Roberts invited Allen into his office and agreed to license the BASIC interpreter for inclusion on all Altair machines. "I couldn't stop grinning," Allen confessed. When he arrived back in Cambridge, bringing with him a working Altair to install in Gates's dorm room, they went out to celebrate. Gates had his usual: a Shirley Temple, ginger ale with maraschino cherry juice.[63]

A month later, Roberts offered Allen a job at MITS as director of software. His colleagues at Honeywell thought he was crazy to consider it. "Your job's safe at Honeywell," they told him. "You can work here for years." But career safety was not an ideal embraced by those eager to lead the computer revolution. So in the spring of 1975, Allen moved to Albuquerque, a city he had only recently learned was not in Arizona.

Gates decided to stay at Harvard, at least for the time being. There he endured what has become a rite of passage, amusing only in retrospect, for many of its most successful students: being hauled before the university's secretive Administrative Board for a disciplinary process, known as being "Ad Boarded." Gates's case arose when auditors from the Defense Department decided to check the use of the PDP-10 that it was funding in Harvard's Aiken Lab. They discovered that one sophomore, W. H. Gates, was using most of the time. After much fretting, Gates prepared a paper defending himself and describing how he had created a version of BASIC using the PDP-10 as an emulator. He ended up being exonerated for his use of the machine, but he was "admonished" for allowing a nonstudent, Paul Allen, to log on with his password. He accepted that minor reprimand and agreed to put his early version of the BASIC interpreter (but not the refined one he and Allen were then working on) into the public domain.[64]

By that time Gates was focusing more on his software partnership with Allen than his course work at Harvard. He finished his sophomore year that spring of 1975, then flew down to Albuquerque for the summer and decided to stay there rather than return for the first semester of his junior year that fall. He went back to Harvard for two more semesters, in the spring and fall of 1976, but then left Harvard for good, two semesters shy of graduating. In June 2007, when he returned to Harvard to get an honorary degree, he began his speech by directing a comment to his father in the audience. "I've been waiting more than 30 years to say this: Dad, I always told you I'd come back and get my degree."[65]

MICRO-SOFT

When Gates arrived in Albuquerque in the summer of 1975, he and Allen were still supplying BASIC for the Altair on a handshake deal with Ed Roberts. Gates insisted on a formal agreement and, after much haggling, agreed to license the software to MITS for ten years, to be bundled with each Altair, for $30 in royalty per copy. Gates was able to win two provisions that would be historically significant. He insisted that he and Allen would retain ownership of the software; MITS would merely have rights to license it. He also required that MITS use its "best efforts" to sublicense the software to other computer makers, splitting the revenues with Gates and Allen. It set a precedent for the deal Gates would make six years later with IBM. "We were able to make sure our software worked on many types of machines," he said. "That allowed us and not the hardware makers to define the market."[66]

Now they needed a name. They kicked around a few ideas, including Allen & Gates, which they decided sounded too much like a law firm. Eventually they picked one that was not particularly exciting or inspiring but did convey that they were writing software for microcomputers. In the final documents for the MITS deal, they referred to themselves as "Paul Allen and Bill Gates doing business as Micro-Soft." A credit line appeared in the source code of what was then their only product: "Micro-Soft BASIC: Paul Allen wrote the non-runtime stuff. Bill Gates wrote the runtime stuff. Monte Davidoff wrote the math package." Within a couple of years, the name was simplified to Microsoft.

After bunking for a while at the Sundowner Motel on a strip of Route 66 known more for prostitutes than programmers, Gates and Allen moved to a cheap furnished apartment. Monte Davidoff, of floating-point math fame, and Chris Larson, a younger student from Lakeside High, moved in, turning the apartment into a frat house doing business as a geek bunker. In the evenings Allen would crank up his Stratocaster guitar and play along with Aerosmith or Jimi Hendrix, and Gates would retaliate by loudly singing Frank Sinatra's "My Way."[67]

Of them all, Gates was the prime example of the innovator's personality. "An innovator is probably a fanatic, somebody who loves what they do, works day and night, may ignore normal things to some degree and therefore be viewed as a bit imbalanced," he said. "Certainly in my teens and 20s, I fit that model."[68] He would work, as he had at Harvard, in bursts that could last up to thirty-six hours, and then curl up on the floor of his office and fall asleep. Said Allen, "He lived in binary states: either bursting with nervous energy on his dozen Cokes a day, or dead to the world."

Gates was also a rebel with little respect for authority, another trait of innovators. To folks like Roberts, a former Air Force officer with five sons who called him "Sir," Gates came across as a brat. "He literally was a spoiled kid, that's what the problem was," said Roberts later. But it was more complex than that. Gates worked hard and lived frugally off his then-meager earnings, but he did not believe in being deferential. The scrawny Gates would go toe to toe with the brawny six-foot-four Roberts and engage in arguments so heated that, as Allen remembered, "you could hear them yelling throughout the plant, and it was a spectacle."

Allen assumed that his partnership with Gates would be fifty-fifty. They had always been a team, and it seemed unnecessary to fight over who had done more. But since their spat over the payroll program in high school, Gates had insisted on being in charge. "It's not right for you to get half," he told Allen. "You had your salary at MITS while I did almost everything on BASIC without one back in Boston. I should get more. I think it should be sixty-forty." Whether or not Gates was right, it was in his nature to insist on such things, and it was in Allen's nature not to. Allen was taken aback but agreed. Worse yet, Gates insisted on revising the split two years later. "I've done most of the work on BASIC, and I gave up a lot to leave Harvard," he told Allen on a walk. "I deserve more than 60 percent." His new demand was that the split be 64-36. Allen was furious. "It exposed the differences between the son of a librarian and the son of a lawyer," he said. "I'd been taught that a deal was a deal and your word was your bond. Bill was more flexible." But again Allen went along.[69]

In fairness to Gates, he was the person who, by then, was actually running the fledgling company. Not only was he writing much of the code, but he also was in charge of sales, making most of the calls himself. He would kick around ideas about product strategy with Allen for hours, but he was the one who made the final decisions on which versions of Fortran or BASIC or COBOL would be built. He was also in charge of business deals with the hardware makers, and he was an even tougher negotiator with them than he had been with Allen. Plus he was in charge of personnel, which meant hiring, firing, and telling people in words of one syllable when their work sucked, which is something Allen would never do. He had the credibility to do so; when there were contests in the office to see who could write a program using the fewest lines of code, Gates usually won.

Allen would sometimes come in late and might even think it was permissible to leave work in time for dinner. But not Gates and his close coterie. "It was hard-core," he recalled. "A small group and me would work late into the night. And I'd sometimes stay all night and then sleep in the office and my secretary would come wake me up if we had a meeting."[70]

Born with a risk-taking gene, Gates would cut loose late at night by driving at terrifying speeds up the mountain roads to an abandoned cement plant. "Sometimes I wondered why Bill drove so fast," Allen said. "I decided it was his way of letting off steam. He'd get so wound up in our work that he needed a way to stop thinking about the business and the code for a while. His breakneck driving wasn't so different from table stakes poker or edge-of-the-envelope waterskiing." Once they had made a little money, Gates splurged on a green Porsche 911, which he would race along the freeway after midnight. At one point he complained to his local dealer that the car's top speed was supposed to be 126 miles per hour, but he could get it up only to 121. Late one night he was caught speeding and got into an argument with the cop about why he wasn't carrying a driver's license. He was thrown into jail. "Got arrested," he said when Allen picked up the phone. He was released in a few hours, but his mug shot from that night became a memorable icon of geek history.[71]

Gates's intensity paid off. It allowed Microsoft to meet software deadlines that seemed insane, beat other competitors to the market for each new product, and charge such a low price that computer manufacturers rarely thought of writing or controlling their own software.

SOFTWARE WANTS TO BE FREE

In June 1975, the month that Gates moved to Albuquerque, Roberts decided to send the Altair on the road as if it were a carnival show exhibit. His goal was to spread the word about the Altair's wonders and create fan clubs in towns across America. He tricked out a Dodge camper van, dubbed it the MITS Mobile, and sent it on a sixty-town tour up the coast of California then down to the Southeast, hitting such hot spots as Little Rock, Baton Rouge, Macon, Huntsville, and Knoxville.

Gates, who went along for part of the ride, thought it was a neat marketing ploy. "They bought this big blue van and they went around the country and created computer clubs everyplace they went," he marveled.[72] He was at the shows in Texas, and Allen joined them when they got to Alabama. At the Huntsville Holiday Inn, sixty people, a mix of hippyish hobbyists and crew-cut engineers, paid $10 to attend, then about four times the cost of a movie. The presentation lasted three hours. At the end of a display of a lunar landing game, doubters peered under the table suspecting that there were cables to some bigger minicomputer hidden underneath. "But once they saw it was real," Allen recalled, "the engineers became almost giddy with enthusiasm."[73]

One of the stops was at Rickeys Hyatt House hotel in Palo Alto on June 5. There a fateful encounter occurred after Microsoft BASIC was demonstrated to a group of hobbyists, including many from the recently formed Homebrew Computer Club. "The room was packed with amateurs and experimenters eager to find out about this new electronic toy," the Homebrew's newsletter reported.[74] Some of them were also eager to act on the hacker credo that software should be free. This was not surprising given the social and cultural attitudes, so different from the entrepreneurial zeal in Albuquerque, that had flowed together in the early 1970s leading up to the formation of the Homebrew Club.

Many of the Homebrew members who met the MITS Mobile had built an Altair and been waiting impatiently to get hold of the BASIC program that Gates and Allen had produced. Some had already sent checks to MITS for it. So they were thrilled to see that the Altairs on display were running a version of it. Indulging in the imperative of hackers, one of the members, Dan Sokol, "borrowed" the punched paper tape that had the program and used a DEC PDP-11 to make copies.[75] At the next Homebrew meeting, there was a cardboard box filled with dozens of BASIC tapes for members to take.* There was one stipulation: you had to make a few copies to replenish the communal box. "Remember to bring back more copies than you take," Lee Felsenstein joked. It was his signature line for any software sharing.[76] Thus did Microsoft BASIC spread freely.

This, not surprisingly, infuriated Gates. He wrote a passionate open letter, displaying all the tact of a nineteen-year-old, which served as the opening shot in the war over the protection of intellectual property in the age of personal computers:

> An Open Letter to Hobbyists . . .
>
> Almost a year ago, Paul Allen and myself, expecting the hobby market to expand, hired Monte Davidoff and developed Altair BASIC. Though the initial work took only two months, the three of us have spent most of the last year documenting, improving and adding features to BASIC. Now we have 4K, 8K, EXTENDED, ROM and DISK BASIC. The value of the computer time we have used exceeds $40,000.
>
> The feedback we have gotten from the hundreds of people who

---

*Reading a draft version of this book online, Steve Wozniak said that Dan Sokol made only eight copies, because they were hard and time-consuming to make. But John Markoff, who reported this incident in *What the Dormouse Said,* shared with me (and Woz and Felsenstein) the transcript of his interview with Dan Sokol, who said he used a PDP-11 with a high-speed tape reader and punch. Every night he would make copies, and he estimated he made seventy-five in all.

say they are using BASIC has all been positive. Two surprising things are apparent, however, 1) Most of these "users" never bought BASIC (less than 10% of all Altair owners have bought BASIC), and 2) The amount of royalties we have received from sales to hobbyists makes the time spent on Altair BASIC worth less than $2 an hour.

Why is this? As the majority of hobbyists must be aware, most of you steal your software. Hardware must be paid for, but software is something to share. Who cares if the people who worked on it get paid?

Is this fair? One thing you don't do by stealing software is get back at MITS for some problem you may have had. . . . One thing you do do is prevent good software from being written. Who can afford to do professional work for nothing? What hobbyist can put 3-man years into programming, finding all bugs, documenting his product and distribute for free? The fact is, no one besides us has invested a lot of money in hobby software. We have written 6800 BASIC, and are writing 8080 APL and 6800 APL, but there is very little incentive to make this software available to hobbyists. Most directly, the thing you do is theft. . . .

I would appreciate letters from anyone who wants to pay up, or has a suggestion or comment. Just write to me at 1180 Alvarado SE, #114, Albuquerque, New Mexico, 87108. Nothing would please me more than being able to hire ten programmers and deluge the hobby market with good software.

Bill Gates
General Partner, Micro-Soft

The letter was printed in the Homebrew Computer Club newsletter and also the Altair user group's *Computer Notes* and the *People's Computer Company*.[77] It stirred up a frenzy. "I got a lot of shit," Gates admitted. Of the three hundred letters he received, only five had a voluntary payment. Most of the rest heaped abuse on him.[78]

Basically, Gates was correct. The creation of software was just as valuable as the creation of hardware. Those who made software deserved to be compensated. If they weren't, people would quit writing software. By resisting the hacker ethos that anything that could be copied should be free, Gates helped ensure the growth of the new industry.

Still, there was a certain audacity to the letter. Gates was, after all, a serial stealer of computer time, and he had manipulated passwords to hack into accounts from eighth grade through his sophomore year at Harvard. Indeed, when he claimed in his letter that he and Allen had used more than $40,000 worth of computer time to make BASIC, he omitted the fact that he had never actually paid for that time and that much of it was on Harvard's military-supplied computer, funded by American taxpayers. The editor of one hobbyist newsletter wrote, "Rumors have been circulating through the hobby computer community that imply that development of the BASIC referred to in Bill Gates's letter was done on a Harvard University computer provided at least in part with government funds and that there was some question as to the propriety if not the legality of selling the results."[79]

Also, though Gates did not appreciate it at the time, the widespread pirating of Microsoft BASIC helped his fledgling company in the long run. By spreading so fast, Microsoft BASIC became a standard, and other computer makers had to license it. When National Semiconductor came out with a new microprocessor, for example, it needed a BASIC and decided to license Microsoft's because everyone was using it. "We made Microsoft the standard," said Felsenstein, "and he called us thieves for doing so."[80]

At the end of 1978 Gates and Allen moved their company from Albuquerque back home to the Seattle area. Just before they left, one of the twelve staffers won a free photo shoot from a local studio, so they posed for what would become a historic photograph, with Allen and most of the others looking like refugees from a hippie commune and Gates sitting up front looking like a Cub Scout. On his drive up the California coast, Gates was slapped with three speeding tickets, two from the same policeman.[81]

APPLE

Among those in Gordon French's garage at the first meeting of the Homebrew Computer Club was a socially awkward young hardware engineer named Steve Wozniak, who had dropped out of college and was working at Hewlett-Packard's calculator division in the Silicon Valley town of Cupertino. A friend had shown him the flyer—"Are you building your own computer?"—and he worked up the courage to attend. "That night turned out to be one of the most important nights of my life," he declared.[82]

Wozniak's father was a Lockheed engineer who loved explaining electronics. "One of my first memories is his taking me to his workplace on a weekend and showing me a few electronic parts, putting them on a table with me so I got to play with them," Wozniak recalled. There were usually stray transistors and resistors lying around the house, and when Steve would ask, "What's that?" his father would start from the beginning and explain how electrons and protons worked. "He pulled out a blackboard from time to time, and he would answer anything and make diagrams for it," Wozniak said. "He taught me how to make an *and* gate and an *or* gate out of parts he got—parts called diodes and resistors. And he showed me how they needed a transistor in between to amplify the signal and connect the output of one gate to the input of the other. To this very moment, that is the way every single digital device on the planet works at its most basic level." It was a striking example of the imprint a parent can make, especially back in the days when parents knew how radios worked and could show their kids how to test the vacuum tubes and replace the one that had burned out.

Wozniak made a crystal radio using scraped pennies when he was in second grade, a multihouse intercom system for the kids in his neighborhood when he was in fifth grade, a Hallicrafters shortwave radio when he was in sixth grade (he and his dad earned ham licenses together), and later that year taught himself how to apply Boolean algebra to electronic circuit design and demonstrate it with a machine that never lost at tic-tac-toe.

By the time he was in high school, Wozniak was applying his electronic wizardry to pranks. In one case he built a metronome attached to stripped batteries that looked like a bomb. When his principal discovered it ticking in a locker, he rushed it onto the playground away from the kids and called the bomb squad. Wozniak had to spend one night in the local house of detention, where he taught his fellow inmates to remove the wires on the ceiling fan and touch them to the iron bars in order to shock the jailer when he came to open the door. Although he had learned to code well, he was at heart a hardware engineer, unlike more refined software jockeys such as Gates. At one point he built a roulette-like game where players put their fingers in slots and when the ball landed one of them got shocked. "Hardware guys will play this game, but software guys are always way too chicken," he said.

Like others, he combined a love of technology with a hippie outlook, although he could not quite pull off the counterculture lifestyle. "I would wear this little Indian headband, and I wore my hair really long and grew a beard," he recalled. "From the neck up, I looked like Jesus Christ. But from the neck down, I still wore the clothes of a regular kid, a kid engineer. Pants. Collared shirt. I never did have the weird hippie clothes."

For fun, he would study the manuals of the office computers made by Hewlett-Packard and DEC and then try to redesign them using fewer chips. "I have no idea why this became the pastime of my life," he admitted. "I did it all alone in my room with my door shut. It was like a private hobby." It was not an activity that made him the life of the party, so he became pretty much a loner, but that talent to save chips served him well when he decided to build a computer of his own. He did so using only twenty chips, compared to the hundreds in most real computers. A friend who lived down the block joined him for the soldering, and because they drank so much Cragmont cream soda, it was dubbed the Cream Soda Computer. There was no screen or keyboard; instructions were fed in by punch card, and answers were conveyed by flashing lights on the front.

The friend introduced Wozniak to a kid who lived a few blocks away and shared their interest in electronics. Steve Jobs was almost five years younger and still at Homestead High, which Wozniak had attended. They sat on the sidewalk swapping tales about pranks they had pulled, Bob Dylan songs they liked, and electronic designs they had made. "Typically, it was really hard for me to explain to people the kind of design stuff I worked on, but Steve got it right away," Wozniak said. "I liked him. He was kind of skinny and wiry and full of energy." Jobs was similarly impressed. "Woz was the first person I'd met who knew more electronics than I did," he later said, stretching his own expertise.

Their greatest escapade, which laid the foundation for the computer partnership they would form, involved what was known as a Blue Box. In the fall of 1971, Wozniak read an article in *Esquire* describing how "phone phreaks" had created a device that emitted just the right tone chirps to fool the Bell System and cadge free long-distance calls. Before he even finished reading the article, he called Jobs, who was just beginning his senior year at Homestead High, and read parts of it aloud to him. It was a Sunday, but they knew how to sneak into a library at Stanford that might have the *Bell System Technical Journal*, which the *Esquire* article said included all the frequencies for the signal tones. After rummaging through the stacks, Wozniak finally found the journal. "I was practically shaking, with goose bumps and everything," he recalled. "It was such a Eureka moment." They drove to Sunnyvale Electronics to buy the parts they needed, soldered them together, and tested it with a frequency counter that Jobs had made as a school project. But it was an analog device, and they couldn't get it to produce tones that were precise and consistent enough.

Wozniak realized he would need to build a digital version, using a circuit with transistors. That fall was one of his infrequent semesters of dropping into college, and he was spending it at Berkeley. With help from a music student in his dorm, he had one built by Thanksgiving. "I have never designed a circuit I was prouder of," he said. "I still think it was incredible." They tested it by calling the Vatican, with Wozniak pretending to be Henry Kissinger needing to speak to the pope; it took a while, but the officials at the Vatican finally realized it was a prank before they woke up the pontiff.

Wozniak had devised an ingenious gadget, but by partnering with Jobs he was able to do much more: create a commercial enterprise. "Hey, let's sell these," Jobs suggested one day. It was a pattern that would lead to one of the most storied partnerships of the digital age, up there with Allen & Gates and Noyce & Moore. Wozniak would come up with some clever feat of engineering, and Jobs would find a way to polish and package it and sell it at a premium. "I got together the rest of the components, like the casing and power supply and keypads, and figured out how we could price it," Jobs said of the Blue Box. Using $40 worth of parts for each Blue Box, they produced a hundred that they sold for $150 apiece. The escapade ended after they got ripped off at gunpoint trying to sell one in a pizza parlor, but from the seeds of the adventure a company would be born. "If it hadn't been for the Blue Boxes, there wouldn't have been an Apple," Jobs later reflected. "Woz and I learned how to work together." Wozniak agreed: "It gave us a taste of what we could do with my engineering skills and his vision."

Jobs spent the following year dropping in and out of Reed College and then seeking spiritual enlightenment on a pilgrimage to India. When he returned in the fall of 1974, he went to work at Atari under Nolan Bushnell and Al Alcorn. Atari, flush with its success with *Pong*, was on a hiring spree. "Have fun; make money," declared one of the ads it took out in the *San Jose Mercury*. Jobs showed up dressed in his hippie garb and said he wouldn't leave the lobby until he was hired. At Alcorn's urging, Bushnell decided to take a chance on him. Thus the torch was passed from the most creative entrepreneur of video games to the man who would become the most creative entrepreneur of personal computers.

Despite his newly acquired Zen sensibilities, Jobs was inclined to inform his coworkers that they were "dumb shits" whose ideas sucked. Yet somehow he also managed to be compelling and inspiring. He sometimes wore a saffron robe, went barefoot, and believed that his strict diet of only fruits and vegetables meant that he need not use deodorant or shower often. As Bushnell recounted, "this was a mistaken theory." So he put Jobs on the night shift, when almost no one else was around. "Steve was prickly, but I kind of liked him. So I asked him to go on the night shift. It was a way to save him."

Jobs would later say that he learned some important lessons at Atari, the most profound being the need to keep interfaces friendly and intuitive. Instructions should be insanely simple: "Insert quarter, avoid Klingons." Devices should not need manuals. "That simplicity rubbed off on him and made him a very focused product person," said Ron Wayne, who worked with Jobs at Atari. In addition, Bushnell was able to help mold Jobs into an entrepreneur. "There is something indefinable in an entrepreneur, and I saw that in Steve," Bushnell recalled. "He was interested not just in engineering, but also the business aspects. I taught him that if you act like you can do something, then it will work. I told him, pretend to be completely in control and people will assume that you are."

Wozniak liked to come by Atari most evenings, after he finished work at Hewlett-Packard, to hang with Jobs and play the auto racing video game, *Gran Trak 10*, that Atari had finally developed. "My favorite game ever," he called it. In his spare time, he pieced together a home version of *Pong* that he could play on his TV set. He was able to program it to blast the word *Hell* or *Damn* whenever a player missed hitting the ball. One night he showed it to Alcorn, who came up with a scheme. He assigned Jobs to engineer a one-player version of *Pong*, to be called *Breakout*, in which a user could volley the ball against a brick wall, dislocating bricks to win points. Alcorn guessed, correctly, that Jobs would convince Wozniak to do the circuit design. Jobs was not a great engineer, but he was good at getting people to do things. "I looked at it as a two-for-one thing," Bushnell explained. "Woz was a better engineer." He was also a lovable and naïve teddy bear of a guy, who was as eager to help Jobs make a video game as Tom Sawyer's friends were to whitewash his fence. "This was the most wonderful offer in my life, to actually design a game that people would use," he recalled.

As Woz stayed up all night churning out elements of the design, Jobs sat on a bench to his left wire-wrapping the chips. Woz thought the task would take weeks, but in an early example of Jobs exerting

Steve Jobs (1955–2011) and
Steve Wozniak (1950–) in 1976.

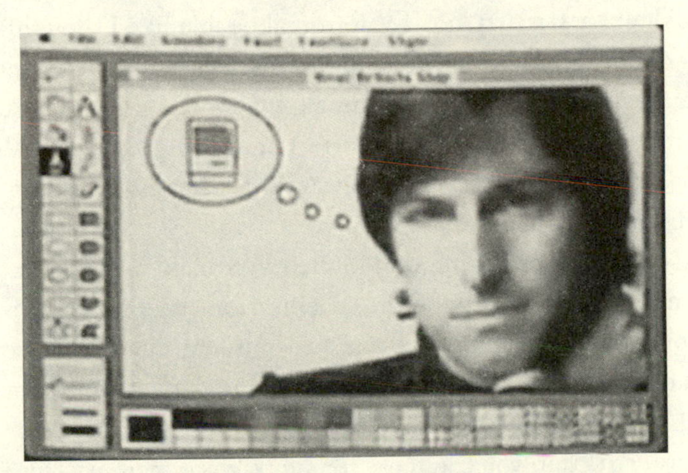

Jobs graphic on the original Macintosh in 1984.

Richard Stallman (1953–).

Linus Torvalds (1969–).

what colleagues called his reality distortion field, he was able to stare unblinkingly at Woz and convince him he could do the job in four days.

The March 1975 first gathering of the Homebrew Computer Club came just after Wozniak had finished designing *Breakout*. At the outset of the meeting, he felt out of place. He had been making calculators and home television game displays, but most of the excitement at that meeting centered on the new Altair computer, which didn't initially interest him. Shy at the best of times, he withdrew into a corner. He later described the scene: "Someone there was holding up the magazine *Popular Electronics*, which had a picture of a computer on the front of it called the Altair. It turned out all these people were really Altair enthusiasts, not TV terminal people like I thought." They went around the room introducing themselves, and when Wozniak's turn came he said, "I'm Steve Wozniak, I work at Hewlett-Packard on calculators and I designed a video terminal." He added that he also liked video games and pay movie systems for hotels, according to the minutes taken by Moore.

But there was one thing that piqued Wozniak's interest. A person at the meeting passed around the specification sheet for the new Intel microprocessor. "That night, I checked out the microprocessor data sheet and I saw it had an instruction for adding a location in memory to the A register," he recalled. "I thought, Wait a minute. Then it had another instruction you could use for subtracting memory from the A register. Whoa. Well, maybe this doesn't mean anything to you, but I knew exactly what these instructions meant, and it was the most exciting thing to discover ever."

Wozniak had been designing a terminal with a video monitor and a keyboard. He had planned for it to be a "dumb" terminal; it would have no computing power of its own, and instead it would connect via a phone line to a time-shared computer somewhere else. But when he saw the specs for the microprocessor—a chip that had a central processing unit on it—he had an insight: he could use a microprocessor to put some of the computing power into the terminal he was building. It would be a great leap from the Altair: a computer and a keyboard and a screen all integrated! "This whole vision of a personal computer just popped into my head," he said. "That night, I started to sketch out on paper what would later become known as the Apple I."

After a day of work on calculator design at HP, Wozniak would go home for a quick dinner and then return to his cubicle to work on his computer. At 10 p.m. on Sunday, June 29, 1975, a historic milestone occurred: Wozniak tapped a few keys on his keyboard, the signal was processed by a microprocessor, and letters appeared on the screen. "I was shocked," he confessed. "It was the first time in history anyone had typed a character on a keyboard and seen it show up on the screen right in front of them." That was not precisely true, but it was indeed the first time a keyboard and a monitor had been integrated with a personal computer designed for hobbyists.

The mission of the Homebrew Computer Club was to share ideas freely. That put it in the crosshairs of Bill Gates, but Wozniak embraced the communal ethos: "I so believed in the club's mission to further computing that I Xeroxed maybe a hundred copies of my complete design and gave it to anyone who wanted it." He was too shy, initially, to stand in front of the group and make a formal presentation, but he was so proud of his design that he loved standing in the back, showing it off to any who gathered around, and handing out the schematics. "I wanted to give it away for free to other people."

Jobs thought differently, just as he had with the Blue Box. And as it turned out, his desire to package and sell an easy-to-use computer—and his instinct for how to do it—changed the realm of personal computers just as much as Wozniak's clever circuit design did. Indeed, Wozniak would have been relegated to minor mentions in the Homebrew newsletter had Jobs not insisted that they create a company to commercialize it.

Jobs began calling chip makers such as Intel to get free samples. "I mean, he knew how to talk to a sales representative," Wozniak marveled. "I could never have done that. I'm too shy." Jobs also started accompanying Wozniak to Homebrew meetings, carrying the television set and conducting the demonstrations, and he came up with a plan to sell circuit boards preprinted with Wozniak's design. It was typical of their partnership. "Every time I'd design something great, Steve would find a way to make money for us," said Wozniak. "It never crossed my mind to sell computers. It was Steve who said, 'Let's hold them in the air and sell a few.'" Jobs sold his Volkswagen Bus and Wozniak sold his HP calculator to raise funding for their endeavor.

They made an odd but powerful partnership: Woz was an angelic naïf who looked like a panda, Jobs a demon-driven mesmerizer who looked like a whippet. Gates had bullied Allen into giving him more than half of their partnership. In the case of Apple, it was Wozniak's father, an engineer who respected engineers and disdained marketers and managers, who insisted that his son, who had made the designs, be given more than 50 percent of the partnership. He confronted Jobs when he came by the Wozniak house: "You don't deserve shit. You haven't produced anything." Jobs began to cry and told Steve Wozniak that he was willing to call off the partnership. "If we're not 50-50," Jobs said, "you can have the whole thing." Wozniak, however, understood what Jobs contributed to their partnership, and it was worth at least 50 percent. If he had been on his own, Wozniak might not have progressed beyond handing out free schematics.

After they demonstrated the computer at a Homebrew meeting, Jobs was approached by Paul Terrell, the owner of a small chain of computer stores called The Byte Shop. After they talked, Terrell said, "Keep in touch," handing Jobs his card. The next day Jobs walked into his store barefoot and announced, "I'm keeping in touch." By the time Jobs had finished his pitch, Terrell had agreed to order fifty of what became known as the Apple I computer. But he wanted them fully assembled, not just printed boards with a pile of components. It was another step in the evolution of personal computers. They would not be just for solder-gun-wielding hobbyists anymore.

Jobs understood this trend. When it came time to build the Apple II, he did not spend much time studying microprocessor specs. Instead he went to Macy's at the Stanford mall and studied the Cuisinart. He decided that the next personal computer should be like an appliance: all fit together with a sleek case and no assembly required. From the power supply to the software, from the keyboard to the monitor, everything should be tightly integrated. "My vision was to create the first fully packaged computer," he explained. "We were no longer aiming for the handful of hobbyists who liked to assemble their own computers, who knew how to buy transformers and keyboards. For every one of them there were a thousand people who would want the machine to be ready to run."

By early 1977 a few other hobbyist computer companies had bubbled up from the Homebrew and other such cauldrons. Lee Felsenstein, the club's master of ceremonies, had launched Processor Technology and come out with a computer called Sol. Other companies included Cromemco, Vector Graphic, Southwest Technical Products, Commodore, and IMSAI. But the Apple II was the first personal computer to be simple and fully integrated, from the hardware to the software. It went on sale in June 1977 for $1,298, and within three years 100,000 of them were sold.

The rise of Apple marked a decline of hobbyist culture. For decades young innovators such as Kilby and Noyce had been introduced to electronics by knowing how to distinguish different transistors, resistors, capacitors, and diodes, then wire-wrapping or soldering them into breadboards to create circuits that became ham radios, rocket controllers, amplifiers, and oscilloscopes. But in 1971 microprocessors began making complex circuit boards obsolete, and Japanese electronics companies began mass-producing products that were cheaper than homemade ones. Sales of do-it-yourself kits withered away. Hardware hackers such as Wozniak ceded primacy to software coders such as Gates. With the Apple II and then, more notably, the Macintosh in 1984, Apple pioneered the practice of creating machines that users were not supposed to open and fiddle with their innards.

The Apple II also established a doctrine that would become a religious creed for Steve Jobs: his company's hardware was tightly integrated with its operating system software. He was a perfectionist who liked to control the user experience end to end. He didn't want to let you buy an Apple machine and run someone else's clunky operating system on it, nor buy Apple's operating system and put it on someone else's junky hardware.

That integrated model did not become standard practice. The launch of the Apple II woke up the big computer companies, most notably IBM, and prompted an alternative to emerge. IBM—more specifically IBM as it was outmaneuvered by Bill Gates—would embrace an approach in which the personal computer's hardware and its operating system were made by different companies. As a result, software would become king, and, except at Apple, most computer hardware would become a commodity.

DAN BRICKLIN AND VISICALC

For personal computers to be useful, and for practical people to justify buying them, they had to become tools rather than merely toys. Even the Apple II might have been a passing fad, once the excitement of the hobbyists receded, if users had not been able to apply it to a practical task. Thus there arose a demand for what became known as application software, programs that could apply a personal computer's processing power to a specific chore.

The most influential pioneer in that field was Dan Bricklin, who conceived the first financial spreadsheet program, VisiCalc.[83] Bricklin was an MIT computer science graduate who spent a few years developing word-processing software at Digital Equipment Corporation and then enrolled at Harvard Business School. Sitting in a lecture one day in the spring of 1978, he watched as the professor created the columns and rows for a financial model on the blackboard. When he found an error or wanted to modify a value in one cell, the professor had to use his eraser and change the values in many of the other cells.[84]

Bricklin had seen Doug Engelbart demonstrate his oNLine System, made famous at the Mother of All Demos, which featured a graphical display and a mouse for pointing and clicking. Bricklin began envisioning an electronic spreadsheet that would use a mouse and simple point-drag-and-click interface. That summer, while riding a bike on Martha's Vineyard, he decided to turn the idea into a product. He was well suited for such an endeavor. He was a software engineer with the instincts of a product person; he had a feel for what users would want. His parents were entrepreneurs, and he was excited by the prospect of starting a business. And he was a good team player, who knew how to find the right partners. "I had the right combination of experience and knowledge to develop software that met a need people had," he observed.[85]

So he teamed up with a friend he had met at MIT, Bob Frankston, another software engineer whose father was an entrepreneur. "The ability for Dan and me to work as a team was crucial," Frankston said. Although Bricklin could have written the program alone, instead he sketched it out and had Frankston develop it. "It gave him the freedom to focus on what the program should do rather than how to do it," Frankston said of their collaboration.[86]

The first decision they made was to develop the program for use on a personal computer rather than on a DEC business computer. They chose the Apple II because Wozniak had made its architecture open and transparent enough that the functions needed by software developers were easily accessible.

They created the prototype over a weekend on an Apple II they borrowed from someone who would, in effect, become a third collaborator, Dan Fylstra. A recent graduate of Harvard Business School, Fylstra had launched a software publishing company, which focused on games such as chess, that he ran out of his Cambridge apartment. In order for a software industry to develop separately from the hardware industry, it was necessary to have publishers who knew how to promote and distribute products.

Because both Bricklin and Frankston had good business sense and a feel for consumer desires, they focused on making VisiCalc a *product*, not just a program. They used friends and professors as focus groups to make sure the interface was intuitive and easy to use. "The goal was to give the user a conceptual model that was unsurprising," Frankston explained. "It was called the principle of least surprise. We were illusionists synthesizing an experience."[87]

Among those who helped turn VisiCalc into a business phenomenon was Ben Rosen, then an analyst with Morgan Stanley who later turned his influential newsletter and conferences into a business of his own and then started a venture capital firm in Manhattan. In May 1979 Fylstra demonstrated an early version of VisiCalc at Rosen's Personal Computer Forum in his hometown of New Orleans. In his newsletter, Rosen enthused, "VisiCalc comes alive visually. . . . In minutes, people who have never used a computer are writing and using programs." He ended with a prediction that came true: "VisiCalc could someday become the software tail that wags (and sells) the personal computer dog."

VisiCalc catapulted the Apple II to triumph, because for a year there were no versions for other personal computers. "That's what really drove the Apple II to the success it achieved," Jobs later said.[88] It was quickly followed by word-processing software, such as Apple Writer and EasyWriter. Thus did VisiCalc not only stimulate the market for personal computers, but it helped to create an entire new profit-driven industry, that of publishing proprietary application software.

THE IBM OPERATING SYSTEM

During the 1970s IBM dominated the mainframe market with its 360 series. But it was beaten by DEC and Wang in the market for refrigerator-size minicomputers, and it looked like it might be left behind in personal computers as well. "IBM bringing out a personal computer would be like teaching an elephant to tap dance," one expert declared.[89]

The company's top management seemed to agree. So they considered instead just licensing the Atari 800 home computer and slapping IBM's name on it. But when that option was debated at a July 1980 meeting, IBM's CEO Frank Carey dismissed it. Surely the world's greatest computer company could create a personal computer of its own, he said. Doing anything new at the company, he complained, seemed to require three hundred people working three years.

That is when Bill Lowe, who was the director of IBM's development lab in Boca Raton, Florida, piped up. "No, sir, you're wrong," he stated. "We can get a project out in a year."[90] His cockiness got him assigned the task of overseeing the project, code-named Acorn, to create an IBM personal computer.

Lowe's new team was led by Don Estridge, who chose Jack Sams, a gentle southerner who was a twenty-year veteran of IBM, to be in charge of piecing together the software. Given the one-year deadline, Sams knew that he would have to license software from outside vendors rather than have it written in-house. So on July 21, 1980, he placed a call to Bill Gates and asked to see him right away. When Gates invited him to fly to Seattle the following week, Sams replied that he was already heading for the airport and wanted to see Gates the next day. Sensing a big fish hungry to be hooked, Gates was thrilled.

A few weeks earlier, Gates had recruited his Harvard dorm mate Steve Ballmer to Microsoft as the business manager, and he asked Ballmer to join him at the IBM meeting. "You're the only other guy here who can wear a suit," Gates pointed out.[91] When Sams arrived, Gates was also wearing a suit, but he did not quite fill it. "This young fellow came out to take us back, and I thought he was the office boy," recalled Sams, who was dressed in the IBM standard blue suit and white shirt. But he and the rest of his team were soon dazzled by Gates's brilliance.

At first the IBM folks wanted to talk about licensing Microsoft BASIC, but Gates turned the conversation into an intense discussion about where technology was heading. By the end of a few hours, they were talking about licensing all of the programming languages Microsoft had or could produce, including Fortran and COBOL in addition to BASIC. "We told IBM, 'Okay, you can have everything we make,' even though we hadn't even made it yet," Gates recalled.[92]

The IBM team returned a few weeks later. There was one essential piece of software, in addition to these programming languages, that IBM was missing. It needed an operating system, the software program that would serve as the foundation for all of the other programs. An operating system handles the basic instructions that other software uses, including such chores as deciding where data should be stored, how memory and processing resources should be allocated, and how applications software interacts with the computer's hardware.

Microsoft did not yet make an operating system. It was instead working with one called CP/M (for Control Program for Microcomputers) that was owned by Gary Kildall, a childhood friend of Gates who had recently moved to Monterey, California. So with Sams sitting in his office, Gates picked up the phone and called Kildall. "I'm sending some guys down," he said, describing what the IBM executives were seeking. "Treat them right, they're important guys."[93]

Kildall didn't. Gates later referred to it as "the day Gary decided to go flying." Instead of meeting the IBM visitors, Kildall chose to pilot his private plane, as he loved to do, and keep a previously scheduled appointment in San Francisco. He left it to his wife to meet with the four dark-suited men of the IBM team in the quirky Victorian house that served as Kildall's company headquarters. When they presented her with a long nondisclosure agreement, she refused to sign it. After much haggling, the IBM folks walked out in disgust. "We popped out our letter that said please don't tell anybody we're here, and we don't want to hear anything confidential, and she read it and said I can't sign this," Sams recalled. "We spent the whole day in Pacific Grove debating with them and with our attorneys and her attorneys and everybody else about whether or not she could even talk to us about talking to us, and then we left." Kildall's little company had just blown its chance to become the dominant player in computer software.[94]

Sams flew back to Seattle to see Gates and asked him to figure out another way to conjure up an operating system. Fortunately, Paul Allen knew someone in Seattle who could help: Tim Paterson, who worked for a small firm called Seattle Computer Products. A few months earlier Paterson had become frustrated that Kildall's CP/M was not available for Intel's newest microprocessors, so he adapted it into an operating system that he dubbed QDOS, for Quick and Dirty Operating System.[95]

By then Gates had come to the realization that one operating system, most likely the one chosen by IBM, would end up being the standard operating system that most personal computers would use. He also figured out that whoever owned that operating system would be in the catbird seat. So instead of sending the IBM folks to see Paterson, Gates and his team said that they would handle things on their own. Ballmer later recalled, "We just told IBM, 'Look, we'll go and get this operating system from this small local company, we'll take care of it, we'll fix it up.'"

Paterson's firm was struggling to make ends meet, so Allen was able to negotiate a savvy deal with his friend. After initially acquiring just a nonexclusive license, Allen went back when an IBM deal looked likely and bought Paterson's software outright, without telling him why. "We ended up working out a deal to buy the operating system from him, for whatever usage we wanted, for fifty thousand dollars," Allen recalled.[96] For that pittance Microsoft acquired the software that, after they spruced it up, would allow it to dominate the software industry for more than three decades.

But Gates almost balked. He was uncharacteristically worried that Microsoft, which was wildly overcommitted with other projects, might not have the capacity to gussy up QDOS into an IBM-worthy operating system. Microsoft had only forty ragtag employees, some of whom slept on the floor and took sponge baths in the morning, and it was led by a twenty-four-year-old who could still be mistaken for an office boy. On a Sunday at the end of September 1980, two months after IBM first came calling, Gates gathered his top team to make the go or no-go decision. It was Kay Nishi, a young computer entrepreneur from Japan with a Gatesian intensity, who was the most adamant. "Gotta do it! Gotta do it!" he squealed repeatedly as he bounced around the room. Gates decided he was right.[97]

Gates and Ballmer took an overnight flight to Boca Raton to negotiate the deal. Their 1980 revenues were $7.5 million, compared to IBM's $30 billion, but Gates was gunning for an agreement that would allow Microsoft to keep ownership of an operating system that IBM would turn into a global standard. In its deal with Paterson's company, Microsoft had bought DOS outright, "for whatever usage," rather than merely licensing it. That was smart, but what was even smarter was not letting IBM force Microsoft to make the same arrangement.

When they landed at the Miami airport, they went to a bathroom to change into suits, and Gates realized he had forgotten a tie. In an unusual display of fastidiousness, he insisted that they stop at a Burdine's department store on the drive to Boca in order to buy one. It did not have the full desired effect on the crisp-suited IBM executives waiting to greet him. One of the software engineers recalled that Gates looked like a "kid that had chased somebody around the block and stolen a suit off him and the suit was way too big for him. His collar stuck up and he looked like some punk kid, and I said, 'Who the hell is this?'"[98]

Once Gates began his presentation, however, they quit focusing on his disheveled appearance. He wowed the IBM team with his mastery of details, both technical and legal, and projected calm confidence when insisting on terms. It was largely an act. When he arrived back in Seattle, Gates went into his office, lay on the floor, and agonized aloud to Ballmer about all of his doubts.

After a month of back-and-forth, a thirty-two-page deal was struck in early November 1980. "Steve and I knew that contract by heart," Gates declared.[99] "We didn't get paid that much. The total was something like $186,000." At least initially. But it had the two provisions that Gates knew would alter the balance of power in the computer industry. The first was that IBM's license to use the operating system, which it would call PC-DOS, would be nonexclusive. Gates could license the same operating system to other personal computer makers under the name MS-DOS. Second, Microsoft would keep control of the source code. This meant that IBM couldn't modify or evolve the software into something that became proprietary to its machines. Only Microsoft could make changes, and then it could license each new version to any company it wanted. "We knew there were going to be clones of the IBM PC," Gates said. "We structured that original contract to allow them. It was a key point in our negotiations."[100]

The deal was similar to the one Gates had made with MITS, when he retained the right to license BASIC to other computer makers as well. That approach allowed Microsoft's BASIC and then, more important, its operating system to become an industry standard, one that Microsoft controlled. "In fact, our tagline in our ad had been 'We set the standard,'" Gates recalled with a laugh. "But when we did in fact set the standard, our antitrust lawyer told us to get rid of that. It's one of those slogans you can use only when it's not true."*[101]

---

*The lawyers were right to be worried. Microsoft later was involved in a protracted antitrust suit brought by the Justice Department, which charged that it had improperly

Gates boasted to his mother about the importance of his deal with IBM, hoping that it would prove that he had been right to drop out of Harvard. Mary Gates happened to be on the board of the United Way with IBM's president John Opel, who was about to take over from Frank Cary as CEO. One day she was flying with Opel on his plane to a meeting, and she mentioned the connection. "Oh, my little boy's doing this project, he's actually working with your company." Opel seemed unaware of Microsoft. So when she came back, she warned Bill, "Look, I mentioned your project and how you dropped out of school and all this stuff to Opel, and he doesn't know who you are, so maybe your project's not as important as you think." A few weeks later, the Boca Raton executives went to IBM headquarters to brief Opel on their progress. "We have a dependency on Intel for the chip, and Sears and ComputerLand are going to do the distribution," the team leader explained. "But probably our biggest dependency is actually a pretty small software company up in Seattle run by a guy named Bill Gates." To which Opel responded, "Oh, you mean Mary Gates's son? Oh, yeah, she's great."[102]

Producing all the software for IBM was a struggle, as Gates predicted, but the ragtag Microsoft crew worked around the clock for nine months to get it done. For one last time, Gates and Allen were a team again, sitting side by side through the night, coding with the shared intensity they had displayed at Lakeside and Harvard. "The one tiff Paul and I had was when he wanted to go see a space shuttle launch and I didn't, because we were late," said Gates. Allen ended up going. "It was the first one," he said. "And we flew back right after the launch. We weren't gone even 36 hours."

By writing the operating system, the two of them helped determine the look and feel of the personal computer. "Paul and I decided every stupid little thing about the PC," Gates said. "The keyboard layout, how the cassette port worked, how the sound port worked,

---

leveraged its dominance of the operating system market to seek advantage in browsers and other products. The case was eventually settled after Microsoft agreed to modify some of its practices.

how the graphics port worked."[103] The result reflected, alas, Gates's nerdy design taste. Other than causing a cohort of users to learn where the backslash key was, there was little good that could be said about human-machine interfaces that relied on prompts such as "c:\>" and files with clunky names such as AUTOEXEC.BAT and CONFIG.SYS.

Years later, at an event at Harvard, the private equity investor David Rubenstein asked Gates why he had saddled the world with the Control+Alt+Delete startup sequence: "Why, when I want to turn on my software and computer, do I need to have three fingers? Whose idea was that?" Gates began to explain that IBM's keyboard designers had failed to provide an easy way to signal the hardware to bring up the operating system, then he stopped himself and sheepishly smiled. "It was a mistake," he admitted.[104] Hard-core coders sometimes forget that simplicity is the soul of beauty.

The IBM PC was unveiled, with a list price of $1,565, at New York's Waldorf Astoria in August 1981. Gates and his team were not invited to the event. "The weirdest thing of all," Gates said, "was when we asked to come to the big official launch, IBM denied us."[105] In IBM's thinking, Microsoft was merely a vendor.

Gates got the last laugh. Thanks to the deal he made, Microsoft was able to turn the IBM PC and its clones into interchangeable commodities that would be reduced to competing on price and doomed to having tiny profit margins. In an interview appearing in the first issue of *PC* magazine a few months later, Gates pointed out that soon all personal computers would be using the same standardized microprocessors. "Hardware in effect will become a lot less interesting," he said. "The total job will be in the software."[106]

THE GRAPHICAL USER INTERFACE

Steve Jobs and his team at Apple bought a new IBM PC as soon as it came out. They wanted to check out what the competition looked like. The consensus was, to use Jobs's phrase, "It sucked." This was not simply a reflection of Jobs's instinctive arrogance, although it was partly that. It was a reaction to the fact that the machine, with its surly c:\> prompts and boxy design, was boring. It didn't occur to Jobs that corporate technology managers might not be yearning for excitement at the office and knew they couldn't get in trouble for choosing a boring brand like IBM over a plucky one like Apple. Bill Gates happened to be at Apple headquarters for a meeting on the day that the IBM PC was announced. "They didn't seem to care," he said. "It took them a year to realize what had happened."[107]

Jobs was aroused by competition, especially when he thought it sucked. He saw himself as an enlightened Zen warrior, fighting the forces of ugliness and evil. He had Apple take out an ad in the *Wall Street Journal*, which he helped to write. The headline: "Welcome, IBM. Seriously."

One reason Jobs was dismissive was that he had already seen the future and was embarked on inventing it. On visits to Xerox PARC, he was shown many of the ideas that Alan Kay, Doug Engelbart, and their colleagues had developed, most notably the graphical user interface (GUI, pronounced GOO-ee), which featured a desktop metaphor with windows, icons, and a mouse that served as a pointer. The creativity of the Xerox PARC team combined with the design and marketing genius of Jobs would make the GUI the next great leap in facilitating the human-machine interaction that Bush, Licklider, and Engelbart had envisioned.

Jobs's two main visits with his team to Xerox PARC were in December 1979. Jef Raskin, an Apple engineer who was designing a friendly computer that would eventually become the Macintosh, had already seen what Xerox was doing and wanted to convince Jobs to look into it. One problem was that Jobs found Raskin insufferable—the technical terminology he used for Raskin was "a shithead who sucks"—but eventually Jobs made the pilgrimage. He had worked out a deal with Xerox that allowed the Apple folks to study the technology in return for allowing Xerox to make a million-dollar investment in Apple.

Jobs was certainly not the first outsider to see what Xerox PARC had wrought. Its researchers had given hundreds of demonstrations to visitors, and they had already distributed more than a thousand Xerox Altos, the expensive computer developed by Lampson, Thacker, and

Kay that used a graphical user interface and other PARC innovations. But Jobs was the first to become obsessed with the idea of incorporating PARC's interface ideas into a simple, inexpensive, personal computer. Once again, the greatest innovation would come not from the people who created the breakthroughs but from the people who applied them usefully.

On Jobs's first visit, the Xerox PARC engineers, led by Adele Goldberg, who worked with Alan Kay, were reserved. They did not show Jobs much. But he threw a tantrum—"Let's stop this bullshit!" he kept shouting—and finally was given, at the behest of Xerox's top management, a fuller show. Jobs bounced around the room as his engineers studied each pixel on the screen. "You're sitting on a goldmine," he shouted. "I can't believe Xerox is not taking advantage of this."

There were three major innovations on display. The first was Ethernet, the technologies developed by Bob Metcalfe for creating local area networks. Like Gates and other pioneers of personal computers, Jobs was not very interested—certainly not as interested as he should have been—in networking technology. He was focused on the ability of computers to empower individuals rather than to facilitate collaboration. The second innovation was object-oriented programming. That, likewise, did not grab Jobs, who was not a programmer.

What caught his attention was the graphical user interface featuring a desktop metaphor that was as intuitive and friendly as a neighborhood playground. It had cute icons for documents and folders and other things you might want, including a trash can, and a mouse-controlled cursor that made them easy to click. Not only did Jobs love it, but he could see ways to improve it, make it simpler and more elegant.

The GUI was made possible by bitmapping, another innovation pioneered at Xerox PARC. Until then, most computers, including the Apple II, would merely generate numerals or letters on the screen, usually in a ghastly green against a black background. Bitmapping allowed each and every pixel on the screen to be controlled by the computer—turned off or on and in any color. That permitted all sorts of wonderful displays, fonts, designs, and graphics. With his feel for design, familiarity with fonts, and love of calligraphy, Jobs was blown away by bitmapping. "It was like a veil being lifted from my eyes," he recalled. "I could see what the future of computing was destined to be."

As Jobs drove back to Apple's office in Cupertino, at a speed that would have awed even Gates, he told his colleague Bill Atkinson that they had to incorporate—and improve upon—Xerox's graphical interface in future Apple computers, such as the forthcoming Lisa and Macintosh. "This is it!" he shouted. "We've got to do it!" It was a way to bring computers to the people.[108]

Later, when he was challenged about pilfering Xerox's ideas, Jobs quoted Picasso: "Good artists copy, great artists steal." He added, "And we have always been shameless about stealing great ideas." He also crowed that Xerox had fumbled its idea. "They were copier-heads who had no clue about what a computer could do," he said of Xerox's management. "They just grabbed defeat from the greatest victory in the computer industry. Xerox could have owned the entire computer industry."[109]

In fact, neither explanation does Jobs and Apple justice. As the case of the forgotten Iowa inventor John Atanasoff shows, conception is just the first step. What really matters is execution. Jobs and his team took Xerox's ideas, improved them, implemented them, and marketed them. Xerox had the chance to do that, and they in fact tried to, with a machine called the Xerox Star. It was clunky and kludgy and costly, and it flopped. The Apple team simplified the mouse so it had only one button, gave it the power to move documents and other items around the screen, allowed file extensions to be changed just by dragging a document and "dropping" it into a folder, created pull-down menus, and allowed the illusion of documents piling on top of each other and overlapping.

Apple launched Lisa in January 1983 and then, more successfully, Macintosh a year later. Jobs knew when he unveiled the Mac that it would propel the personal computer revolution by being a machine that was friendly enough to take home. At the dramatic product launch, he walked across a dark stage to pull the new computer out of a cloth bag. The theme from *Chariots of Fire* began to play, and the word *MACINTOSH* scrolled horizontally across the screen, then underneath it the words *insanely great!* appeared in elegant script, as if being slowly written by hand. There was a moment of awed silence in the auditorium, then a few gasps. Most had never seen, or even imagined, something so spectacular. The screen then flicked through displays of different fonts, documents, charts, drawings, a chess game, spreadsheet, and a rendering of Jobs with a thought bubble containing a Macintosh by his head. The ovation lasted for five minutes.[110]

The Macintosh launch was accompanied by a memorable ad, "1984," that showed a young heroine outracing the authoritarian police to throw a hammer into a screen, destroying Big Brother. It was Jobs the rebel taking on IBM. And Apple now had an advantage: it had perfected and implemented a graphical user interface, the great new leap in human-machine interaction, while IBM and its operating system supplier Microsoft were still using curt command lines with c:\> prompts.

WINDOWS

In the early 1980s, before the introduction of the Macintosh, Microsoft had a good relationship with Apple. In fact, on the day that IBM launched its PC in August 1981, Gates was visiting Jobs at Apple, which was a regular occurrence since Microsoft was making most of its revenue writing software for the Apple II. Gates was still the supplicant in the relationship. In 1981 Apple had $334 million in revenue, compared to Microsoft's $15 million. Jobs wanted Microsoft to write new versions of its software for the Macintosh, which was still a secret development project. So at their August 1981 meeting, he confided his plans to Gates.

Gates thought that the idea of the Macintosh—an inexpensive computer for the masses with a simple graphical user interface—sounded, as he put it, "super neat." He was willing, indeed eager, to have Microsoft write application software for it. So he invited Jobs up to Seattle. In his presentation there to the Microsoft engineers, Jobs was at his charismatic best. With a bit of metaphorical license, he spun his vision of a factory in California that would take in sand, the raw material of silicon, and churn out an "information appliance" that was so simple it would need no manual. The Microsoft folks code-named the project "Sand." They even reverse-engineered it into an acronym: Steve's Amazing New Device.[111]

Jobs had one major worry about Microsoft: he didn't want it to copy the graphical user interface. With his feel for what would wow average consumers, he knew that the desktop metaphor with point-and-click navigation would be, if done right, the breakthrough that would make computers truly personal. At a design conference in Aspen in 1981, he waxed eloquently about how friendly computer screens would become by using "metaphors that people already understand such as that of documents on a desktop." His fear that Gates would steal the idea was somewhat ironic, since Jobs himself had filched the concept from Xerox. But to Jobs's way of thinking, he had made a business deal for the rights to appropriate Xerox's idea. Plus he had improved it.

So Jobs wrote into his contract with Microsoft a clause that he believed would give Apple at least a year's head start in having a graphical user interface. It decreed that for a certain period Microsoft would not produce for any company other than Apple any software that "utilizes a mouse or tracking ball" or had a point-and-click graphical interface. But Jobs's reality distortion field got the better of him. Because he was so intent on getting Macintosh on the market by late 1982, he became convinced that it would happen. So he agreed that the prohibition would last until the end of 1983. As it turned out, Macintosh did not ship until January 1984.

In September 1981 Microsoft secretly began designing a new operating system, intended to replace DOS, based on the desktop metaphor with windows, icons, mouse, and pointer. It hired from Xerox PARC Charles Simonyi, a software engineer who had worked alongside Alan Kay in creating graphical programs for the Xerox Alto. In February 1982 the *Seattle Times* ran a picture of Gates and Allen that, as a sharp-eyed reader may have noted, had a whiteboard in the background with a few sketches and the words *Window manager* on top. By that summer, just as Jobs began to realize that the release date for the Macintosh would slip until at least late 1983, he became paranoid.

His fears were heightened when his close pal Andy Hertzfeld, an engineer on the Macintosh team, reported that his contact at Microsoft had begun asking detailed questions about how bitmapping was executed. "I told Steve that I suspected that Microsoft was going to clone the Mac," Hertzfeld recalled.[112]

Jobs's fears were realized in November 1983, two months before the Macintosh was launched, when Gates held a press conference at the Palace Hotel in Manhattan. He announced that Microsoft was developing a new operating system that would be available for IBM PCs and their clones, featuring a graphical user interface. It would be called Windows.

Gates was within his rights. His restrictive agreement with Apple expired at the end of 1983, and Microsoft did not plan to ship Windows until well after that. (As it turned out, Microsoft took so long to finish even a shoddy version 1.0 that Windows would not end up shipping until November 1985.) Nevertheless, Jobs was livid, which was not a pretty sight. "Get Gates down here immediately," he ordered one of his managers. Gates complied, but he was unintimidated. "He called me down to get pissed off at me," Gates recalled. "I went down to Cupertino, like a command performance. I told him, 'We're doing Windows.' I said to him, 'We're betting our company on graphics interface.'" In a conference room filled with awed Apple employees, Jobs shouted back, "You're ripping us off! I trusted you, and now you're stealing from us!"[113] Gates had a habit of getting calmer and cooler whenever Jobs worked himself into a frenzy. At the end of Jobs's tirade, Gates looked at him and, in his squeaky voice, replied with what became a classic zinger: "Well, Steve, I think there's more than one way of looking at it. I think it's more like we both had this rich neighbor named Xerox and I broke into his house to steal the TV set and found out that you had already stolen it."[114]

Jobs remained angry and resentful for the rest of his life. "They just ripped us off completely, because Gates has no shame," he said almost thirty years later, shortly before he died. Upon hearing this, Gates responded, "If he believes that, he really has entered into one of his own reality distortion fields."[115]

The courts ended up ruling that Gates was legally correct. A decision by a federal appeals court noted that "GUIs were developed as a user-friendly way for ordinary mortals to communicate with the Apple computer . . . based on a desktop metaphor with windows, icons and pull-down menus which can be manipulated on the screen with a hand-held device called a mouse." But it ruled, "Apple cannot get patent-like protection for the idea of a graphical user interface, or the idea of a desktop metaphor." Protecting a look-and-feel innovation was almost impossible.

Whatever the legalities were, Jobs had a right to be angry. Apple had been more innovative, imaginative, elegant in execution, and brilliant in design. Microsoft's GUI was shoddy, with tiled windows that could not overlap with each other and graphics that looked like they had been designed by drunkards in a Siberian basement.

Nevertheless, Windows eventually clawed its way to dominance, not because its design was better but because its business model was better. The market share commanded by Microsoft Windows reached 80 percent by 1990 and kept rising, to 95 percent by 2000. For Jobs, Microsoft's success represented an aesthetic flaw in the way the universe worked. "The only problem with Microsoft is they just have no taste, they have absolutely no taste," he later said. "I don't mean that in a small way. I mean that in a big way, in the sense that they don't think of original ideas and they don't bring much culture into their product."[116]

The primary reason for Microsoft's success was that it was willing and eager to license its operating system to any hardware maker. Apple, by contrast, opted for an integrated approach. Its hardware came only with its software and vice versa. Jobs was an artist, a perfectionist, and thus a control freak who wanted to be in charge of the user experience from beginning to end. Apple's approach led to more beautiful products, a higher profit margin, and a more sublime user experience. Microsoft's approach led to a wider choice of hardware. It also turned out to be a better path for gaining market share.

RICHARD STALLMAN, LINUS TORVALDS, AND THE FREE AND OPEN-SOURCE SOFTWARE MOVEMENTS

In late 1983, just as Jobs was preparing to unveil the Macintosh and Gates was announcing Windows, another approach to the creation of software emerged. It was pushed by one of the diehard denizens of the MIT Artificial Intelligence Lab and Tech Model Railroad Club, Richard Stallman, a truth-possessed hacker with the looks of an Old Testament prophet. With even greater moral fervor than the Homebrew Computer Club members who copied tapes of Microsoft BASIC, Stallman believed that software should be collaboratively created and freely shared.[117]

At first glance, this did not seem like an approach that would provide incentives for people to produce great software. The joy of free sharing wasn't what motivated Gates, Jobs, and Bricklin. But because there was a collaborative and communitarian ethic that permeated hacker culture, the free and open-source software movements ended up being powerful forces.

Born in 1953, Richard Stallman was intensely interested in math as a child growing up in Manhattan, and he conquered calculus on his own as a young boy. "Mathematics has something in common with poetry," he later said. "It's made out of these true relationships, true steps, true deductions, so it has this beauty about it." Unlike his classmates, he was deeply averse to competition. When his high school teacher divided the students into two teams for a quiz contest, Stallman refused to answer any questions. "I resisted the notion of competing," he explained. "I saw that I was being manipulated and my classmates were falling prey to this manipulation. They all wanted to beat the other people, who were just as much their friends as were the people on their own team. They started demanding that I answer the questions so we could win. But I resisted the pressure because I had no preference for one team or the other."[118]

Stallman went to Harvard, where he became a legend even among the math wizards, and during the summers and after he graduated he worked at the MIT Artificial Intelligence Lab, two subway stops away in Cambridge. There he added to the train track layout at the

Tech Model Railroad Club, wrote a PDP-11 simulator to run on the PDP-10, and grew enamored with the collaborative culture. "I became part of a software-sharing community that had existed for many years," he recalled. "Whenever people from another university or a company wanted to port and use a program, we gladly let them. You could always ask to see the source code."[119]

Like a good hacker, Stallman defied restrictions and locked doors. With his fellow students, he devised multiple ways to break into offices where there were forbidden terminals; his own specialty was climbing through the false ceilings, pushing aside a tile, and lowering a long strip of magnetic tape tipped with wads of sticky duct tape to open door handles. When MIT instituted a database of users and a system of strong passwords, Stallman resisted, and he rallied his colleagues to do so as well: "I thought that was disgusting, so I didn't fill out the form and I created a null-set password." At one point a professor warned that the university might delete his directory of files. That would be unfortunate for everyone, Stallman replied, since some of the system's resources were in his directory.[120]

Unfortunately for Stallman, the hacker camaraderie at MIT began to dissipate in the early 1980s. The lab bought a new time-sharing computer with a software system that was proprietary. "You had to sign a nondisclosure agreement even to get an executable copy," Stallman lamented. "This meant that the first step in using a computer was to promise not to help your neighbor. A cooperating community was forbidden."[121]

Instead of rebelling, many of his colleagues joined for-profit software firms, including a spinoff from the MIT lab called Symbolics, where they made a lot of money by not sharing freely. Stallman, who sometimes slept in his office and looked like he shopped in a thrift store, did not share their money-seeking motivations and regarded them as traitors. The final straw came when Xerox donated a new laser printer and Stallman wanted to institute a software hack so that it would warn users on the network when it jammed. He asked someone to provide the printer's source code, but he refused, saying he had signed a nondisclosure agreement. Stallman was morally outraged.

All of these events turned Stallman into even more of a Jeremiah, railing against idolatry and preaching from a book of lamentations. "Some people do compare me with an Old Testament prophet, and the reason is Old Testament prophets said certain social practices were wrong," he asserted. "They wouldn't compromise on moral issues."[122] Neither would Stallman. Proprietary software was "evil," he said, because "it required people to agree not to share and that made society ugly." The way to resist and defeat the forces of evil, he decided, was to create free software.

So in 1982, repelled by the selfishness that seemed to pervade Reagan-era society as well as software entrepreneurs, Stallman embarked on a mission to create an operating system that was free and completely nonproprietary. In order to prevent MIT from making a claim to any rights to it, he quit his job at the Artificial Intelligence Lab, though he was allowed by his indulgent supervisor to keep his key and continue using the lab's resources. The operating system Stallman decided to develop was one that would be similar to and compatible with UNIX, which had been developed at Bell Labs in 1971 and was the standard for most universities and hackers. With a coder's subtle humor, Stallman created a recursive acronym for his new operating system, GNU, which stood for GNU's Not UNIX.

In the March 1985 issue of *Dr. Dobb's Journal*, a publication that sprang out of the Homebrew Computer Club and *People's Computer Company*, Stallman issued a manifesto: "I consider that the Golden Rule requires that if I like a program I must share it with other people who like it. Software sellers want to divide the users and conquer them, making each user agree not to share with others. I refuse to break solidarity with other users in this way. . . . Once GNU is written, everyone will be able to obtain good system software free, just like air."[123]

Stallman's free software movement was imperfectly named. Its goal was not to insist that all software come free of charge but that it be liberated from any restrictions. "When we call software 'free,' we mean that it respects the users' essential freedoms: the freedom to run it, to study and change it, and to redistribute copies with or without changes," he repeatedly had to explain. "This is a matter of freedom, not price, so think of 'free speech,' not 'free beer.'"

For Stallman, the free software movement was not merely a way to develop peer-produced software; it was a moral imperative for making a good society. The principles that it promoted were, he said, "essential not just for the individual users' sake, but for society as a whole because they promote social solidarity—that is, sharing and cooperation."[124]

To enshrine and certify his creed, Stallman came up with a GNU General Public License and also the concept, suggested by a friend, of "copyleft," which is the flipside of asserting a copyright. The essence of the General Public License, Stallman said, is that it gives "everyone permission to run the program, copy the program, modify the program, and distribute modified versions—but not permission to add restrictions of their own."[125]

Stallman personally wrote the first components for the GNU operating system, including a text editor, a compiler, and many other tools. But it became increasingly clear that one key element was missing. "What about the kernel?" *Byte* magazine asked in a 1986 interview. The central module of an operating system, a kernel manages the requests from software programs and turns them into instructions for the computer's central processing unit. "I'm finishing the compiler before I go to work on the kernel," Stallman answered. "I am also going to have to rewrite the file system."[126]

For a variety of reasons, he found it difficult to complete a kernel for GNU. Then, in 1991, one became available not from Stallman or his Free Software Foundation, but from a most unexpected source: a twenty-one-year-old toothy and boyish Swedish-speaking Finn at the University of Helsinki named Linus Torvalds.

Linus Torvalds's father was a Communist Party member and TV journalist, his mother a student radical and then print journalist, but as a child in Helsinki he became more interested in technology than in politics.[127] He described himself as "good at math, good at physics, and with no social graces whatsoever, and this was before being a nerd was considered a good thing."[128] Especially in Finland.

When Torvalds was eleven, his grandfather, a professor of statistics, gave him a used Commodore Vic 20, one of the first personal computers. Using BASIC, Torvalds began writing his own programs, including one that amused his younger sister by writing "Sara is the best" over and over. "One of the biggest joys," he said, "was learning that computers are like mathematics: You get to make up your own world with its own rules."

Tuning out his father's urgings to learn to play basketball, Torvalds focused instead on learning to write programs in machine language, the numerical instructions executed directly by a computer's central processing unit, exposing him to the joy of being "intimate with a machine." He later felt lucky to have learned assembly language and machine code on a very basic device: "Computers were actually better for kids when they were less sophisticated, when dweebie youngsters like me could tinker under the hood."[129] Like car engines, computers eventually became harder to take apart and put back together.

After enrolling in the University of Helsinki in 1988 and serving his year in the Finnish Army, Torvalds bought an IBM clone with an Intel 386 processor. Unimpressed with its MS-DOS, which Gates and company had produced, he decided that he wanted to install UNIX, which he had learned to like on the university's mainframes. But UNIX cost $5,000 per copy and wasn't configured to run on a home computer. Torvalds set out to remedy that.

He read a book on operating systems by a computer science professor in Amsterdam, Andrew Tanenbaum, who had developed MINIX, a small clone of UNIX for teaching purposes. Deciding that he would replace the MS-DOS with MINIX on his new PC, Torvalds paid the $169 license fee ("I thought it was outrageous"), installed the sixteen floppy disks, and then started to supplement and modify MINIX to suit his tastes.

Torvalds's first addition was a terminal emulation program so that he could dial into the university's mainframe. He wrote the program from scratch in assembly language, "at the bare hardware level," so he didn't need to depend on MINIX. During the late spring of 1991, he hunkered down to code just as the sun reappeared from its winter hibernation. Everyone was emerging into the outdoors, except him. "I was spending most of my time in a bathrobe, huddled over my unattractive new computer, with thick black window shades shielding me from the sunlight."

Once he got a rudimentary terminal emulator working, he wanted to be able to download and upload files, so he built a disk driver and file system driver. "By the time I did this it was clear the project was on its way to becoming an operating system," he recalled. In other words, he was embarking on building a software package that could serve as a kernel for a UNIX-like operating system. "One moment I'm in my threadbare robe hacking away on a terminal emulator with extra functions. The next moment I realize it's accumulating so many functions that it has metamorphosed into a new operating system in the works." He figured out the hundreds of "system calls" that UNIX could do to get the computer to perform basic operations such as Open and Close, Read and Write, and then wrote programs to implement them in his own way. He was still living in his mother's apartment, often fighting with his sister Sara, who had a normal social life, because his modem hogged their phone line. "Nobody could call us," she complained.[130]

Torvalds initially planned to name his new software "Freax," to evoke "free" and "freaks" and "UNIX." But the person who ran the FTP site he was using didn't like the name, so Torvalds resorted to calling it "Linux," which he pronounced, similarly to the way he pronounced his first name, "LEE-nucks."[131] "I never wanted to use that name because I felt, OK, that's a little too egotistical," he said. But he later conceded that there was a part of his ego that enjoyed getting acclaim after so many years of living in the body of a reclusive nerd, and he was glad he went along with the name.[132]

In the early fall of 1991, when the Helsinki sun started disappearing again, Torvalds emerged with the shell of his system, which contained ten thousand lines of code.* Instead of trying to market what he had produced, he decided simply to offer it publicly. He had

*By 2009 the Debian version 5.0 of GNU/Linux had 324 million source lines of code, and one study estimated that it would have cost about $8 billion to develop by conventional means (http://gsyc.es/~frivas/paper.pdf).

recently gone with a friend to hear a lecture by Stallman, who had become an itinerant global preacher for the doctrine of free software. Torvalds didn't actually get religion or embrace the dogma: "It probably didn't make a huge impact on my life at that point. I was interested in the technology, not the politics—I had enough politics at home."[133] But he did see the practical advantages of the open approach. Almost by instinct rather than as a philosophical choice, he felt Linux should be freely shared with the hope that anyone who used it might help improve it.

On October 5, 1991, he posted a cheeky message on the MINIX discussion newsgroup. "Do you pine for the nice days of minix-1.1, when men were men and wrote their own device drivers?" he began. "I'm working on a free version of a minix-lookalike for AT-386 computers. It has finally reached the stage where it's even usable (though may not be depending on what you want), and I am willing to put out the sources for wider distribution."[134]

"It wasn't much of a decision to post it," he recalled. "It was how I was accustomed to exchanging programs." In the computer world, there was (and still is) a strong culture of shareware, in which people voluntarily sent in a few dollars to someone whose program they downloaded. "I was getting emails from people asking me if I would like them to send me thirty bucks or so," Torvalds said. He had racked up $5,000 in student loans and was still paying $50 a month for the installment loan on his computer. But instead of seeking donations he asked for postcards, and they started flooding in from people all over the world who were using Linux. "Sara typically picked up the mail, and she was suddenly impressed that her combative older brother was somehow hearing from new friends so far away," Torvalds recalled. "It was her first tip-off that I was doing anything potentially useful during those many hours when I had the phone line engaged."

Torvalds's decision to eschew payments came from a mix of reasons, as he later explained, including a desire to live up to his family heritage:

> I felt I was following in the footsteps of centuries of scientists and other academics who built their work on the foundations of

> others. . . . I also wanted feedback (okay, and praise). It didn't make sense to charge people who could potentially help improve my work. I suppose I would have approached it differently if I had not been raised in Finland, where anyone exhibiting the slightest sign of greediness is viewed with suspicion, if not envy. And yes, I undoubtedly would have approached the whole no-money thing a lot differently if I had not been brought up under the influence of a diehard academic grandfather and a diehard communist father.

"Greed is never good," Torvalds declared. His approach helped turn him into a folk hero, suitable for veneration at conferences and on magazine covers as the anti-Gates. Charmingly, he was self-aware enough to know that he relished such acclaim and that this made him a little bit more egotistical than his admirers realized. "I've never been the selfless, ego-free, techno-lovechild the hallucinating press insists I am," he admitted.[135]

Torvalds decided to use the GNU General Public License, not because he fully embraced the free-sharing ideology of Stallman (or for that matter his own parents) but because he thought that letting hackers around the world get their hands on the source code would lead to an open collaborative effort that would make it a truly awesome piece of software. "My reasons for putting Linux out there were pretty selfish," he said. "I didn't want the headache of trying to deal with parts of the operating system that I saw as the crap work. I wanted help."[136]

His instinct was right. His release of his Linux kernel led to a tsunami of peer-to-peer volunteer collaboration that became a model of the shared production that propelled digital-age innovation.[137] By the fall of 1992, a year after its release, Linux's newsgroup on the Internet had tens of thousands of users. Selfless collaborators added improvements such as a Windows-like graphical interface and tools to facilitate the networking of computers. Whenever there was a bug, someone somewhere stepped in to fix it. In his book *The Cathedral and the Bazaar*, Eric Raymond, one of the seminal theorists of the open software movement, propounded what he called "Linus's Law": "Given enough eyeballs, all bugs are shallow."[138]

Peer-to-peer sharing and commons-based collaboration were nothing new. An entire field of evolutionary biology has arisen around the question of why humans, and members of some other species, cooperate in what seem to be altruistic ways. The tradition of forming voluntary associations, found in all societies, was especially strong in early America, evidenced in cooperative ventures ranging from quilting bees to barn raisings. "In no country in the world has the principle of association been more successfully used, or more unsparingly applied to a multitude of different objects, than in America," Alexis de Tocqueville wrote.[139] Benjamin Franklin in his *Autobiography* propounded an entire civic creed, with the motto "To pour forth benefits for the common good is divine," to explain his formation of voluntary associations to create a hospital, militia, street-sweeping corps, fire brigade, lending library, night-watch patrol, and many other community endeavors.

The hacker corps that grew up around GNU and Linux showed that emotional incentives, beyond financial rewards, can motivate voluntary collaboration. "Money is not the greatest of motivators," Torvalds said. "Folks do their best work when they are driven by passion. When they are having fun. This is as true for playwrights and sculptors and entrepreneurs as it is for software engineers." There is also, intended or not, some self-interest involved. "Hackers are also motivated, in large part, by the esteem they can gain in the eyes of their peers by making solid contributions. . . . Everybody wants to impress their peers, improve their reputation, elevate their social status. Open source development gives programmers the chance."

Gates's "Letter to Hobbyists," complaining about the unauthorized sharing of Microsoft BASIC, asked in a chiding way, "Who can afford to do professional work for nothing?" Torvalds found that an odd outlook. He and Gates were from two very different cultures, the communist-tinged radical academia of Helsinki versus the corporate elite of Seattle. Gates may have ended up with the bigger house, but Torvalds reaped antiestablishment adulation. "Journalists seemed to love the fact that, while Gates lived in a high-tech lakeside mansion, I was tripping over my daughter's playthings in a three-bedroom ranch house with bad plumbing in boring Santa Clara," he said with ironic self-awareness. "And that I drove a boring Pontiac. And answered my own phone. Who *wouldn't* love me?"

Torvalds was able to master the digital-age art of being an accepted leader of a massive, decentralized, nonhierarchical collaboration, something that Jimmy Wales at Wikipedia was doing at around the same time. The first rule for such a situation is to make decisions like an engineer, based on technical merit rather than personal considerations. "It was a way of getting people to trust me," Torvalds explained. "When people trust you, they take your advice." He also realized that leaders in a voluntary collaborative have to encourage others to follow their passion, not boss them around. "The best and most effective way to lead is by letting people do things because they *want* to do them, not because you want them to." Such a leader knows how to empower groups to self-organize. When it's done right, a governance structure by consensus naturally emerges, as happened both with Linux and Wikipedia. "What astonishes so many people is that the open source model actually works," Torvalds said. "People know who has been active and who they can trust, and it just happens. No voting. No orders. No recounts."[140]

The combination of GNU with Linux represented, at least in concept, the triumph of Richard Stallman's crusade. But moral prophets rarely indulge in victory celebrations. Stallman was a purist. Torvalds wasn't. The Linux kernel he eventually distributed contained some binary blobs with proprietary features. That could be remedied; indeed Stallman's Free Software Foundation created a version that was completely free and nonproprietary. But there was a deeper and more emotional issue for Stallman. He complained that referring to the operating system as "Linux," which almost everybody did, was misleading. Linux was the name of the kernel. The system as a whole should be called GNU/Linux, he insisted, sometimes angrily. One person who was at a software expo recounted how Stallman had reacted when a nervous fourteen-year-old boy asked him about Linux. "You ripped into that boy and tore him a brand new asshole, and I watched as his face fell and his devotion to you and our cause crumpled in a heap," the onlooker later berated Stallman.[141]

Stallman also insisted that the goal should be to create what he called *free software*, a phrase that reflected a moral imperative to share. He objected to the phrase that Torvalds and Eric Raymond began to use, *open-source software*, which emphasized the pragmatic goal of getting people to collaborate in order to create software more effectively. In practice, most free software is also open-source and vice versa; they are usually thrown together under the rubric of *free and open-source software*. But to Stallman it mattered not only how you made your software but also your motivations. Otherwise the movement might be susceptible to compromise and corruption.

The disputes went beyond mere substance and became, in some ways, ideological. Stallman was possessed by a moral clarity and unyielding aura, and he lamented that "anyone encouraging idealism today faces a great obstacle: the prevailing ideology encourages people to dismiss idealism as 'impractical.'"[142] Torvalds, on the contrary, was unabashedly practical, like an engineer. "I led the pragmatists," he said. "I have always thought that idealistic people are interesting, but kind of boring and scary."[143]

Torvalds admitted to "not exactly being a huge fan" of Stallman, explaining, "I don't like single-issue people, nor do I think that people who turn the world into black and white are very nice or ultimately very useful. The fact is, there aren't just two sides to any issue, there's almost always a range of responses, and 'it depends' is almost always the right answer in any big question."[144] He also believed that it should be permissible to make money from open-source software. "Open source is about letting everybody play. Why should business, which fuels so much of society's technological advancement, be excluded?"[145] Software may want to be free, but the people who write it may want to feed their kids and reward their investors.

These disputes should not overshadow the astonishing accomplishment that Stallman and Torvalds and their thousands of collaborators wrought. The combination of GNU and Linux created an operating system that has been ported to more hardware platforms, ranging from the world's ten biggest supercomputers to embedded systems in mobile phones, than any other operating system. "Linux is subversive," wrote Eric Raymond. "Who would have thought that a world-class operating system could coalesce as if by magic out of part-time hacking by several thousand developers scattered all over the planet, connected only by the tenuous strands of the Internet?"[146] Not only did it become a great operating system; it became a model for commons-based peer production in other realms, from Mozilla's Firefox browser to Wikipedia's content.

By the 1990s there were many models for software development. There was the Apple approach, in which the hardware and the operating system software were tightly bundled, as with the Macintosh and iPhone and every iProduct in between. It made for a seamless user experience. There was the Microsoft approach, in which the operating system was unbundled from the hardware. That allowed more user choices. In addition, there were the free and open-source approaches, which allowed the software to be completely unfettered and modifiable by any user. Each model had its advantages, each had its incentives for creativity, and each had its prophets and disciples. But the approach that worked best was having all three models coexisting, along with various combinations of open and closed, bundled and unbundled, proprietary and free. Windows and Mac, UNIX and Linux, iOS and Android: a variety of approaches competed over the decades, spurring each other on—and providing a check against any one model becoming so dominant that it stifled innovation.

Larry Brilliant (1944–) and Stewart Brand on Brand's houseboat in 2010.

William von Meister (1942–1995).

Steve Case (1958–).

CHAPTER TEN

ONLINE

The Internet and the personal computer were both born in the 1970s, but they grew up apart from one another. This was odd, and all the more so when they continued to develop on separate tracks for more than a decade. This was partly because there was a difference in mind-set between those who embraced the joys of networking and those who got giddy at the thought of a personal computer of their very own. Unlike the utopians of the Community Memory project who loved forming virtual communities, many early fans of personal computers wanted to geek out alone on their own machines, at least initially.

There was also a more tangible reason that personal computers arose in a way that was disconnected from the rise of networks. The ARPANET of the 1970s was not open to ordinary folks. In 1981 Lawrence Landweber at the University of Wisconsin pulled together a consortium of universities that were not connected to the ARPANET to create another network based on TCP/IP protocols, which was called CSNET. "Networking was available only to a small fraction of the U.S. computer research community at the time," he said.[1] CSNET became the forerunner of a network funded by the National Science Foundation, NSFNET. But even after these were all woven together into the Internet in the early 1980s, it was hard for an average person using a personal computer at home to get access. You generally had to be affiliated with a university or research institution to jack in.

So for almost fifteen years, beginning in the early 1970s, the growth of the Internet and the boom in home computers proceeded in parallel. They didn't intertwine until the late 1980s, when it became possible for ordinary people at home or in the office to dial up and go online. This would launch a new phase of the Digital Revolution, one that would fulfill the vision of Bush, Licklider, and Engelbart that computers would augment human intelligence by being tools both for personal creativity and for collaborating.

EMAIL AND BULLETIN BOARDS

"The street finds its own uses for things," William Gibson wrote in "Burning Chrome," his 1982 cyberpunk story. Thus it was that the researchers who had access to the ARPANET found their own use for it. It was supposed to be a network for time-sharing computer resources. In that it was a modest failure. Instead, like many technologies, it shot to success by becoming a medium for communications and social networking. One truth about the digital age is that the desire to communicate, connect, collaborate, and form community tends to create killer apps. And in 1972 the ARPANET got its first. It was email.

Electronic mail was already used by researchers who were on the same time-sharing computer. A program called SNDMSG allowed a user of a big central computer to send a message to the personal folder of another user who was sharing the same computer. In late 1971 Ray Tomlinson, an MIT engineer working at BBN, decided to concoct a cool hack that would allow such messages to be sent to folders on other mainframes. He did it by combining SNDMSG with an experimental file transfer program called CPYNET, which could exchange files between distant computers on the ARPANET. Then he came up with something that was even more ingenious: in order to instruct a message to go to the file folder of a user at a different site, he used the @ sign on his keyboard to create the addressing system that we all use now, username@hostname. Thus Tomlinson created not only email but the iconic symbol of the connected world.[2]

The ARPANET allowed researchers at one center to tap into the computing resources somewhere else, but that rarely happened. Instead email became the main method for collaborating. ARPA's director, Stephen Lukasik, became one of the first email addicts, thus causing all researchers who needed to deal with him to follow suit. He commissioned a study in 1973 which found that, less than two years after it was invented, email accounted for 75 percent of the traffic on the ARPANET. "The largest single surprise of the ARPANET program has been the incredible popularity and success of network mail," a BBN report concluded a few years later. It should not have been a surprise. The desire to socially network not only drives innovations, it co-opts them.

Email did more than facilitate the exchange of messages between two computer users. It led to the creation of virtual communities, ones that, as predicted in 1968 by Licklider and Taylor, were "selected more by commonality of interests and goals than by accidents of proximity."

The earliest virtual communities began with email chains that were distributed to large self-selected groups of subscribers. They became known as mailing lists. The first major list, in 1975, was SF-Lovers, for science fiction fans. The ARPA managers initially wanted to shut it down out of fear that some senator might not be amused by the use of military money to support a sci-fi virtual hangout, but the moderators of the group successfully argued that it was a valuable training exercise in juggling large information exchanges.

Soon other methods of forming online communities arose. Some used the backbone of the Internet; others were more jury-rigged. In February 1978 two members of the Chicago Area Computer Hobbyists' Exchange, Ward Christensen and Randy Suess, found themselves snowed in by a huge blizzard. They spent the time developing the first computer Bulletin Board System, which allowed hackers and hobbyists and self-appointed "sysops" (system operators) to set up their own online forums and offer files, pirated software, information, and message posting. Anyone who had a way to get online could join in.

The following year, students at Duke University and the University of North Carolina, which were not yet connected to the Internet, developed another system, hosted on personal computers, which featured threaded message-and-reply discussion forums. It became known as "Usenet," and the categories of postings on it were called "newsgroups." By 1984 there were close to a thousand Usenet terminals at colleges and institutes around the country.

Even with these new bulletin boards and newsgroups, most average PC owners could not easily join virtual communities. Users needed a way to connect, which wasn't easy from home or even most offices. But then, in the early 1980s, an innovation came along, part technological and part legal, that seemed small but had a huge impact.

MODEMS

The little device that finally created a connection between home computers and global networks was called a modem. It could modulate and demodulate (hence the name) an analog signal, like that carried by a telephone circuit, in order to transmit and receive digital information. It thus allowed ordinary people to connect their computers to others online by using phone lines. The online revolution could now begin.

It was slow in coming because AT&T had a near-monopoly over the nation's phone system, even controlling the equipment you could use in your home. You couldn't connect anything to your phone line, or even to your phone, unless Ma Bell leased it to you or approved it. Although AT&T offered some modems in the 1950s, they were clunky and costly and designed mainly for industrial or military use, rather than being conducive to homebrew hobbyists creating virtual communities.

Then came the *Hush-A-Phone* case. It involved a simple plastic mouthpiece that could be snapped onto a phone to amplify your voice while making it harder for those nearby to overhear you. It had been around for twenty years, causing no harm, but then an AT&T lawyer spotted one in a shopwindow, and the company decided to sue on the absurd ground that any external device, including a little plastic cone, could damage its network. It showed how far the company would go to protect its monopoly.

Fortunately, AT&T's effort backfired. A federal appeals court dismissed the company's claim, and the barriers to jacking into its network began to crumble. It was still illegal to connect a modem into the phone system electronically, but you could do so mechanically, such as by taking your phone's handset and cradling it into the suction cups of an acoustical coupler. By the early 1970s there were a few modems of this type, including the Pennywhistle, designed for the hobbyist crowd by Lee Felsenstein, that could send and receive digital signals at three hundred bits per second.*

The next step came when, a headstrong Texas cowboy won, after a twelve-year legal battle he financed by selling off his cattle, the right for his customers to use a radio-enabled extension phone he had invented. It took a few years for all of the regulations to be worked out, but by 1975 the Federal Communications Commission opened the way for consumers to attach electronic devices to the network.

The rules were stringent, due to AT&T lobbying, so electronic modems were initially expensive. But in 1981 the Hayes Smartmodem came on the market. It could be plugged directly into a phone line and connected to a computer, with no need for a clunky acoustic coupler. Pioneering hobbyists and cyberpunks, along with ordinary home computer users, could type in the phone number of an online service provider, hold their breath while they waited for the staticky screech that indicated a data connection had been made, and then tap into the virtual communities that formed around bulletin boards, newsgroups, mailing lists, and other online hangouts.

THE WELL

In almost every decade of the Digital Revolution, the amused and amusing Stewart Brand found a way to stand at the locus where technology overlapped with community and the counterculture. He had

---

*An Ethernet or WiFi today can transmit data at a billion bps, which is more than 3 million times faster.

produced the techno-psychedelic show at Ken Kesey's Trips Festival, reported on *Spacewar* and Xerox PARC for *Rolling Stone*, aided and abetted Doug Engelbart's Mother of All Demos, and founded the *Whole Earth Catalog*. So in the fall of 1984, just as modems were becoming easily available and personal computers were becoming user-friendly, it was not surprising that Brand helped to conjure up the idea for the prototypic online community, The WELL.

It began when Brand was visited by another of the playfully earnest and creative denizens of the idealistic techno-counterculture, Larry Brilliant. A physician and epidemiologist, Brilliant had a compulsion to change the world and have fun while doing so. He had served as the doctor for an American Indian occupation of Alcatraz, sought enlightenment at a Himalayan ashram with the famed guru Neem Karoli Baba (where he first crossed paths with Steve Jobs), enlisted in the World Health Organization's campaign to eliminate smallpox, and with support from Jobs and the counterculture luminaries Ram Dass and Wavy Gravy founded the Seva Foundation, which focused on curing blindness in poor communities around the world.

When one of the helicopters used by the Seva Foundation in Nepal had mechanical problems, Brilliant used a computer conferencing system and an Apple II that Jobs had donated to organize online a repair mission. The potential power of online discussion groups impressed him. When he went to teach at the University of Michigan, he helped to build a company around a computer conferencing system that had been created on the university's network. Known as PicoSpan, it allowed users to post comments on different topics and strung them into threads for all to read. Brilliant's idealism, techno-utopianism, and entrepreneurialism flowed together. He used the conferencing system to bring medical expertise to Asian hamlets and organize missions when something went wrong.

When Brilliant went to a conference in San Diego, he called his old friend Stewart Brand for lunch. They met at a beachside restaurant near where Brand planned to spend the day skinny-dipping. Brilliant had two interwoven goals: to popularize the PicoSpan conferencing software and to create an online intellectual commune.

He pitched Brand on a partnership in which Brilliant would put up $200,000 in capital, buy a computer, and provide the software. "Stewart would then manage the system and extend it throughout his network of smart, interesting people," Brilliant explained.[3] "My idea was to use this new technology as a way of discussing everything in the *Whole Earth Catalog*. There can be a social network around Swiss Army knives or solar stoves or anything."[4]

Brand turned the idea into something grander: creating the world's most stimulating online community where people could discuss anything they wanted. "Let's just have a conversation and get the smartest people in the world," he suggested, "and let them figure out whatever they want to talk about."[5] Brand came up with a name, The WELL, and reverse-engineered an acronym for it: the Whole Earth 'Lectronic Link. A playful apostrophe, he later said, was "always worth having in a name."[6]

Brand championed a concept, abandoned by many later virtual communities, that was critical to making The WELL a seminal service. The participants could not be totally anonymous; they could use a handle or pseudonym, but they had to provide their real name when they joined, and other members could know who they were. Brand's credo, which popped up on the opening screen, was "You own your own words." You were accountable for what you posted.

Like the Internet itself, The WELL became a system designed by its users. By 1987 the topics of its online forums, known as conferences, ranged from the Grateful Dead (the most popular) to UNIX programming, from art to parenting, aliens to software design. There was minimal hierarchy or control, so it evolved in a collaborative way. That made it both an addictive experience and a fascinating social experiment. Whole books were written about it, including ones by the influential tech chroniclers Howard Rheingold and Katie Hafner. "Just being on The Well, talking with people you might not consider befriending in any other context, was its own seduction," Hafner wrote.[7] In his book Rheingold explained, "It's like having the corner bar, complete with old buddies and delightful newcomers and new tools waiting to take home and fresh graffiti and letters, except instead of putting on my coat, shutting down the computer, and walking down to the corner, I just invoke my telecom program and there they are."[8] When Rheingold discovered that his two-year-old daughter had a tick in her scalp, he found out how to treat it from a doctor on The WELL before his own physician had called him back.

Online conversations could be intense. A discussion leader named Tom Mandel, who became a central character in Hafner's book and also helped me and my colleagues at *Time* manage our online forums, regularly engaged in fiery exchanges, known as flame wars, with other members. "I expressed opinions about everything," he recalled. "I even started an altercation that dragged half of West Coast cyberspace into an electronic brawl and got myself banished from the WELL."[9] But when he revealed he was dying of cancer, they rallied around him emotionally. "I'm sad, terribly sad, I cannot tell you how sad and grief stricken I am that I cannot stay to play and argue with you much longer," he wrote in one of his last posts.[10]

The WELL was a model of the type of intimate, thoughtful community that the Internet used to feature. It still remains, after three decades, a tight-knit community, but it was long ago overtaken in popularity by more commercial online services and then by less communal discussion venues. The widespread retreat into anonymity online has undermined Brand's creed that people should be accountable for what they say, thus making many online comments less thoughtful and discussions less intimate. As the Internet goes through different cycles—it has been a platform for time-sharing, community, publishing, blogging, and social networking—there may come a time when the natural yearning that humans have for forging trusted communities, akin to corner bars, will reassert itself, and The WELL or startups that replicate its spirit will become the next hot innovation. Sometimes innovation involves recovering what has been lost.

AMERICA ONLINE

William von Meister was an early example of the new frontiersmen who would drive digital innovation beginning in the late 1970s. Like Ed Roberts of Altair, von Meister was a supercharged serial entrepreneur. Fueled by the proliferation of venture capitalists, this breed of innovators threw off ideas like sparks, got an adrenaline rush from risk taking, and touted new technologies with the zeal of evangelists. Von Meister was both an exemplar and a caricature. Unlike Noyce and Gates and Jobs, he did not set out to build companies but instead to launch them and see where they landed. Rather than being afraid of failure, he was energized by it, and his ilk made forgiving failure a feature of the Internet age. A magnificent rogue, he started nine companies in ten years, most of which either crashed or ejected him. But through his serial failures, he helped to define the archetype of the Internet entrepreneur and, in the process, invent the online business.[11]

Von Meister's mother was an Austrian countess and his father, a godson of Kaiser Wilhelm II, ran the U.S. division of the German zeppelin company that operated the *Hindenburg* until its 1937 explosion, and then ran a division of a chemical company until he was indicted for fraud. His style rubbed off on young Bill, born in 1942, who seemed hell-bent on matching his father's flameouts in flamboyance if not in severity. Growing up in a whitewashed brick mansion known as Blue Chimneys on a twenty-eight-acre estate in New Jersey, he loved escaping to the attic to operate his ham radio and build electronic gadgets. Among the devices he made was a radio transmitter that his father kept in his car and used to signal when he was nearing home from work so that the household staff could prepare his tea.

After a desultory academic career that consisted of dropping into and out of colleges in Washington, DC, von Meister joined Western Union. He made money with a bunch of side ventures, including salvaging some of the company's discarded equipment, and then launched a service that allowed people to dictate important letters to call centers for overnight delivery. It was successful, but in what became a pattern, von Meister was forced out for spending wildly and not paying any attention to operations.*

Von Meister was one of the original breed of media entrepreneurs—think Ted Turner rather than Mark Zuckerberg—who lived larger than

*Western Union later bought the business and turned it into its Mailgram service.

life and mixed craziness with shrewdness so thoroughly that they became almost indistinguishable. He had a taste for flashy women and fine red wine, race cars and private planes, single-malt Scotch and contraband cigars. "Bill von Meister was not just a serial entrepreneur, he was a pathological entrepreneur," according to Michael Schrage, who covered him for the *Washington Post*. "Bill von Meister's ideas, on average, when you look back at them, don't seem stupid. But at the time they seemed outlandish. The big risk was he was such a loon that his looniness would get confused with the idea, because they're so intertwined."[12]

Von Meister continued to prove adept at coming up with new notions and raising money from venture capitalists, though not at running anything. Among his startups: a bulk telephone routing service for businesses, a restaurant in suburban Washington called McLean Lunch and Radiator that allowed customers to make free long-distance calls from phones at their table, and a service called Infocast that sent information to computers by piggybacking digital data on FM radio signals. Then in 1978, when he had become bored or unwelcome at these ventures, he combined his interests in phones, computers, and information networks to create a service that he called The Source.

The Source linked home computers via telephone lines into a network that offered bulletin boards, message exchanges, news stories, horoscopes, restaurant guides, wine rankings, shopping, weather, airline schedules, and stock quotes. In other words, it was one of the first consumer-oriented online services. (The other was CompuServe, a business-oriented time-sharing network that in 1979 was just venturing into the consumer dial-up market.) "It can take your personal computer anywhere in the world," proclaimed an early marketing brochure. Von Meister told the *Washington Post* that it would become a "utility" that would provide information "like water comes out of a faucet." In addition to piping information into the home, The Source focused on creating community: forums and chat rooms and private file-sharing areas where users could post their own writings for others to download. At the official launch of the service in July 1979 at

Manhattan's Plaza Hotel, the sci-fi writer and pitchman Isaac Asimov proclaimed, "This is the beginning of the Information Age!"[13]

As usual, von Meister was soon mismanaging the company and squandering money, causing him to be ousted after a year by his principal funder, who said, "Billy von Meister is a terrific entrepreneur but he didn't know how to stop entrepreneuring." The Source was eventually sold to *Reader's Digest*, which later sold it to CompuServe. But despite being short-lived, it pioneered the online era by showing that consumers wanted not just information piped to them but also the chance to connect with friends and generate their own content to be shared.

Von Meister's next idea, also slightly ahead of its time, was a home music store that would sell streaming music through cable TV networks. Record stores and recording companies ganged up to block his access to songs, so the idea-a-minute von Meister switched his focus to video games. It was an even riper target; at the time, there were 14 million Atari home game consoles. Thus was born Control Video Corporation (CVC). Von Meister's new service allowed users to download games for purchase or rent. He dubbed the service GameLine, and he began to bundle with it some of the information services that had been in The Source. "We're going to turn the videogame jockey into an information junkie," he proclaimed.[14]

GameLine and CVC set up shop in a strip mall on the way to Washington's Dulles Airport. Von Meister selected a board of directors that symbolized the official passing of the torch to a new breed of Internet pioneers. Among its members were Larry Roberts and Len Kleinrock, architects of the original ARPANET. Another was the pathbreaking venture capitalist Frank Caufield of what had become Silicon Valley's most influential financial firm, Kleiner Perkins Caufield & Byers. Representing the investment bank Hambrecht & Quist was Dan Case, a smooth and energetic young Rhodes Scholar from Hawaii and Princeton.

Dan Case joined von Meister in Las Vegas for the January 1983 Consumer Electronics Show, where CVC's GameLine was hoping to make a splash. Von Meister, ever the showman, paid for a hot air balloon shaped like a joystick and emblazoned with the name GameLine to float above the town, and he rented a sprawling suite at the Tropicana Hotel, which he festooned with hired showgirls.[15] Case relished the scene. Hovering in the corner was his younger brother, Steve, who was more reticent and, with his enigmatic smile and equable face, harder to read.

Born in 1958 and raised in Hawaii, with a placid temperament that made it seem as if he had been nurtured by dolphins, Steve Case had a pacific façade. Called by some "the Wall" because his face rarely flickered with emotion, he was shy but not insecure. To some people who didn't really know him, that made him appear aloof or arrogant, which he wasn't. As he grew up, he taught himself to joke and trade friendly insults in a flat and nasal tone, like a newbie at a fraternity. But beneath the banter he was deeply thoughtful and earnest.

In high school, Dan and Steve turned their bedrooms into offices from which they ran a series of businesses that, among other things, sold greeting cards and distributed magazines. "The first lesson of Case entrepreneurship," Steve recalled, "was I came up with the idea and he provided the funding and then owned half the company."[16]

Steve went to Williams College, where the famed historian James MacGregor Burns drily noted, "He was among my median students."[17] He spent more time thinking about starting businesses than studying for class. "I remember a professor pulling me aside and suggesting I should defer my business interests and focus on my studies as college represented a once-in-a-lifetime opportunity," Case recalled. "Needless to say, I disagreed." He took only one computer class and hated it "because this was the punch-card era and you'd write a program and then have to wait hours to get the results."[18] The lesson he learned was that computers needed to be made more accessible and interactive.

One aspect of computers he liked was the notion of using them to tap into networks. "The far-away connections seemed magical," he told the journalist Kara Swisher. "It struck me as the most completely obvious use for them, and the rest was just for computer wonks."[19] After reading *The Third Wave* by the futurist Alvin Toffler, he became riveted by the concept of "the electronic frontier," in which technology would connect people to each other and all of the world's information.[20]

In early 1980 he applied for a job at the J. Walter Thompson advertising agency. "I firmly believe that technological advances in communications are on the verge of significantly altering our way of life," he wrote in his application letter. "Innovations in telecommunications (especially two-way cable systems) will result in our television sets (big-screen, of course!) becoming an information line, newspaper, school, computer, referendum machine, and catalog."[21] He didn't get the job, and he was also initially turned down by Proctor & Gamble. But he talked his way into having a second interview at P&G, traveling to Cincinnati at his own expense, and ended up as a junior brand manager in a group that handled a soon-defunct hair conditioner towelette called Abound. There Case learned the trick of giving away free samples in order to launch a new product. "That was in part the inspiration behind AOL's free disk trial strategy a decade later," he said.[22] After two years he left to work at PepsiCo's Pizza Hut division. "The reason I did that was because it was highly entrepreneurial. It was a company run by the franchisees, almost the opposite of Procter & Gamble, which is more of a top-down, process-oriented company where all the key decisions were made in Cincinnati."[23]

As a young bachelor based in Wichita, Kansas, where there was not a whole lot to do in the evenings, he became a fan of The Source. It was a perfect refuge for someone with his mix of shyness and desire for connection. He learned two lessons: that people like to be part of communities and that technology needs to be simple if it is going to appeal to the masses. When he first tried to log on to The Source, he had trouble getting his Kaypro portable computer configured. "It was like climbing Mount Everest, and my first thought was to figure out why it had to be so hard," he recalled. "But when I finally logged in and found myself linked to all over the country from this sorry little apartment in Wichita, it was exhilarating."[24]

On the side, Case formed his own small marketing company. He was at heart an entrepreneur in an era when most other college kids sought jobs at big companies. He rented a maildrop with an upscale address in San Francisco, got it printed on stationery, and had his business correspondence forwarded to his little apartment in Wichita. His passion was to help companies that wanted to pioneer the electronic frontier, so when his brother Dan joined Hambrecht & Quist in 1981, he began sending Steve business plans for interesting companies. One of them was for von Meister's Control Video Corporation. During a Colorado ski vacation in December 1982, they discussed whether Dan should invest, and they also decided to go together to the Consumer Electronics Show in Las Vegas the following month.[25]

The irrepressible von Meister and the repressible Steve spent a long dinner in Las Vegas talking about ways to market GameLine. Perhaps because they had shared interests but different personalities, they hit it off. During a drunken conversation in the bathroom halfway through the dinner, von Meister asked Dan whether it would be all right for him to hire young Steve. Dan allowed that it would be just fine. Steve started at CVC as a part-time consultant, then was hired full-time in September 1983 and moved to Washington, DC. "I thought the GameLine idea had real promise," Case said. "But I also felt that even if it failed, the lessons I'd learn by working alongside Bill would be a valuable education. And that certainly proved to be accurate."[26]

Within a few months, CVC was on the brink of bankruptcy. Von Meister had still not learned to be a prudent manager, and the Atari game market had deflated. When told the sales numbers at a board meeting that year, the venture capitalist Frank Caufield responded, "You would've thought they would've shoplifted more than that." So Caufield insisted that a disciplined manager be brought in. The person he tapped was a close friend and classmate from West Point, Jim Kimsey, who had the gruff exterior of a Special Forces commando cloaking the personable heart of a bartender.

Kimsey was not the obvious person to whip into shape an interactive digital service; he was far more familiar with guns and whiskey glasses than keyboards. But he had the mix of tenacity and rebelliousness that makes for a good entrepreneur. Born in 1939, he grew up in Washington, DC, and in his senior year was kicked out of the town's top Catholic school, Gonzaga High, for being disruptive. Nevertheless, he was eventually able to wrangle an appointment to West Point, where he was suited to an atmosphere that celebrated, channeled, and controlled aggression. Upon graduation, he was deployed to the Dominican Republic, then served two tours in Vietnam in the late 1960s. While there as a major with the Airborne Rangers, he took charge of building an orphanage for a hundred Vietnamese kids. Had it not been for his tendency to mouth off to those higher in the chain of command, he may have made the military a career.[27]

Instead he went back to Washington in 1970, bought an office building downtown, rented out much of it to brokerage firms, and on the ground floor opened a bar called The Exchange that had a working ticker-tape machine. He soon opened other popular singles bars, with names like Madhatter and Bullfeathers, while embarking on additional real estate ventures. Part of his routine was going on adventure trips with his West Point pal Frank Caufield and their sons. It was on a 1983 rafting trip that Caufield recruited him to CVC as a minder for von Meister and, eventually, as CEO.

Faced with sluggish sales, Kimsey fired most of the staff except for Steve Case, whom he promoted to vice president of marketing. Kimsey had a colorful saloonkeeper's way with words, especially scatological ones. "My job is to make chicken salad out of chicken shit," he declared. And he was fond of the old joke about a young boy who merrily digs through a pile of horse manure and, when asked why, declares, "There must be a pony somewhere in this shit."

It was an odd triumvirate: the undisciplined idea generator von Meister, the coolly strategic Case, and the rough-edged commando Kimsey. While von Meister played showman and Kimsey played backslapping barkeep, Case hovered in the corner observing and coming up with new ideas. Together they showed once again how a diverse team can promote innovation. Ken Novack, an outside counsel, later observed, "It was no accident that they created this business together."[28]

Case and von Meister had long been interested in building computer networks that could connect ordinary users. When CBS, Sears, and IBM joined forces in 1984 to launch such a service that became known as Prodigy, other computer makers realized that there might be a real market. Commodore came to CVC and asked it to create an online service. So Kimsey reconfigured CVC into a company called Quantum, which launched a service named Q-Link for Commodore users in November 1985.

For $10 a month, Q-Link had everything that von Meister—who was then being eased out of the company—and Case had envisioned: news, games, weather, horoscopes, reviews, stocks, soap opera updates, a shopping mall, and more, along with the regular crashes and downtime that became endemic in the online world. But most important, Q-Link had an area filled with active bulletin boards and live chat rooms, dubbed People Connection, which enabled members to form communities.

Within two months, by the beginning of 1986, Q-Link had ten thousand members. But growth began to taper off, largely because Commodore's computer sales were slumping in the face of new competition from Apple and others. "We have to take control of our destiny," Kimsey told Case.[29] It was clear that for Quantum to succeed, it had to create its Link online services for other computer makers, most notably Apple.

With the tenacity that came with his patient personality, Case targeted the executives at Apple. Even after its brilliantly controlling cofounder Steve Jobs had been forced out of the company, at least for the time being, Apple was difficult to partner with. So Case moved across the country to Cupertino and took an apartment near Apple's headquarters. From there he waged his siege. There were many possible units within Apple he could try to conquer, and he was eventually able to get a little desk inside the company. Despite his reputation for being aloof, he had a whimsical sense of humor; on his desk, he put up a sign that said "Steve Held Hostage"* along with the number of days he had been there.[30] In 1987, after three months of daily campaigning, he was successful: Apple's customer service department agreed to strike a deal with Quantum for a service called AppleLink.

---

*A reference to the phrase used during the 1980 drama in which Americans were held hostage in Iran.

When it launched a year later, the first live chat forum featured Apple's lovable cofounder Steve Wozniak.

Case went on to make a similar deal with Tandy to launch PC-Link. But he soon realized that his strategy of creating separate private-label services for different computer makers needed to be revised. Users of one service could not connect with those on another. In addition, the computer makers were controlling Quantum's products, marketing, and future. "Look, we can no longer rely on these partnerships," Case told his team. "We really need to stand on our own two feet and kind of have our own brand."[31]

This became a more urgent problem—but also an opportunity—when relations with Apple frayed. "The powers that be at Apple decided they were uncomfortable with a third-party company using the Apple brand name," Case said. "Apple's decision to pull the rug out on us led to the need to rebrand."[32] Case and Kimsey decided to combine the users of all three of their services into one integrated online service with a brand name all its own. The software approach pioneered by Bill Gates would apply to the online realm as well: online services would be unbundled from the hardware and would work on all computer platforms.

Now they needed to come up with a name. There were many suggestions, such as Crossroads and Quantum 2000, but they all sounded like religious retreats or mutual funds. Case came up with America Online, which caused many of his colleagues to gag. It was hokey and awkwardly patriotic. But Case liked it. He knew, just as Jobs had when he named his company Apple, that it was important to be, as he later said, "simple, unintimidating, and even a bit sappy."[33] With no marketing dollars, Case needed a name that clearly described what the service did. And the name America Online accomplished that.

AOL, as it became known, was like going online with training wheels. It was unintimidating and easy to use. Case applied the two lessons he had learned at Proctor & Gamble: make a product simple and launch it with free samples. America was carpet-bombed with software disks offering two months of free service. A voice-over actor named Elwood Edwards, who was the husband of an early employee of AOL, recorded perky greetings—"Welcome!" and "You've got mail!"—that made the service seem friendly. So America went online.

As Case understood, the secret sauce was not games or published content; it was a yearning for connection. "Our big bet, even back in 1985, was what we called community," he recounted. "Now people refer to it as social media. We thought the killer app of the Internet was going to be people. People interacting with people they already knew in new ways that were more convenient, but also people interacting with people they didn't yet know, but should know because they had some kind of shared interest."[34] Among AOL's primary offerings were chat rooms, instant messaging, buddy lists, and text messaging. As on The Source, there was news, sports, weather, and horoscopes. But social networking was the focus. "Everything else—commerce and entertainment and financial services—was secondary," Case said. "We thought community trumped content."[35]

Particularly popular were the chat rooms, where people with similar interests—computers, sex, soap operas—could gather. They could even go off into "private rooms" to talk by mutual consent or, at the other extreme, visit one of the "auditoriums" that might feature a session with a celebrity. AOL's users were not called customers or subscribers; they were *members*. AOL thrived because it helped to create a social network. CompuServe and Prodigy, which began primarily as information and shopping services, did the same with tools such as CompuServe's CB Simulator, which replicated in text the wacky pleasure of talking on a citizens-band radio.

Kimsey the bar owner could never quite get why healthy people would spend their Saturday nights in chat rooms and on bulletin boards. "Admit it, don't you think it's all horseshit?" he would ask Case half jokingly.[36] Case would shake his head. He knew that there was a pony in it.

AL GORE AND THE ETERNAL SEPTEMBER

Online services such as AOL developed independently of the Internet. An entanglement of laws, regulations, traditions, and practices made it impossible for commercial companies to offer direct Internet

access to ordinary folks who were not connected to an educational or research institution. "It now seems really silly, but up until 1992, it was illegal to connect a commercial service like AOL to the Internet," Steve Case said.[37]

But beginning in 1993, the barrier was lowered and the Internet was made accessible to everyone. This disrupted the online services, which until then had been walled gardens where members were coddled in a controlled environment. It also transformed the Internet by producing a flood of new users. But most important was that it began to connect the strands of the Digital Revolution in the way that Bush, Licklider, and Engelbart had envisioned. Computers and communications networks and repositories of digital information were woven together and put at the fingertips of every individual.

It began in earnest when AOL, following the lead of a smaller competitor named Delphi, opened a portal in September 1993 to allow its members access to the newsgroups and bulletin boards of the Internet. In Internet lore, the deluge was called, especially by contemptuous veteran netizens, the Eternal September. The name referred to the fact that every September a new wave of freshmen would enter universities and, from their campus networks, get access to the Internet. Their postings tended to be annoying at first, but within weeks most had acquired enough netiquette to assimilate into the Internet culture. The opened floodgates of 1993, however, produced a never-ending flow of newbies, overwhelming the social norms and clubbiness of the net. "September 1993 will go down in net.history as the September that never ended," an Internet hand named Dave Fischer posted in January 1994.[38] A newsgroup sprang up named alt.aol-sucks, where old-timers posted their diatribes. The AOL interlopers, read one, "couldn't get a clue if they stood in a clue field in clue mating season, dressed as a clue, and drenched with clue pheromones."[39] In fact, the Eternal September's democratization of the Internet was a good thing, but it took a while for veterans to appreciate this.

This opening up of the Internet, which paved the way for an astonishing era of innovation, did not happen by chance. It was the result of

government policies, carefully crafted in a thoughtful and bipartisan atmosphere, that assured America's lead in building an information-age economy. The most influential person in this process, which may come as a surprise to those who know of his role only as a punch line to jokes, was Senator Al Gore Jr. of Tennessee.

Gore's father was also a senator. "I remember driving with my dad from Carthage to Nashville listening to him say how we needed better than these two-lane roads," the younger Gore recalled. "They won't handle our needs."[40] Gore Sr. helped craft the bipartisan legislation for the interstate highway program, and his son took that as an inspiration for helping to promote what he dubbed the "Information Superhighway."

In 1986 Gore launched a congressional study that looked at a variety of topics, including creating supercomputer centers, interconnecting the various research networks, increasing their bandwidth, and opening them up to more users. It was chaired by the ARPANET pioneer Len Kleinrock. Gore followed up with detailed hearings that led to the High Performance Computing Act of 1991, known as the Gore Act, and the Scientific and Advanced Technology Act of 1992. These allowed commercial networks, such as AOL, to connect with the research network run by the National Science Foundation, and hence to the Internet itself. After he was elected vice president in 1992, Gore pushed the National Information Infrastructure Act of 1993, which made the Internet widely available to the general public and moved it into the commercial sphere so that its growth could be funded by private as well as government investment.

When I told people I was writing a book about the people who helped invent computers and the Internet, the most predictable quip I got, especially from those who knew little about Internet history, was "Oh, you mean Al Gore?" Then they would laugh. It's a mark of our political discourse that one of the significant nonpartisan achievements on behalf of American innovation got turned into a punch line because of something that Gore never quite said—that he "invented" the Internet. When he was asked by CNN's Wolf Blitzer in March 1999 to list his qualifications to be a candidate for president, he cited, among other things, "During my service in the United States

Congress, I took the initiative in creating the Internet."[41] It was inelegantly phrased, as answers on cable news shows often are, but he never used the word *invented*.

Vint Cerf and Bob Kahn, two of the people who did in fact invent the Internet's protocols, spoke up on Gore's behalf. "No one in public life has been more intellectually engaged in helping to create the climate for a thriving Internet than the Vice President," they wrote.[42] Even Republican Newt Gingrich defended him, observing, "It's something Gore had worked on a long time. . . . Gore is not the Father of the Internet, but in all fairness, Gore is the person who, in the Congress, most systematically worked to make sure that we got to an Internet."[43]

The takedown of Gore was the harbinger of a new era of rising partisanship accompanied by a lack of faith in what government could do. That's why it's useful to reflect on what led to the Eternal September of 1993. Over the course of more than three decades, the federal government, working with private industry and research universities, had designed and built a massive infrastructure project, like the interstate highway system but vastly more complex, and then threw it open to ordinary citizens and commercial enterprises. It was funded primarily by public dollars, but it paid off thousands of times over by seeding a new economy and an era of economic growth.

Tim Berners-Lee (1955–).

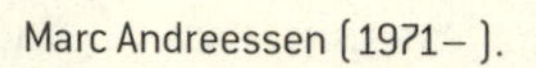

Marc Andreessen (1971–).

Justin Hall (1974–) and Howard Rheingold (1947–) in 1995.

CHAPTER ELEVEN

THE WEB

There was a limit to how popular the Internet could be, at least among ordinary computer users, even after the advent of modems and the rise of online services made it possible for almost anyone to get connected. It was a murky jungle with no maps, filled with clusters of weird foliage with names like alt.config and Wide Area Information Servers that could intimidate all but the most intrepid pathfinder.

But just when the online services began opening up to the Internet in the early 1990s, a new method of posting and finding content miraculously appeared, as if it had burst into life from an underground atom smasher, which in fact was close to what happened. It made the carefully packaged online services obsolete, and it fulfilled—indeed far surpassed—the utopian dreams of Bush, Licklider, and Engelbart. More than most innovations of the digital age it was invented primarily by one man, who gave it a name that managed to be, as he was personally, both expansive and simple: the World Wide Web.

TIM BERNERS-LEE

As a kid growing up on the edge of London in the 1960s, Tim Berners-Lee came to a fundamental insight about computers: they were very good at crunching step by step through programs, but they were not very good at making random associations and clever links, the way that an imaginative human could.

This is not something that most kids ponder, but both of Berners-Lee's parents were computer scientists. They worked as programmers on the Ferranti Mark I, the commercial version of the Manchester University stored-program computer. One evening at home his father, who had been asked by his boss to draft a speech on how to make computers more intuitive, talked about some books on the human brain that he was reading. His son recalled, "The idea stayed with me that computers could become much more powerful if they could be programmed to link otherwise unconnected information."[1] They also talked about Alan Turing's concept of a universal machine. "It made me realize that the limitations on what you could do with a computer were just the limitations of your imagination."[2]

Berners-Lee was born in 1955, the same year as Bill Gates and Steve Jobs, and he considered it a lucky time to be interested in electronics. Kids of that era found it easy to get hold of basic equipment and components that they could play with. "Things came along at the right time," he explained. "Anytime we understood one technology, then industry produced something more powerful that we could afford with our pocket money."[3]

In primary school, Berners-Lee and a friend hung around hobby shops, where they used their allowance to buy electromagnets and make their own relays and switches. "You'd have an electromagnet banged into a bit of wood," he recalled. "When you turned it on, it would attract a bit of tin and that would complete a circuit." From that they developed a deep understanding of what a bit was, how it could be stored, and the things that could be done with a circuit. Just when they were outgrowing simple electromagnetic switches, transistors became common enough that he and his friends could buy a bag of a hundred pretty cheaply. "We learned how to test transistors and use them to replace the relays we had built."[4] In doing so, he could visualize clearly what each component was doing by comparing them to the old electromagnetic switches they superseded. He used them to make audio sounds for his train set and to create circuits that controlled when the train should slow down.

"We began to imagine quite complicated logical circuits, but those became impractical because you would have to use too many transistors," he said. But just as he ran into that problem, microchips became available at the local electronics store. "You buy these little bags of microchips with your pocket money and you'd realize you could make the core of a computer."[5] Not only that, but you could *understand* the core of the computer because you had progressed from simple switches to transistors to microchips and knew how each worked.

One summer just before he went off to Oxford, Berners-Lee had a job in a lumber yard. When he was dumping a pile of sawdust into a Dumpster, he spied an old calculator, partly mechanical and partly electronic, with rows of buttons. He salvaged it, wired it up with some of his switches and transistors, and soon had it working as a rudimentary computer. At a repair shop he bought a broken television set and used the monitor to serve as a display, after figuring out how the circuit of vacuum tubes worked.[6]

During his Oxford years, microprocessors became available. So, just as Wozniak and Jobs had done, he and his friends designed boards that they tried to sell. They were not as successful as the Steves, partly because, as Berners-Lee later said, "we didn't have the same ripe community and cultural mix around us like there was at the Homebrew and in Silicon Valley."[7] Innovation emerges in places with the right primordial soup, which was true of the Bay Area but not of Oxfordshire in the 1970s.

His step-by-step hands-on education, starting with electromagnetic switches and progressing to microprocessors, gave him a deep understanding of electronics. "Once you've made something with wire and nails, when someone says a chip or circuit has a relay you feel confident using it because you know you could make one," he said. "Now kids get a MacBook and regard it as an appliance. They treat it like a refrigerator and expect it to be filled with good things, but they don't know how it works. They don't fully understand what I knew, and my parents knew, which was what you could do with a computer was limited only by your imagination."[8]

There was a second childhood memory that lingered: that of a Victorian-era almanac and advice book in his family home with the magical and musty title *Enquire Within Upon Everything*. The introduction proclaimed, "Whether You Wish to Model a Flower in Wax; to Study the Rules of Etiquette; to Serve a Relish for Breakfast or Supper; to Plan a Dinner for a Large Party or a Small One; to Cure a Headache; to Make a Will; to Get Married; to Bury a Relative; Whatever You May Wish to Do, Make, or to Enjoy, Provided Your Desire has Relation to the Necessities of Domestic Life, I Hope You will not Fail to 'Enquire Within.'"[9] It was, in some ways, the *Whole Earth Catalog* of the nineteenth century, and it was filled with random information and connections, all well indexed. "Enquirers are referred to the index at the end," the title page instructed. By 1894 it had gone through eighty-nine editions and sold 1,188,000 copies. "The book served as a portal to a world of information, everything from how to remove clothing stains to tips on investing money," Berners-Lee observed. "Not a perfect analogy for the Web, but a primitive starting point."[10]

Another concept that Berners-Lee had been chewing on since childhood was how the human brain makes random associations—the smell of coffee conjures up the dress a friend wore when you last had coffee with her—whereas a machine can make only the associations that it has been programmed to make. He was also interested in how people work together. "You got half the solution in your brain, and I got half in my brain," he explained. "If we are sitting around a table, I'll start a sentence and you might help finish it, and that's the way we all brainstorm. Scribble stuff on whiteboard, and we edit each other's stuff. How can we do that when we are separated?"[11]

All of these elements, from *Enquire Within* to the brain's ability to make random associations and to collaborate with others, were jangling around in Berners-Lee's head when he graduated from Oxford. Later he would realize a truth about innovation: New ideas occur when a lot of random notions churn together until they coalesce. He described the process this way: "Half-formed ideas, they float around. They come from different places, and the mind has got this wonderful way of somehow just shoveling them around until one day they fit. They may fit not so well, and then we go for a bike ride or something, and it's better."[12]

For Berners-Lee, his own innovative concepts began to coalesce when he took a consulting job at CERN, the mammoth supercollider and particle physics lab near Geneva. He needed a way to catalogue the connections among the ten thousand or so researchers, their projects, and their computer systems. Both the computers and the people spoke many different languages and tended to make ad hoc links to one another. Berners-Lee needed to keep track of them, so he wrote a program to help him do so. He noticed that when people explained to him the various relationships at CERN, they tended to scribble diagrams with a lot of arrows. So he devised a method to replicate these in his program. He would type in the name of a person or project and then create links that would show which were related. Thus it was that Berners-Lee created a computer program that he named, after the Victorian almanac of his childhood, Enquire.

"I liked Enquire," he wrote, "because it stored information without using structures like matrices or trees."[13] Such structures are hierarchical and rigid, whereas the human mind makes more random leaps. As he worked on Enquire, he developed a grander vision for what it could become. "Suppose all the information stored on computers everywhere were linked. There would be a single global information space. A web of information would form."[14] What he imagined, although he didn't know it at the time, was Vannevar Bush's memex machine—which could store documents, cross-reference them, retrieve them—writ global.

But before he got very far in creating Enquire, his consultancy at CERN came to an end. He left behind his computer and his eight-inch floppy disk containing all of the code, and it was promptly lost and forgotten. For a few years he worked in England for a company that made software for publishing documents. But he got bored and applied for a fellowship at CERN. In September 1984 he arrived back there to work with the group that was responsible for gathering the results of all of the experiments being done at the institute.

CERN was a cauldron of diverse peoples and computer systems using dozens of languages, both verbal and digital. All had to share information. "In this connected diversity," Berners-Lee recalled, "CERN was a microcosm of the rest of the world."[15] In such a setting, he found himself returning to his childhood ruminations about how people with different perspectives work together to turn each other's half-formed notions into new ideas. "I've always been interested in how people work together. I was working with a lot of people at other institutes and universities, and they had to collaborate. If they had been in the same room, they would have written all over the blackboard. I was looking for a system that would allow people to brainstorm and to keep track of the institutional memory of a project."[16]

Such a system, he felt, would connect people from afar so that they could complete each other's sentences and add useful ingredients to each other's half-formed notions. "I wanted it to be something which would allow us to work together, design things together," he said. "The really interesting part of the design is when we have lots of people all over the planet who have part of it in their heads. They have parts of the cure for AIDS, part of an understanding of cancer."[17] The goal was to facilitate team creativity—the brainstorming that occurs when people sit around fleshing out each other's ideas—when the players are not in the same place.

So Berners-Lee reconstructed his Enquire program and began thinking about ways to expand it. "I wanted to access different kinds of information, such as a researcher's technical papers, the manual for different software modules, minutes of meetings, hastily scribbled notes, and so on."[18] Actually, he wanted to do much more than that. He had the placid exterior of a congenital coder, but lurking underneath he harbored the whimsical curiosity of a child who stayed up late reading *Enquire Within Upon Everything*. Rather than merely devising a data management system, he yearned to create a collaborative playground. "I wanted to build a creative space," he later said, "something like a sandpit where everyone could play together."[19]

He hit upon a simple maneuver that would allow him to make the connections he wanted: *hypertext*. Now familiar to any Web surfer, hypertext is a word or phrase that is coded so that when clicked it sends the reader to another document or piece of content. Envisioned by Bush in his description of a memex machine, it was named in 1963 by the tech visionary Ted Nelson, who dreamed up a brilliantly ambitious project called Xanadu, never brought to fruition, in which all pieces of information would be published with two-way hypertext links to and from related information.

Hypertext was a way to allow the connections that were at the core of Berners-Lee's Enquire program to proliferate like rabbits; anyone could link to documents on other computers, even those with different operating systems, without asking permission. "An Enquire program capable of external hypertext links was the difference between imprisonment and freedom," he exulted. "New webs could be made to bind different computers together." There would be no central node, no command hub. If you knew the web address of a document, you could link to it. That way the system of links could spread and sprawl, "riding on top of the Internet," as Berners-Lee put it.[20] Once again, an innovation was created by weaving together two previous innovations: in this case, hypertext and the Internet.

Using a NeXT computer, the handsome hybrid of a workstation and personal computer that Jobs created after being ousted from Apple, Berners-Lee adapted a protocol that he had been working on, called a Remote Procedure Call, that allowed a program running on one computer to call up a subroutine that was on another computer. Then he drew up a set of principles for naming each document. Initially he called these Universal Document Identifiers. The folks at the Internet Engineering Task Force in charge of approving standards balked at what they said was his "arrogance" in calling his scheme *universal*. So he agreed to change it to *uniform*. In fact he was pushed into changing all three words, turning it into Uniform Resource Locators—those URLs, such as http://www.cern.ch, that we now use every day.[21] By the end of 1990 he had created a suite of tools that allowed his network to come to life: a Hypertext Transfer Protocol (HTTP) to allow hypertext to be exchanged online, a Hypertext Markup Language (HTML) for creating pages, a rudimentary browser to serve as the application software that retrieved and displayed information, and server software that could respond to requests from the network.

In March 1989 Berners-Lee had his design in place and officially submitted a funding proposal to the top managers at CERN. "The hope would be to allow a pool of information to develop which could grow and evolve," he wrote. "A 'web' of notes with links between them is far more useful than a fixed hierarchical system."[22] Unfortunately, his proposal elicited as much bafflement as enthusiasm. "Vague, but exciting," his boss, Mike Sendall, wrote atop the memo. "When I read Tim's proposal," he later admitted, "I could not figure out what it was, but I thought it was great."[23] Once again, a brilliant inventor found himself in need of a collaborator to turn a concept into a reality.

More than most digital-age innovations, the conception of the Web was driven primarily by one person. But Berners-Lee did need a partner in bringing it to fruition. Fortunately, he was able to find one in Robert Cailliau, a Belgian engineer at CERN, who had been toying with similar ideas and was willing to join forces. "In the marriage of hypertext and the Internet," said Berners-Lee, "Robert was best man."

With his personable demeanor and bureaucratic skills, Cailliau was the perfect person to be the evangelist for the project within CERN and the project manager who got things done. A fastidious dresser who methodically scheduled his haircuts, he was "the kind of engineer who can be driven mad by the incompatibility of power plugs in different countries," according to Berners-Lee.[24] They formed a partnership often seen in innovative teams: the visionary product designer paired with the diligent project manager. Cailliau, who loved planning and organizational work, cleared the way, he said, for Berners-Lee to "bury his head in the bits and develop his software." One day Cailliau tried to go over a project plan with Berners-Lee and realized, "He just did not understand the concept!"[25] Because of Cailliau, he didn't have to.

Cailliau's first contribution was to sharpen the funding proposal that Berners-Lee had submitted to CERN administrators by making it less vague while keeping it exciting. He began with its title, "Information Management." Cailliau insisted that they figure out a catchier name for the project, which shouldn't be too hard. Berners-Lee had a few ideas. The first was Mine of Information, but that abbreviated to MOI, French for *me*, which sounded a bit egocentric. The second idea was The Information Mine, but that abbreviated to TIM, which was even more so. Cailliau rejected the approach, often used at CERN, of plucking the name of some Greek god or Egyptian pharaoh. Then Berners-Lee came up with something that was direct and descriptive. "Let's call it the World Wide Web," he said. It was the metaphor he had used in his original proposal. Cailliau balked. "We can't call it that, because the abbreviation WWW sounds longer than the full name!"[26] The initials have three times the syllables as the name itself. But Berners-Lee could be quietly stubborn. "It sounds good," he declared. So the title of the proposal was changed to "WorldWideWeb: Proposal for a HyperText Project." Thus the Web was named.

Once the project was officially embraced, the CERN administrators wanted to patent it. When Cailliau raised the issue, Berners-Lee objected. He wanted the Web to spread and evolve as quickly as possible, and that meant it should be free and open. At one point he looked at Cailliau and asked accusingly, "Robert, do you want to be rich?" As Cailliau recalled, his initial reaction was "Well, it helps, no?"[27] That was the incorrect response. "He apparently didn't care about that," Cailliau realized. "Tim's not in it for the money. He accepts a much wider range of hotel-room facilities than a CEO would."[28]

Instead Berners-Lee insisted that the Web protocols should be made available freely, shared openly, and put forever in the public domain. After all, the whole point of the Web, and the essence of its design, was to promote sharing and collaboration. CERN issued a document declaring that it "relinquishes all intellectual property rights to this code, both source and binary form, and permission is granted for anyone to use, duplicate, modify, and redistribute it."[29] Eventually CERN joined forces with Richard Stallman and adopted his GNU General Public License. The result was one of the grandest free and open-source projects in history.

That approach reflected Berners-Lee's self-effacing style. He was averse to any hint of personal aggrandizement. Its wellsprings also came from someplace deeper within him: a moral outlook based on peer sharing and respect, something he found in the Unitarian Universalist Church that he adopted. As he said of his fellow Unitarians, "They meet in churches instead of wired hotels, and discuss justice, peace, conflict, and morality rather than protocols and data formats, but in other ways the peer respect is very similar to that of the Internet Engineering Task Force. . . . The design of the Internet and the Web is a search for a set of rules which will allow computers to work together in harmony, and our spiritual and social quest is for a set of rules which allow people to work together in harmony."[30]

Despite the hoopla that accompanies many product announcements—think Bell Labs unveiling the transistor or Steve Jobs the Macintosh—some of the most momentous innovations tiptoe quietly onto history's stage. On August 6, 1991, Berners-Lee was glancing through the Internet's alt.hypertext newsgroup and ran across this question: "Is anyone aware of research or development efforts in . . . hypertext links enabling retrieval from multiple heterogeneous sources?" His answer, "from: timbl@info.cern.ch at 2:56 pm," became the first public announcement of the Web. "The WorldWideWeb project aims to allow links to be made to any information anywhere," he began. "If you're interested in using the code, mail me."[31]

With his low-key personality and even lower-key posting, Berners-Lee did not fathom what a profound idea he had unleashed. *Any information anywhere.* "I spent a lot of time trying to make sure people could put anything on the web," he said more than two decades later. "I had no idea that people would put literally everything on it."[32] Yes, everything. *Enquire Within Upon Everything.*

MARC ANDREESSEN AND MOSAIC

For people to summon forth sites on the Web, they needed a piece of client software on their own computers that became known as a browser. Berners-Lee wrote one that could both read and edit documents; his hope was that the Web would become a place where users could collaborate. But his browser worked only on NeXT computers, of which there were few, and he had neither the time nor the resources to create other browser versions. So he enlisted a young intern at CERN, an undergraduate named Nicola Pellow who was majoring in math at Leicester Polytechnic, to write the first all-purpose browser for UNIX and Microsoft operating systems. It was rudimentary, but it worked. "It was to be the vehicle that allowed the Web to take its first tentative step on to the world stage, but Pellow was unfazed," Cailliau recalled. "She was given the task and she simply sat down to do it, little realizing the enormity of what she was about to unleash."[33] Then she went back to Leicester Polytechnic.

Berners-Lee began urging others to improve on Pellow's work: "We energetically suggested to everyone everywhere that the creation of browsers would make useful projects."[34] By the fall of 1991 there were a half-dozen experimental versions, and the Web quickly spread to other research centers in Europe.

That December it made the leap across the Atlantic. Paul Kunz, a particle physicist at the Stanford Linear Accelerator Center, was visiting CERN, and Berners-Lee recruited him to the world of the Web. "He twisted my arm and insisted that I come see him," according to Kunz, who worried that he was in for a boring demonstration of information management. "But then he showed me something that opened my eyes."[35] It was a Web browser on Berners-Lee's NeXT calling up information from an IBM machine somewhere else. Kunz brought the software back with him, and http://slacvm.slac.stanford.edu/ became the first Web server in the United States.

The World Wide Web hit orbital velocity in 1993. The year began with fifty Web servers in the world, and by October there were five hundred. One reason was that the primary alternative to the Web for accessing information on the Internet was a sending and fetching protocol developed at the University of Minnesota called Gopher,* and word leaked out that the developers were planning to charge a fee for use of the server software. A more important impetus was the creation of the first easy-to-install Web browser with graphic capabilities, named Mosaic. It was developed at the National Center for

*Like the Web's HTTP, Gopher was an Internet (TCP/IP) application layer protocol. It primarily facilitated a menu-based navigation for finding and distributing documents (usually text-based) online. The links were done by the servers rather than embedded in the documents. It was named after the university's mascot and was also a pun on "go for."

Supercomputing Applications (NCSA) at the University of Illinois at Urbana-Champaign, which had been funded by the Gore Act.

The man, or overgrown kid, most responsible for Mosaic was a gentle but intense undergraduate named Marc Andreessen, a corn-fed six-foot-four jolly giant born in Iowa in 1971 and raised in Wisconsin. Andreessen was a fan of the pioneers of the Internet, and their writings inspired him: "When I got a copy of Vannevar Bush's 'As We May Think,' I said to myself, 'Yep, there it is! He figured it out!' Bush envisioned the Internet as fully as you could, given that you didn't have digital computers. He and Charles Babbage are in the same league." Another hero was Doug Engelbart. "His lab was node four on the Internet, which was like having the fourth telephone in the world. He had the amazing foresight to understand what the Internet would be before it got built."[36]

When Andreessen saw the Web demonstrated in November 1992, he was blown away. So he enlisted an NCSA staffer, Eric Bina, a first-class programmer, to partner with him in building a more exciting browser. They loved Berners-Lee's concepts, but they thought CERN's implementation software was drab and devoid of cool features. "If someone were to build the right browser and server, that would be really interesting," Andreessen told Bina. "We can run with this and really make it work."[37]

For two months they engaged in a programming binge that rivaled those of Bill Gates and Paul Allen. For three or four days straight they would code around the clock—Andreessen fueled by milk and cookies, Bina by Skittles and Mountain Dew—and then crash for a full day to recover. They were a great team: Bina was a methodical programmer, Andreessen a product-driven visionary.[38]

On January 23, 1993, with just a little more fanfare than Berners-Lee had indulged in when launching the Web, marca@ncsa.uiuc.edu announced Mosaic on the www-talk Internet newsgroup. "By the power vested in me by nobody in particular," Andreessen began, "alpha/beta version 0.5 of NCSA's Motif-based networked information systems and World Wide Web browser, X Mosaic, is hereby released." Berners-Lee, who was initially pleased, posted a response two days later: "Brilliant! Every new browser is sexier than the last." He added it to the growing list of browsers available for download from info.cern.ch.[39]

Mosaic was popular because it could be installed simply and enabled images to be embedded in Web pages. But it became even more popular because Andreessen knew one of the secrets of digital-age entrepreneurs: he fanatically heeded user feedback and spent time on Internet newsgroups soaking up suggestions and complaints. Then he persistently released updated versions. "It was amazing to launch a product and get immediate feedback," he enthused. "What I got out of that feedback loop was an instant sense of what was working and what wasn't."[40]

Andreessen's focus on continual improvement impressed Berners-Lee: "You'd send him a bug report and then two hours later he'd mail you a fix."[41] Years later, as a venture capitalist, Andreessen made a rule of favoring startups whose founders focused on running code and customer service rather than charts and presentations. "The former are the ones who become the trillion-dollar companies," he said.[42]

There was something about Andreessen's browser, however, that disappointed and then began to annoy Berners-Lee. It was beautiful, even dazzling, but Andreessen's emphasis was on enabling rich media for publishing eye-catching pages, and Berners-Lee felt that the focus should instead be on providing tools that would facilitate serious collaboration. So in March 1993, after a meeting in Chicago, he drove "across the seemingly interminable cornfields" of central Illinois to visit Andreessen and Bina at NCSA.

It was not a pleasant session. "All of my earlier meetings with browser developers had been meetings of minds," Berners-Lee recalled. "But this one had a strange tension to it." He felt that the Mosaic developers, who had their own public relations staff and were garnering a lot of publicity, were "attempting to portray themselves as the center of Web development and to basically rename the Web as Mosaic."[43] They seemed to be trying to own the Web, he thought, and perhaps profit from it.*

*A year later, Andreessen would join with the serially successful entrepreneur Jim Clark to launch a company called Netscape that produced a commercial version of the Mosaic browser.

Andreessen found Berners-Lee's recollection amusing. "When Tim came, it was more of a state visit than a working session. The Web had already become a brush fire, and he was uncomfortable that he was no longer controlling it." Berners-Lee's opposition to embedding images struck him as quaint and purist. "He only wanted text," Andreessen remembered. "He specifically didn't want magazines. He had a very pure vision. He basically wanted it used for scientific papers. His view was that images are the first step on the road to hell. And the road to hell is multimedia content and magazines, garishness and games and consumer stuff." Because he was customer-focused, Andreessen thought that this was academic hogwash. "I'm a Midwestern tinkerer type. If people want images, they get images. Bring it on."[44]

Berners-Lee's more fundamental criticism was that by focusing on fancy display features, such as multimedia and ornamental fonts, Andreessen was ignoring a capability that should have been in the browser: editing tools that would allow users to interact with and contribute to the content on a Web page. The emphasis on display rather than editing tools nudged the Web into becoming a publishing platform for people who had servers rather than a place for collaboration and shared creativity. "I was disappointed that Marc didn't put editing tools in Mosaic," Berners-Lee said. "If there had been more of an attitude of using the Web as a collaborative medium rather than a publishing medium, then I think it would be much more powerful today."[45]

Early versions of Mosaic did have a "collaborate" button, which allowed users to download a document, work on it, and repost it. But the browser was not a full-fledged editor, and Andreessen felt it was impractical to turn it into one. "I was amazed at this near-universal disdain for creating an editor," complained Berners-Lee. "Without a hypertext editor, people would not have the tools to really use the Web as an intimate collaborative medium. Browsers would let them find and share information, but they could not work together intuitively."[46] To some extent, he was right. Despite the astonishing success of the Web, the world would have been a more interesting place if the Web had been bred as a more collaborative medium.

Berners-Lee also paid a visit to Ted Nelson, who lived on a houseboat in Sausalito in the shadow of the Golden Gate Bridge.

Twenty-five years earlier, Nelson had pioneered the concept of a hypertext network with his proposed Xanadu project. It was a pleasant meeting, but Nelson was annoyed that the Web lacked key elements of Xanadu.[47] He believed that a hypertext network should have two-way links, which would require the approval of both the person creating the link and the person whose page was being linked to. Such a system would have the side benefit of enabling micropayments to content producers. "HTML is precisely what we were trying to prevent—ever-breaking links, links going outward only, quotes you can't follow to their origins, no version management, no rights management," Nelson later lamented.[48]

Had Nelson's system of two-way links prevailed, it would have been possible to meter the use of links and allow small automatic payments to accrue to those who produced the content that was used. The entire business of publishing and journalism and blogging would have turned out differently. Producers of digital content could have been compensated in an easy, frictionless manner, permitting a variety of revenue models, including ones that did not depend on being beholden solely to advertisers. Instead the Web became a realm where aggregators could make more money than content producers. Journalists at both big media companies and little blogging sites had fewer options for getting paid. As Jaron Lanier, the author of *Who Owns the Future?*, has argued, "The whole business of using advertising to fund communication on the Internet is inherently self-destructive. If you have universal backlinks, you have a basis for micropayments from somebody's information that's useful to somebody else."[49] But a system of two-way links and micropayments would have required some central coordination and made it hard for the Web to spread wildly, so Berners-Lee resisted the idea.

As the Web was taking off in 1993–94, I was the editor of new media for Time Inc., in charge of the magazine company's Internet strategy. Initially we had made deals with the dial-up online services, such as AOL, CompuServe, and Prodigy. We supplied our content, marketed their services to our subscribers, and moderated chat rooms and bulletin boards that built up communities of members. For that we were able to command between one and two million dollars in annual royalties.

When the open Internet became an alternative to these proprietary online services, it seemed to offer an opportunity to take control of our own destiny and subscribers. At the April 1994 National Magazine Awards lunch, I had a conversation with Louis Rossetto, the editor and founder of *Wired*, about which of the emerging Internet protocols and finding tools—Gopher, Archie, FTP, the Web—might be best to use. He suggested that the best option was the Web because of the neat graphic capabilities being built into browsers such as Mosaic. In October 1994 both *HotWired* and a collection of Time Inc. websites launched.

At Time Inc. we experimented with using our established brands—*Time*, *People*, *Life*, *Fortune*, *Sports Illustrated*—as well as creating a new portal named Pathfinder. We also conjured up new brands, ranging from the *Virtual Garden* to the *Netly News*. Initially we planned to charge a small fee or subscription, but Madison Avenue ad buyers were so enthralled by the new medium that they flocked to our building offering to buy the banner ads we had developed for our sites. Thus we and other journalism enterprises decided that it was best to make our content free and garner as many eyeballs as we could for eager advertisers.

It turned out not to be a sustainable business model.[50] The number of websites, and thus the supply of slots for ads, went up exponentially every few months, but the total amount of advertising dollars remained relatively flat. That meant advertising rates eventually tumbled. It was also not an ethically healthy model; it encouraged journalists to cater primarily to the desires of their advertisers rather than the needs of their readers. By then, however, consumers had been conditioned to believe that content should be free. It took two decades to start trying to put that genie back in the bottle.

In the late 1990s Berners-Lee tried to develop a micropayments system for the Web through the World Wide Web Consortium (W3C), which he headed. The idea was to devise a way to embed in a Web page the information needed to handle a small payment, which would allow different "electronic wallet" services to be created by banks or entrepreneurs. It was never implemented, partly because of the changing complexity of banking regulations. "When we started, the first thing we tried to do was enable small payments to people who posted content," Andreessen explained. "But we didn't have the resources at the University of Illinois to implement that. The credit card systems and banking system made it impossible. We tried hard, but it was so painful to deal with those guys. It was cosmically painful."[51]

In 2013 Berners-Lee began reviving some of the activities of the W3C's Micropayments Markup Working Group. "We are looking at micropayment protocols again," he said. "It would make the Web a very different place. It might be really enabling. Certainly the ability to pay for a good article or song could support more people who write things or make music."[52] Andreessen said he hoped that Bitcoin,* a digital currency and peer-to-peer payment system created in 2009, might turn out to be a model for better payment systems. "If I had a time machine and could go back to 1993, one thing I'd do for sure would be to build in Bitcoin or some similar form of cryptocurrency."[53]

We at Time Inc. and other media companies made one other mistake, I think: we abandoned our focus on creating community after we settled into the Web in the mid-1990s. On our AOL and CompuServe sites, much of our effort had been dedicated to creating communities with our users. One of the early denizens of The WELL, Tom Mandel, was hired to moderate *Time*'s bulletin boards and emcee our chat rooms. Posting articles from the magazine was secondary to creating a sense of social connection and community among our users. When we migrated to the Web in 1994, we initially tried to replicate that approach. We created bulletin boards and chat groups on Pathfinder and pushed our engineers to replicate AOL's simple discussion threads.

---

*Bitcoin and other cryptocurrencies incorporate mathematically coded encryption techniques and other principles of cryptography to create a secure currency that is not centrally controlled.

But as time went on, we began to pay more attention to publishing our own stories online rather than creating user communities or enabling user-generated content. We and other media companies repurposed our print publications into Web pages to be passively consumed by our readers, and we relegated the discussions to a string of reader comments at the bottom of the page. These were often unmoderated rants and blather that few people, including us, ever read. Unlike the Usenet newsgroups or The WELL or AOL, the focus was not on discussions and communities and content created by users. Instead, the Web became a publishing platform featuring old wine—the type of content you could find in print publications—being poured into new bottles. It was like the early days of television, when the offerings were nothing more than radio shows with pictures. Thus we failed to thrive.

Fortunately, the street finds its own uses for things, and new forms of media soon arose to take advantage of the new technology. Led by the growth of blogs and wikis, both of which emerged in the mid-1990s, a revitalized Web 2.0 arose that allowed users to collaborate, interact, form communities, and generate their own content.

JUSTIN HALL AND HOW WEB LOGS BECAME BLOGS

As a freshman at Swarthmore College in December 1993, Justin Hall picked up a stray copy of the *New York Times* in the student lounge and read a story by John Markoff about the Mosaic browser. "Think of it as a map to the buried treasures of the Information Age," it began. "A new software program available free to companies and individuals is helping even novice computer users find their way around the global Internet, the network of networks that is rich in information but can be baffling to navigate."[54] A willowy computer geek with an impish smile and blond hair flowing over his shoulders, Hall seemed to be a cross between Huck Finn and a Tolkien elf. Having spent his childhood in Chicago dialing into computer bulletin boards, he immediately downloaded the browser and began surfing. "The whole concept blew me away," he remembered.[55]

Hall quickly realized something: "Nearly all of the online publishing efforts were amateur, people who didn't have anything to say." So he decided to create a website, using an Apple PowerBook and MacHTTP software he downloaded for free, that would amuse himself and others who shared his cheeky outlook and teenage obsessions. "I could put my writings and words up electronically, make them look pretty, and engage the web with links."[56] He got his site up in mid-January 1994, and a few days later, to his delight, strangers from around the Web began to stumble across it.

His first home page had a tone of mischievous intimacy. It included a photo of Hall mugging behind Colonel Oliver North, another of Cary Grant taking acid, and a sincere shout-out to "Al Gore, the information tollroad's first official pedestrian." The tone was conversational. "Howdy," the home page declared. "This is twenty-first century computing. Is it worth our patience? I'm publishing this, and I guess you're readin' this, in part to figure that out, huh?"

At the time, there were no Web directories or search engines, other than very staid ones like the W3 Catalog from the University of Geneva and a "What's New" page from NCSA at the University of Illinois. So Hall invented one for his site, which he elegantly titled "Here's a Menu of Cool Shit." Shortly thereafter, in an homage to Dostoevsky, he renamed it "Justin's Links from the Underground." It included links to the Electronic Frontier Foundation, the World Bank, and websites created by beer connoisseurs, fans of the rave music scene, and a guy at the University of Pennsylvania named Ranjit Bhatnagar who had created a similar Web page. "Believe me, the author is a very cool guy," Hall noted. He also included a list of bootleg concert recordings, featuring Jane's Addiction and Porno for Pyros. "Leave me a note if you are interested in these, or if you have any of your own," he wrote. Not surprisingly, given the fixations of Justin and his users, there were also many sections devoted to erotica, including pages called "Survey of Sexuality on the Sprawl" and "Pointers to Pages o' Purveyed Prurience." He helpfully reminded his users, "Don't forget to wipe the semen off your keyboards!"

Justin's Links from the Underground became the spiky pathfinder for a proliferation of directories, such as Yahoo and then Lycos and Excite, that began to blossom later that year. But in addition to providing a portal to the wonderland of the Web, Hall created something weirdly beguiling that turned out to be even more significant: a running Web log of his personal activities, random thoughts, deep musings, and intimate encounters. It became the first wholly new form of content to be created for, and take advantage of, personal computer networks. His Web log included poignant poems about his father's suicide, musings about his diverse sexual desires, pictures of his penis, endearingly edgy stories about his stepfather, and other effusions that darted back and forth across the line of Too Much Information. In short, he became the founding scamp of blogging.

"I was on the literary magazine in high school," he said, "and I had published some very personal things." That became the recipe for his and many future blogs: stay casual, get personal, be provocative. He posted a picture of himself standing nude onstage that he had been prevented from using in his high school yearbook, along with the tale of the girl editors "giggling while checking out the black-and-white photo of my tweeter." He later told the story of an evening of painful intercourse with a girl, after which his foreskin had swelled; it was illustrated with many close-ups of his genital situation. In so doing, he helped innovate a sensibility for a new age. "I always tried to provoke, and nudity was part of the provocation," he explained, "so I have a long tradition of doing things that would make my mom blush."[57]

Hall's willingness to push the boundaries of Too Much Information became a hallmark of blogging. It was cheekiness raised to a moral attitude. "TMI is like the deep lab data from all of our human experiments," he later explained. "If you reveal TMI, it can make people feel a little less alone." That was no trivial feat. Indeed, making people feel a little less alone was part of the essence of the Internet.

The case of his swollen foreskin was an example; within a few hours people from around the world posted comments offering their own stories, cures, and assurance that the condition was temporary. A more poignant case came from his postings about his father, an alcoholic who had committed suicide when Justin was eight. "My father was a wry, humanistic, sensitive man," he wrote. "Also an intolerant spiteful bastard." Hall recounted how his father would sing Joan Baez folk songs to him, but also down bottles of vodka and wave guns and berate waitresses. After Hall learned that he was the last person to talk to his father before he killed himself, he posted a poem: "What did we say / I wonder / and / what did it matter? / Could I have changed your mind?" These entries gave rise to a virtual support group. Readers sent in their own stories, and Hall posted them. Sharing led to connections. Emily Ann Merkler wrestled with losing her father to epilepsy. Russell Edward Nelson included scans of his late father's driver's license and other documents. Werner Brandt put up a remembrance page of his father that featured piano songs he had liked. Justin posted them along with his own musings. It became a social network. "The Internet encourages participation," he noted. "By exposing myself on the Web, I hope folks will be inspired to put a little soul in their systems."

A few months after he started his Web log, Hall managed to wrangle, through a tenacious volley of phone calls and emails, an internship for the summer of 1994 at HotWired.com in San Francisco. *Wired* magazine, under its charismatic editor Louis Rossetto, was in the process of creating one of the first magazine websites. Its executive editor was Howard Rheingold, an insightful online sage who had just published *The Virtual Community*, which described the social mores and satisfactions that came from "homesteading on the electronic frontier." Hall became Rheingold's friend and protégé, and together they engaged in a struggle with Rossetto over the soul of the new site.[58]

Rheingold felt that HotWired.com, in contrast to the printed magazine, should be a loosely controlled community, a "global jam session" filled with user-generated material. "I was part of Howard's faction that really felt that community was important and wanted to build user forums and tools that made it easy for people to comment to each other," Hall recalled. One idea they pushed was devising ways that members of the community could develop their own online identities and reputations. "The value is users talking to users," Hall argued to Rossetto. "People are the content."

Rossetto instead felt that HotWired should be a well-crafted and highly designed publishing platform, featuring rich imagery, that would extend the brand of the magazine and create a striking

Wired-like identity online. "We have these great artists and we should feature them," he argued. "We are going to make something beautiful and professional and polished, which is what the Web lacks." Building a lot of tools for user-generated content and comments would be "too much of a sideshow."[59]

The debate was waged in long meetings and impassioned email chains. But Rossetto prevailed, and his outlook, which was shared by many other print-world editors, ended up shaping the evolution of the Web. It became primarily a platform for publishing content rather than for creating virtual communities. "The era of public-access Internet has come to an end," Rossetto declared.[60]

When Hall returned from his extended summer gig at HotWired, he decided to become an evangelist for the other side of the argument, believing that the public-access aspects of the Internet should be celebrated and supported. With less sociological sophistication than Rheingold but more youthful exuberance, he began to preach the redemptive nature of virtual communities and Web logs. "I've been putting my life online, telling stories about the people I know and the things that happen to me when I'm not geeking out," he explained online after a year. "Talking about myself keeps me going."

His manifestos described the appeal of a new public-access medium. "When we tell stories on the Internet, we claim computers for communication and community over crass commercialism," he declared in one of his early postings. As someone who had spent hours on the Internet's early bulletin boards when he was growing up, he wanted to recapture the spirit of the Usenet newsgroups and The WELL.

And so Hall became the Johnny Appleseed of Web logging. On his site, he posted an offer to teach people HTML publishing if they would host him for a night or two, and in the summer of 1996 he traveled by bus across the United States, dropping in on those who took him up on the offer. "He took a medium that had been conceived as a repository for scholarship and scaled it down to personal size," Scott Rosenberg wrote in his history of blogging, *Say Everything*.[61] Yes, but he also helped to do something more: return the Internet and the Web to what they were intended to be, tools for sharing rather than platforms for commercial publishing. Web logging made the Internet more humanizing, which was no small transformation. "The best use of our technology enhances our humanity," Hall insisted. "It lets us shape our narrative and share our story and connect us."[62]

The phenomenon quickly spread. In 1997 John Barger, who produced a fun website called Robot Wisdom, coined the term *weblog*, and two years later a web designer named Peter Merholz jokingly broke the word back into two by saying he was going to use the phrase *we blog*. The word *blog* entered the common parlance.* By 2014 there would be 847 million blogs in the world.

It was a social phenomenon that was not fully appreciated by the traditional wordcrafting elite. It was easy, and not altogether incorrect, to denigrate much of the self-centered blatherings that appeared on blogs and to smirk at those who spent their evenings posting on little-read pages. But as Arianna Huffington pointed out early on when she created her blogging outlet, the *Huffington Post*, people decided to partake in these acts of social discourse because they found them fulfilling.[63] They got the chance to express their ideas, tailor them for public consumption, and get feedback. This was a new opportunity for people who had previously spent evenings passively consuming what was fed to them through their television screens. "Before the Internet came along, most people rarely wrote anything at all for pleasure or intellectual satisfaction after graduating from high school or college," Clive Thompson noted in his book, *Smarter Than You Think*. "This is something that's particularly hard to grasp for professionals whose jobs require incessant writing, like academics, journalists, lawyers or marketers."[64]

In his own sweet way, Justin Hall understood the glory of this. It was what would make the digital age different from the era of television. "By publishing ourselves on the web, we reject the role of passive media marketing recipient," he wrote. "If we all have a place to post our pages—the Howard Rheingold channel, the Rising City High School channel—there's no way the web will end up as banal and

---

*In March 2003 *blog* as both a noun and a verb was admitted into the Oxford English Dictionary.

mediocre as television. There will be as many places to find fresh and engaging content as there are people who yearn to be heard. Good telling of human stories is the best way to keep the Internet and the World Wide Web from becoming a waste vastland."[65]

EV WILLIAMS AND BLOGGER

By 1999 blogs were proliferating. They were no longer mainly the playpen of offbeat exhibitionists like Justin Hall who posted personal journals about their lives and fancies. They had become a platform for freelance pundits, citizen journalists, advocates, activists, and analysts. But there was one problem: to publish and maintain an independent blog required some coding skills and access to a server. Creating user simplicity is one of the keys to successful innovation. For blogging to become a whole new medium that would transform publishing and democratize public discourse, someone had to make it easy, as easy as "Type in this box and then press this button." Enter Ev Williams.

Born in 1972 on a corn and soybean farm on the edge of the hamlet of Clarks, Nebraska (population: 374), Ev Williams grew up as a lanky, shy, and often lonely boy who never got into hunting and football, which made him a bit of an oddity. Instead he played with Legos, built wooden skateboards, took apart bikes, and spent a lot of time on of his family's green tractor, after he had finished his irrigation chores, staring into the distance and daydreaming. "Books and magazines were my outlet to the larger world," he recalled. "My family never really traveled, so I never went anywhere."[66]

He didn't have a computer growing up, but when he went to the University of Nebraska in 1991 he discovered the world of online services and bulletin boards. He began reading all he could about the Internet, even subscribing to a magazine about electronic bulletin boards. After dropping out of college, he decided to start a company to make CD-ROMs explaining the online world for local businessmen. Shot in his basement with a borrowed camera, the videos looked like a no-budget community access show, and they didn't sell. So he wandered off to California and took a job as a junior writer at the tech publisher O'Reilly Media, where he revealed his prickly independence by sending an email to the entire staff refusing to write material for one of the company's products because it "was a piece of shit."

With the instincts of a serial entrepreneur, he was always itching to start his own companies, and at the beginning of 1999 he launched one called Pyra Labs with a savvy woman named Meg Hourihan, whom he'd briefly dated. Unlike others jumping into the dotcom frenzy of that period, they focused on using the Internet for its original purpose: online collaboration. Pyra Labs offered a suite of Web-based applications that allowed teams to share project plans, lists of things to do, and jointly created documents. Williams and Hourihan found that they needed a simple way to share their own random notions and interesting items, so they began posting on a little internal website, which they dubbed "Stuff."

By this time Williams, who had always loved magazines and publications, had gotten into reading blogs. Rather than personal journals such as Hall's, he became a fan of the technology commentators who were pioneering serious Web journalism, such as Dave Winer, who had created one of the first weblogs, Scripting News, and designed an XML syndication format for it.[67]

Williams had his own home page, called EvHead, on which he posted a section of updated notes and comments. Like others who added such logs to their home pages, he had to type each item and update using HTML code. Wanting to streamline the process, he wrote a simple software script that automatically converted his posts into the proper format. It was a little hack that had a transforming effect. "The idea that I could have a thought and I could type in a form and it would be on my website in a matter of seconds completely transformed the experience. It was one of those things that, by automating the process, completely morphed what it was I was doing."[68] He soon began to wonder whether this little side dish could become a product of its own.

One of the basic lessons for innovation is to stay focused. Williams knew that his first company had failed because it tried to do thirty things and succeeded at none. Hourihan, who had been a management consultant, was adamant: Williams's blogger scripting tool was neat, but it was a distraction. It could never be a commercial product. Williams acquiesced, but in March he quietly registered the domain name blogger.com. He couldn't resist. "I have always been a product guy, and am just always thinking about products and thought this would be a cool little idea." In July, when Hourihan was on vacation, he launched Blogger as a separate product, without telling her. He was following another basic lesson for innovation: Don't stay *too* focused.

When Hourihan returned and discovered what had happened, she started shouting and threatened to quit. Pyra had only one other employee besides themselves, and there was no capacity to take on distractions. "She was pissed," Williams recalled. "But we talked her into thinking that it made sense." It did. Blogger attracted enough fans in the ensuing months that Williams, with his laconic and awkward charm, became one of the stars of the March 2000 South by Southwest conference. By the end of the year, Blogger had 100,000 accounts.

What it did not have, however, was revenue. Williams had been offering Blogger for free in the vague hope that it would entice folks into buying the Pyra app. But by the summer of 2000 he had pretty much abandoned Pyra. With the Internet bubble bursting, it was not an easy time to raise money. The relationship between Williams and Hourihan, always a bit fraught, degenerated to the point that shouting matches at the office were a regular occurrence.

In January 2001 the cash crisis came to a head. Desperately in need of new servers, Williams made an appeal to Blogger's users for donations. Close to $17,000 came in, which was enough to buy new hardware but not to pay salaries.[69] Hourihan demanded that Williams step aside as CEO, and when he refused, she quit. "On Monday I resigned from the company I co-founded," she wrote on her blog. "I'm still crying and crying and crying."[70] The other employees, by then six in total, walked out as well.

Williams posted a long entry titled "And Then There Was One" on his own blog. "We are out of money, and I have lost my team. . . . The last two years have been a long, hard, exciting, educational, once-in-a-lifetime, painful, and, ultimately, very rewarding and worthwhile journey for me." Vowing to keep the service alive, even if he had to do it alone, he ended with a postscript: "If anyone wants to share some office space for a while, lemme know. I could use the cost savings (and the company)."[71]

Most people would have quit at that point. There was no money for rent, no one to keep the servers running, no sight of any revenue. He also faced painful personal and legal attacks from his former employees, causing him to rack up lawyer's bills. "The story apparently was that I fired all my friends and I didn't pay them and took over the company," he said. "It was really ugly."[72]

But ingrained in Williams's hardscrabble heritage was the patience of a corn farmer and the stubbornness of an entrepreneur. He had an abnormally high level of immunity to frustration. So he persevered, testing that hazy borderline between persistence and cluelessness, remaining placid as problems bombarded him. He would run the company by himself, from his apartment. He would tend to the servers and the coding himself. "I basically went underground and did nothing but try to keep Blogger going."[73] Revenues were close to zero, but he could bring his costs in line with that. As he wrote in his Web posting, "I'm actually in surprisingly good shape. I'm optimistic. (I'm always optimistic.) And I have many, many ideas. (I always have many ideas.)"[74]

A few people expressed sympathy and offered help, most notably Dan Bricklin, a beloved and collaborative tech leader who had cocreated VisiCalc, the first computer spreadsheet program. "I didn't like the idea of Blogger being lost in the dotcom crash," Bricklin said.[75] After reading Williams's forlorn post, he sent an email asking if there was anything he could do to help. They agreed to meet when Bricklin, who lived in Boston, came to an O'Reilly conference in San Francisco. Over sushi at a nearby restaurant, Bricklin told the tale of how, years earlier, when his own company was foundering, he had run into Mitch Kapor of Lotus. Though competitors, they shared a collaborative hacker ethic, so Kapor offered a deal that helped Bricklin stay personally solvent. Bricklin went on to found a company, Trellix, that made its own website publishing system. Paying forward Kapor's band-of-hackers helpfulness to a semicompetitor, Bricklin worked out a deal for Trellix to license Blogger's software for $40,000, thus keeping it alive. Bricklin was, above all, a nice guy.

Throughout 2001 Williams worked around the clock from his apartment or in borrowed space to keep Blogger running. "Everybody I knew just thought I was crazy," he remembered. The low point came at Christmas when he went to visit his mother, who had moved to Iowa. His site got hacked on Christmas Day. "I was in Iowa trying to assess the damage over a dial-up connection and a tiny laptop. And I didn't have a system administrator or anyone else working for me at the time. I ended up spending most of the day in a Kinko's doing damage control."[76]

Things began to turn around in 2002. He launched Blogger Pro, which users paid for, and with the help of a new partner got a licensing deal in Brazil. The world of blogging was growing exponentially, which made Blogger a hot commodity. In October, with some prodding from Williams's old publishing boss, Tim O'Reilly, Google came calling. It was still mainly a search engine and had no history of buying other companies, but it made an offer to buy Blogger. Williams accepted.

Williams's simple little product helped to democratize publishing. "Push-button publishing for the people" was his mantra. "I love the world of publishing, and I'm fiercely independent minded, both of which came from growing up on a remote farm," he said. "When I found a way to let people publish on the Internet, I knew I could help give power and voice to millions."

At least initially, Blogger was primarily a tool for publishing rather than for interactive discussion. "Instead of promoting dialogue, it let people just get on a soap box," Williams admitted. "The Internet has a community side and a publishing side to it. There are people who obsess about the community part more than I do. I'm more driven by the publishing of knowledge side of it, because I grew up learning about the world from what other people published, and I'm not a huge participant in the community side."[77]

However, most digital tools eventually get commandeered for social purposes, that being the nature of humans. The blogosphere evolved into being a community rather than merely a collection of soap boxes. "It ended up turning into a community, even though we he modified a superb software product developed by 's most enchanting innovators, Bill Atkinson. It was Card, and it allowed users to make their own hyper- nd documents on their computers. Apple had little idea th the software, so at Atkinson's insistence Apple gave ith its computers. It was easy to use, and even kids— —found ways to make HyperCard stacks of linked ames.

m was blown away by HyperCard when he first saw d it cumbersome. So he created a supersimple way of ards and links: a blank box on each card in which you tle or word or phrase. If you wanted to make a link to arry's Video Project or anything else, you simply typed the box. "It was fun to do," he said.[80]

eated an Internet version of his HyperText program, st a few hundred lines of Perl code. The result was a anagement application that allowed users to edit and Web page. Cunningham used the application to build d the Portland Pattern Repository, that allowed soft- s to exchange programming ideas and improve on the thers had posted. "The plan is to have interested parties es about the People, Projects and Patterns that have ay they program," he wrote in an announcement posted 'The writing style is casual, like email. . . . Think of it as st where anyone can be moderator and everything is ot quite a chat, still, conversation is possible."[81]

eded a name. What he had created was a quick Web kWeb sounded lame, as if conjured up by a com- rosoft. Fortunately, there was another word for *quick* om the recesses of his memory. When he was on his Hawaii thirteen years earlier, he remembered, "the agent directed me to take the wiki wiki bus between en he asked what it meant, he was told that *wiki* was word for *quick*, and *wiki wiki* meant superquick. So he pages and the software that ran them WikiWikiWeb,

all had our own blogs, bec
each other," Williams said y
nity there, just as real as any
ally I came to appreciate th

Williams went on to be
ing and micropublishing se
designed to promote collab
alized that he did indeed va
as much as the publishing
connecting and finding a
to the Internet was very d
with a community is always
I founded Blogger, that it w
ing into a community is on
world."[79]

WARD CUNNINGHAM, AND WIKIS

When he launched the We
be used as a collaboration to
Mosaic browser did not gi
they were viewing. It turne
published content. That laps
ging, which encouraged use
dium was invented that we
on the Web. It was called a
modify Web pages—not by
by clicking and typing direc

The application was dev
those congenial Midwest n
making ham radios and get
they fostered. After gradua
tronic equipment company,
track of projects, a task sin
went to CERN.

To do thi
one of Apple
called Hyper
linked cards a
what to do w
it away free
especially kid
pictures and g

Cunningh
it, but he fou
creating new
could type a t
Jane Doe or H
those words i

Then he c
writing it in j
new content
contribute to
a service, call
ware develope
patterns that
write web pa
changed the w
in May 1995.
a moderated
archived. It's n

Now he n
tool, but Qu
mittee at Mi
that popped f
honeymoon i
airport count
terminals." W
the Hawaiian
named his We
wiki for short

Dan Bricklin (1951–) and
Ev Williams (1972–) in 2001.

Jimmy Wales (1966–).

Sergey Brin (1973–) and Larry Page (1973–).

In his original version, the syntax Cunningham used for creating links in a text was to smash words together so that there would be two or more capital letters—as in CapitalLetters—in a term. It became known as CamelCase, and its resonance would later be seen in scores of Internet brands such as AltaVista, MySpace, and YouTube.

WardsWiki (as it became known) allowed anyone to edit and contribute, without even needing a password. Previous versions of each page would be stored, in case someone botched one up, and there would be a "Recent Changes" page so that Cunningham and others could keep track of the edits. But there would be no supervisor or gatekeeper preapproving the changes. It would work, he said with cheery midwestern optimism, because "people are generally good." It was just what Berners-Lee had envisioned, a Web that was read-write rather than read-only. "Wikis were one of the things that allowed collaboration," Berners-Lee said. "Blogs were another." [83]

Like Berners-Lee, Cunningham made his basic software available for anyone to modify and use. Consequently, there were soon scores of wiki sites as well as open-source improvements to his software. But the wiki concept was not widely known beyond software engineers until January 2001, when it was adopted by a struggling Internet entrepreneur who was trying, without much success, to build a free, online encyclopedia.

Jimmy Wales was born in 1966 in Huntsville, Alabama, a town of rednecks and rocket scientists. Six years earlier, in the wake of *Sputnik*, President Eisenhower had personally gone there to open the Marshall Space Flight Center. "Growing up in Huntsville during the height of the space program kind of gave you an optimistic view of the future," Wales observed.[84] "An early memory was of the windows in our house rattling when they were testing the rockets. The space program was basically our hometown sports team, so it was exciting and you felt it was a town of technology and science." [85]

Wales, whose father was a grocery store manager, went to a one-room private school that was started by his mother and grandmother, who taught music. When he was three, his mother bought a *World Book Encyclopedia* from a door-to-door salesman; as he learned to read, it became an object of veneration. It put at his fingertips a cornucopia of knowledge along with maps and illustrations and even a few cellophane layers of transparencies you could lift to explore such things as the muscles, arteries, and digestive system of a dissected frog. But Wales soon discovered that the *World Book* had shortcomings: no matter how much was in it, there were many more things that weren't. And this became more so with time. After a few years, there were all sorts of topics—moon landings and rock festivals and protest marches, Kennedys and kings—that were not included. *World Book* sent out stickers for owners to paste on the pages in order to update the encyclopedia, and Wales was fastidious about doing so. "I joke that I started as a kid revising the encyclopedia by stickering the one my mother bought."[86]

After graduating from Auburn and a halfhearted stab at graduate school, Wales took a job as a research director for a Chicago financial trading firm. But it did not fully engage him. His scholarly attitude was combined with a love for the Internet that had been honed by playing Multi-User Dungeons fantasies, which were essentially crowdsourced games. He founded and moderated an Internet mailing list discussion on Ayn Rand, the Russian-born American writer who espoused an objectivist and libertarian philosophy. He was very open about who could join the discussion forum, frowned on rants and the personal attack known as flaming, and managed comportment with a gentle hand. "I have chosen a 'middle-ground' method of moderation, a sort of behind-the-scenes prodding," he wrote in a posting.[87]

Before the rise of search engines, among the hottest Internet services were Web directories, which featured human-assembled lists and categories of cool sites, and Web rings, which created through a common navigation bar a circle of related sites that were linked to one another. Jumping on these bandwagons, Wales and two friends in 1996 started a venture that they dubbed BOMIS, for Bitter Old Men in Suits, and began casting around for ideas. They launched a panoply of startups that were typical of the dotcom boom of the late 1990s: a used-car ring and directory with pictures, a food-ordering service, a business directory for Chicago, and a sports ring. After Wales relocated to San Diego, he launched a directory and ring that served as "kind of a guy-oriented search engine," featuring pictures of scantily clad women.[88]

The rings showed Wales the value of having users help generate the content, a concept that was reinforced as he watched how the crowds of sports bettors on his site provided a more accurate morning line than any single expert could. He also was impressed by Eric Raymond's *The Cathedral and the Bazaar*, which explained why an open and crowd-generated bazaar was a better model for a website than the carefully controlled top-down construction of a cathedral.[89]

Wales next tried an idea that reflected his childhood love of the *World Book*: an online encyclopedia. He dubbed it Nupedia, and it had two attributes: it would be written by volunteers, and it would be free. It was an idea that had been proposed in 1999 by Richard Stallman, the pioneering advocate of free software.[90] Wales hoped eventually to make money by selling ads. To help develop it, he hired a doctoral student in philosophy, Larry Sanger, whom he first met in online discussion groups. "He was specifically interested in finding a philosopher to lead the project," Sanger recalled.[91]

Sanger and Wales developed a rigorous, seven-step process for creating and approving articles, which included assigning topics to proven experts, whose credentials had been vetted, and then putting the drafts through outside expert reviews, public reviews, professional copy editing, and public copy editing. "We wish editors to be true experts in their fields and (with few exceptions) possess Ph.Ds.," the Nupedia policy guidelines stipulated.[92] "Larry's view was that if we didn't make it more academic than a traditional encyclopedia, people wouldn't believe in it and respect it," Wales explained. "He was wrong, but his view made sense given what we knew at the time."[93] The first article, published in March 2000, was on atonality by a scholar at the Johannes Gutenberg University in Mainz, Germany.

It was a painfully slow process and, worse yet, not a lot of fun. The whole point of writing for free online, as Justin Hall had shown, was that it produced a jolt of joy. After a year, Nupedia had only about a dozen articles published, making it useless as an encyclopedia, and 150 that were still in draft stage, which indicated how unpleasant the process had become. It had been rigorously engineered not to scale.

This hit home to Wales when he decided that he would personally write an article on Robert Merton, an economist who had won the Nobel Prize for creating a mathematical model for markets containing derivatives. Wales had published a paper on option pricing theory, so he was very familiar with Merton's work. "I started to try to write the article and it was very intimidating, because I knew they were going to send my draft out to the most prestigious finance professors they could find," Wales said. "Suddenly I felt like I was back in grad school, and it was very stressful. I realized that the way we had set things up was not going to work."[94]

That was when Wales and Sanger discovered Ward Cunningham's wiki software. Like many digital-age innovations, the application of wiki software to Nupedia in order to create Wikipedia—combining two ideas to create an innovation—was a collaborative process involving thoughts that were already in the air. But in this case a very non-wiki-like dispute erupted over who deserved the most credit.

The way Sanger remembered the story, he was having lunch in early January 2001 at a roadside taco stand near San Diego with a friend named Ben Kovitz, a computer engineer. Kovitz had been using Cunningham's wiki and described it at length. It then dawned on Sanger, he claimed, that a wiki could be used to help solve the problems he was having with Nupedia. "Instantly I was considering whether wiki would work as a more open and simple editorial system for a free, collaborative encyclopedia," Sanger later recounted. "The more I thought about it, without even having seen a wiki, the more it seemed obviously right." In his version of the story, he then convinced Wales to try the wiki approach.[95]

Kovitz, for his part, contended that he was the one who came up with the idea of using wiki software for a crowdsourced encyclopedia and that he had trouble convincing Sanger. "I suggested that instead of just using the wiki with Nupedia's approved staff, he open it up to the general public and let each edit appear on the site immediately, with no review process," Kovitz recounted. "My exact words were to allow 'any fool in the world with Internet access' to freely modify any page on the site." Sanger raised some objections: "Couldn't total idiots put up blatantly false or biased descriptions of things?" Kovitz replied,

"Yes, and other idiots could delete those changes or edit them into something better."[96]

As for Wales's version of the story, he later claimed that he had heard about wikis a month before Sanger's lunch with Kovitz. Wikis had, after all, been around for more than four years and were a topic of discussion among programmers, including one who worked at BOMIS, Jeremy Rosenfeld, a big kid with a bigger grin. "Jeremy showed me Ward's wiki in December 2000 and said it might solve our problem," Wales recalled, adding that when Sanger showed him the same thing, he responded, "Oh, yes, wiki, Jeremy showed me this last month."[97] Sanger challenged that recollection, and a nasty crossfire ensued on Wikipedia's discussion boards. Wales finally tried to de-escalate the sniping with a post telling Sanger, "Gee, settle down," but Sanger continued his battle against Wales in a variety of forums.[98]

The dispute presented a classic case of a historian's challenge when writing about collaborative creativity: each player has a different recollection of who made which contribution, with a natural tendency to inflate his own. We've all seen this propensity many times in our friends, and perhaps even once or twice in ourselves. But it is ironic that such a dispute attended the birth of one of history's most collaborative creations, a site that was founded on the faith that people are willing to contribute without requiring credit.*

More important than determining who deserved credit is appreciating the dynamics that occur when people share ideas. Ben Kovitz, for one, understood this. He was the player who had the most insightful view—call it the "bumblebee at the right time" theory—on the collaborative way that Wikipedia was created. "Some folks, aiming to criticize or belittle Jimmy Wales, have taken to calling me one of the founders of Wikipedia, or even 'the true founder,'" he said. "I suggested the idea, but I was not one of the founders. I was only the bumblebee. I had buzzed around the wiki flower for a while, and then pollinated the free-encyclopedia flower. I have talked with many oth-

---

*Tellingly, and laudably, Wikipedia's entries on its own history and the roles of Wales and Sanger have turned out, after much fighting on the discussion boards, to be balanced and objective.

ers who had the same idea, just not in times or places where it could take root."[99]

That is the way that good ideas often blossom: a bumblebee brings half an idea from one realm, and pollinates another fertile realm filled with half-formed innovations. This is why Web tools are valuable, as are lunches at taco stands.

Cunningham was supportive, indeed delighted when Wales called him up in January 2001 to say he planned to use the wiki software to juice up his encyclopedia project. Cunningham had not sought to patent or copyright either the software or the wiki name, and he was one of those innovators who was happy to see his products become tools that anyone could use or adapt.

At first Wales and Sanger conceived of Wikipedia merely as an adjunct to Nupedia, sort of like a feeder product or farm team. The wiki articles, Sanger assured Nupedia's expert editors, would be relegated to a separate section of the website and not be listed with the regular Nupedia pages. "If a wiki article got to a high level it could be put into the regular Nupedia editorial process," he wrote in a post.[100] Nevertheless, the Nupedia purists pushed back, insisting that Wikipedia be kept completely segregated, so as not to contaminate the wisdom of the experts. The Nupedia Advisory Board tersely declared on its website, "Please note: the editorial processes and policies of Wikipedia and Nupedia are totally separate; Nupedia editors and peer reviewers do not necessarily endorse the Wikipedia project, and Wikipedia contributors do not necessarily endorse the Nupedia project."[101] Though they didn't know it, the pedants of the Nupedia priesthood were doing Wikipedia a huge favor by cutting the cord.

Unfettered, Wikipedia took off. It became to Web content what GNU/Linux was to software: a peer-to-peer commons collaboratively created and maintained by volunteers who worked for the civic satisfactions they found. It was a delightful, counterintuitive concept, perfectly suited to the philosophy, attitude, and technology of the Internet. Anyone could edit a page, and the results would show up instantly. You didn't have to be an expert. You didn't have to fax in a copy of your diploma. You didn't have to be authorized by the Powers

That Be. You didn't even have to be registered or use your real name. Sure, that meant vandals could mess up pages. So could idiots or ideologues. But the software kept track of every version. If a bad edit appeared, the community could simply get rid of it by clicking on a "revert" link. "Imagine a wall where it was easier to remove graffiti than add it" is the way the media scholar Clay Shirky explained the process. "The amount of graffiti on such a wall would depend on the commitment of its defenders."[102] In the case of Wikipedia, its defenders were fiercely committed. Wars have been fought with less intensity than the reversion battles on Wikipedia. And somewhat amazingly, the forces of reason regularly triumphed.

One month after Wikipedia's launch, it had a thousand articles, approximately seventy times the number that Nupedia had after a full year. By September 2001, after eight months in existence, it had ten thousand articles. That month, when the September 11 attacks occurred, Wikipedia showed its nimbleness and usefulness; contributors scrambled to create new pieces on such topics as the World Trade Center and its architect. A year after that, the article total reached forty thousand, more than were in the *World Book* that Wales's mother had bought. By March 2003 the number of articles in the English-language edition had reached 100,000, with close to five hundred active editors working almost every day. At that point, Wales decided to shut Nupedia down.

By then Sanger had been gone for a year. Wales had let him go. They had increasingly clashed on fundamental issues, such as Sanger's desire to give more deference to experts and scholars. In Wales's view, "people who expect deference because they have a PhD and don't want to deal with ordinary people tend to be annoying."[103] Sanger felt, to the contrary, that it was the nonacademic masses who tended to be annoying. "As a community, Wikipedia lacks the habit or tradition of respect for expertise," he wrote in a New Year's Eve 2004 manifesto that was one of many attacks he leveled after he left. "A policy that I attempted to institute in Wikipedia's first year, but for which I did not muster adequate support, was the policy of respecting and deferring politely to experts." Sanger's elitism was rejected not only by Wales but by the Wikipedia community. "Consequently, nearly everyone with much expertise but little patience will avoid editing Wikipedia," Sanger lamented.[104]

Sanger turned out to be wrong. The uncredentialed crowd did not run off the experts. Instead the crowd itself became the expert, and the experts became part of the crowd. Early on in Wikipedia's development, I was researching a book about Albert Einstein and I noticed that the Wikipedia entry on him claimed that he had traveled to Albania in 1935 so that King Zog could help him escape the Nazis by getting him a visa to the United States. This was completely untrue, even though the passage included citations to obscure Albanian websites where this was proudly proclaimed, usually based on some thirdhand series of recollections about what someone's uncle once said a friend had told him. Using both my real name and a Wikipedia handle, I deleted the assertion from the article, only to watch it reappear. On the discussion page, I provided sources for where Einstein actually was during the time in question (Princeton) and what passport he was using (Swiss). But tenacious Albanian partisans kept reinserting the claim. The Einstein-in-Albania tug-of-war lasted weeks. I became worried that the obstinacy of a few passionate advocates could undermine Wikipedia's reliance on the wisdom of crowds. But after a while, the edit wars ended, and the article no longer had Einstein going to Albania. At first I didn't credit that success to the wisdom of crowds, since the push for a fix had come from me and not from the crowd. Then I realized that I, like thousands of others, was in fact a part of the crowd, occasionally adding a tiny bit to its wisdom.

A key principle of Wikipedia was that articles should have a neutral point of view. This succeeded in producing articles that were generally straightforward, even on controversial topics such as global warming and abortion. It also made it easier for people of different viewpoints to collaborate. "Because of the neutrality policy, we have partisans working together on the same articles," Sanger explained. "It's quite remarkable."[105] The community was usually able to use the lodestar of the neutral point of view to create a consensus article offering competing views in a neutral way. It became a model, rarely emulated, of how digital tools can be used to find common ground in a contentious society.

Not only were Wikipedia's articles created collaboratively by the community; so were its operating practices. Wales fostered a loose system of collective management, in which he played guide and gentle prodder but not boss. There were wiki pages where users could jointly formulate and debate the rules. Through this mechanism, guidelines were evolved to deal with such matters as reversion practices, mediation of disputes, the blocking of individual users, and the elevation of a select few to administrator status. All of these rules grew organically from the community rather than being dictated downward by a central authority. Like the Internet itself, power was distributed. "I can't imagine who could have written such detailed guidelines other than a bunch of people working together," Wales reflected. "It's common in Wikipedia that we'll come to a solution that's really well thought out because so many minds have had a crack at improving it."[106]

As it grew organically, with both its content and its governance sprouting from its grassroots, Wikipedia was able to spread like kudzu. At the beginning of 2014, there were editions in 287 languages, ranging from Afrikaans to Žemaitška. The total number of articles was 30 million, with 4.4 million in the English-language edition. In contrast, the *Encyclopedia Britannica*, which quit publishing a print edition in 2010, had eighty thousand articles in its electronic edition, less than 2 percent of the number in Wikipedia. "The cumulative effort of Wikipedia's millions of contributors means you are a click away from figuring out what a myocardial infarction is, or the cause of the Agacher Strip War, or who Spangles Muldoon was," Clay Shirky has written. "This is an unplanned miracle, like 'the market' deciding how much bread goes in the store. Wikipedia, though, is even odder than the market: not only is all that material contributed for free, it is available to you free."[107] The result has been the greatest collaborative knowledge project in history.

So why do people contribute? Harvard Professor Yochai Benkler dubbed Wikipedia, along with open-source software and other free collaborative projects, examples of "commons-based peer production." He explained, "Its central characteristic is that groups of individuals successfully collaborate on large-scale projects following a diverse cluster of motivational drives and social signals, rather than either market prices or managerial commands."[108] These motivations include the psychological reward of interacting with others and the personal gratification of doing a useful task. We all have our little joys, such as collecting stamps or being a stickler for good grammar, knowing Jeff Torborg's college batting average or the order of battle at Trafalgar. These all find a home on Wikipedia.

There is something fundamental, almost primordial at work. Some Wikipedians refer to it as "wiki-crack." It's the rush of dopamine that seems to hit the brain's pleasure center when you make a smart edit and it appears instantly in a Wikipedia article. Until recently, being published was a pleasure afforded only to a select few. Most of us in that category can remember the thrill of seeing our words appear in public for the first time. Wikipedia, like blogs, made that treat available to anyone. You didn't have to be credentialed or anointed by the media elite.

For example, many of Wikipedia's articles on the British aristocracy were largely written by a user known as Lord Emsworth. They were so insightful about the intricacies of the peerage system that some were featured as the "Article of the Day," and Lord Emsworth rose to become a Wikipedia administrator. It turned out that Lord Emsworth, a name taken from P. G. Wodehouse's novels, was actually a sixteen-year-old schoolboy in South Brunswick, New Jersey. On Wikipedia, nobody knows you're a commoner.[109]

Connected to that is the even deeper satisfaction that comes from helping to create the information that we use rather than just passively receiving it. "Involvement of people in the information they read," wrote the Harvard professor Jonathan Zittrain, "is an important end itself."[110] A Wikipedia that we create in common is more meaningful than would be the same Wikipedia handed to us on a platter. Peer production allows people to be engaged.

Jimmy Wales often repeated a simple, inspiring mission for Wikipedia: "Imagine a world in which every single person on the planet is given free access to the sum of all human knowledge. That's what we're doing." It was a huge, audacious, and worthy goal. But it badly understated what Wikipedia did. It was about more than people being "given" free access to knowledge; it was also about empowering them, in a way not seen before in history, to be part of the process of creating and distributing knowledge. Wales came to realize that. "Wikipedia allows people not merely to access other people's knowledge but to share their own," he said. "When you help build something, you own it, you're vested in it. That's far more rewarding than having it handed down to you."[111]

Wikipedia took the world another step closer to the vision propounded by Vannevar Bush in his 1945 essay, "As We May Think," which predicted, "Wholly new forms of encyclopedias will appear, ready made with a mesh of associative trails running through them, ready to be dropped into the memex and there amplified." It also harkened back to Ada Lovelace, who asserted that machines would be able to do almost anything, except think on their own. Wikipedia was not about building a machine that could think on its own. It was instead a dazzling example of human-machine symbiosis, the wisdom of humans and the processing power of computers being woven together like a tapestry. When Wales and his new wife had a daughter in 2011, they named her Ada, after Lady Lovelace.[112]

LARRY PAGE, SERGEY BRIN, AND SEARCH

When Justin Hall created his quirky home page in January 1994, there were only seven hundred websites in the world. By the end of that year there were ten thousand, and by the end of the following year there were 100,000. The combination of personal computers and networks had led to something amazing: anyone could get content from anywhere and distribute their own content everywhere. But for this exploding universe to be useful, it was necessary to find an easy way, a simple human-computer-network interface, that would enable people to find what they needed.

The first attempts to do this were hand-compiled directories. Some were quirky and frivolous, like Hall's Links from the Underground and Paul Phillips's Useless Pages. Others were sober and serious, like Tim Berners-Lee's World Wide Web Virtual Library, NCSA's "What's New" page, and Tim O'Reilly's Global Network

Navigator. Somewhere in between, and taking the concept to a new level, was a site created in early 1994 by two Stanford graduate students that was called, in one of its many early incarnations, Jerry and David's Guide to the Web.

As they were finishing their doctoral dissertations, Jerry Yang and David Filo would procrastinate by playing fantasy league basketball. "We did everything we could to avoid writing our theses," Yang recalled.[113] Yang spent time devising ways to ferret out player stats from servers that used FTP and Gopher, two protocols for distributing documents on the Internet that were popular before the rise of the Web.

When the Mosaic browser was released, Yang turned his attention to the Web, and he and Filo began compiling by hand an ever-expanding directory of sites. It was organized by categories—such as business, education, entertainment, government—each of which had dozens of subcategories. By the end of 1994, they had renamed their guide to the Web "Yahoo!"

There was one obvious problem: with the number of websites increasing tenfold each year, there was no way to keep a directory updated by hand. Fortunately, there was a tool that was already being used to ferret out information that resided on FTP and Gopher sites. It was called a crawler, because it crawled from server to server on the Internet compiling an index. The two most famous were named, like the comic book couple, Archie (for FTP archives) and Veronica (for Gopher). By 1994 a variety of enterprising engineers were creating crawlers that would serve as search tools for the Web. These included the WWW Wanderer built by Matthew Gray at MIT, WebCrawler by Brian Pinkerton at the University of Washington, AltaVista by Louis Monier at the Digital Equipment Corporation, Lycos by Michael Mauldin at Carnegie Mellon University, OpenText by a team from Canada's University of Waterloo, and Excite by six friends from Stanford. All of them used link-hopping robots, or bots, that could dart around the Web like a binge drinker on a pub crawl, scarfing up URLs and information about each site. This would then be tagged, indexed, and placed in a database that could be accessed by a query server.

Filo and Yang did not build their own web crawler; instead they decided to license one to add to their home page. Yahoo! continued to emphasize the importance of its directory, which was compiled by humans. When a user typed in a phrase, the Yahoo! computers would see if it related to an entry in the directory, and if so that handcrafted list of sites would pop up. If not, the query would be handed off to the Web-crawling search engine.

The Yahoo! team believed, mistakenly, that most users would navigate the Web by exploring rather than seeking something specific. "The shift from exploration and discovery to the intent-based search of today was inconceivable," recalled Srinija Srinivasan, Yahoo!'s first editor in chief, who oversaw a newsroom of more than sixty young editors and directory compilers.[114] This reliance on the human factor meant that Yahoo! would be much better than its rivals over the years (and even to the present) in choosing news stories, although not in providing search tools. But there was no way that Srinivasan and her team could keep up with the number of Web pages being created. Despite what she and her colleagues at Yahoo! believed, automated search engines would become the primary method for finding things on the Web, with another pair of Stanford graduate students leading the way.

Larry Page was born and bred in the world of computing.[115] His father was a professor of computer science and artificial intelligence at the University of Michigan, and his mother taught programming there. In 1979, when Larry was six, his father brought home an Exidy Sorcerer, a hobbyist home computer.* "I remember being really excited that we had a computer, because it was a big deal, and it was probably expensive, kind of like buying a car," he said.[116] Larry soon mastered it and was using it for his schoolwork. "I think I was the first kid in my elementary school to turn in a word-processed document."[117]

One of his childhood heroes was Nikola Tesla, the imaginative pioneer of electricity and other inventions who was outmaneuvered

*Created by the Byte Shop's owner Paul Terrell, who had launched the Apple I by ordering the first fifty for his store.

in business by Thomas Edison and died in obscurity. When he was twelve, Page read a biography of Tesla and found the story troubling. "He was one of the greatest inventors, but it's a sad, sad story," he said. "He couldn't commercialize anything, he could barely fund his own research. You'd want to be more like Edison. If you invent something, that doesn't necessarily help anybody. You've got to actually get it into the world; you've got to produce, make money doing it so you can fund it."[118]

Larry's parents used to take him and his brother, Carl, on long road trips, sometimes to computer conferences. "I think I ended up being in almost every state by the time I left for college," he observed. One such trip was to the International Joint Conference on Artificial Intelligence in Vancouver, which was filled with wondrous robots. Because he was under sixteen, Larry was told he couldn't come in, but his father insisted. "He just basically yelled at them. It's one of the few times I'd seen him argue."[119]

Like Steve Jobs and Alan Kay, Larry's other love besides computers was music. He played the saxophone and studied composition. During summers, he went to the famous music camp at Interlochen in northern Michigan. It had a method for crowdsourcing the rank of each kid: at the beginning of camp, kids were assigned chairs in the orchestra, and anyone could challenge the person a seat above them; the two competitors were given a selection of music, and all of the other kids would face the other direction and then vote on who sounded better. "After a while, things would kind of calm down, and everyone would kind of know where they were," he said.[120]

Page's parents not only taught at Michigan but had met as students there, so they were only half joking when they told him that he would go there as well. He did. He made a point of majoring in business as well as computer science, partly because of the cautionary tale of Tesla, who could invent but not commercialize. In addition, he had a role model in his brother, Carl, nine years older, who after college became a cofounder of an early social network company that was later sold to Yahoo! for $413 million.

The college course that made the greatest impression on him, Page said, was one on human-computer interaction taught by Judith

Olson. The goal was to understand how to design interfaces that were easy and intuitive. Page did his research paper on the display of the Eudora mail client, estimating and then testing how long it would take to perform various tasks. He discovered, for example, that command keys actually slowed people down by 0.9 second compared to using a mouse. "I feel like I developed an intuition for how people will interact with a screen, and I realized those things were pretty important," he said. "But they're not well understood, even to this day."[121]

One summer during his college years, Page went to a camp run by a leadership training institute called LeaderShape. It encouraged students to have a "healthy disregard for the impossible." The institute inculcated in him a desire, which he would indulge at Google, to launch projects that others considered to be on the borderline between audacious and insane. In particular, both at Michigan and later, he pushed futuristic ideas for personal transportation systems and driverless cars.[122]

When it came time to go to graduate school, Page was rejected by MIT but accepted by Stanford. That was fortuitous; for someone interested in the intersection of technology and business, Stanford was the place to be. Ever since the Stanford grad Cyril Elwell had founded Federal Telegraph in 1909, tech entrepreneurship was not merely tolerated but expected there, an attitude that was reinforced when the engineering dean Fred Terman built an industrial park on university land in the early 1950s. Even among the faculty, the focus was on startup business plans as much as on scholarly publications. "That's the kind of professor I wanted, one who has one foot in industry and wants to do crazy stuff that's world-breaking," Page said. "A lot of Stanford computer science professors are that way."[123]

At the time, most other elite universities emphasized scholarly research and avoided commercial endeavors. Stanford led the way in regarding the university not just as an academy but as an incubator. Among the companies spawned by Stanford were Hewlett-Packard, Cisco, Yahoo!, and Sun Microsystems. Page, who would eventually add the biggest name of all to that list, believed that this outlook actually improved the research. "I think the productivity of pure research was a lot higher, because it had a real-world grounding," he maintained. "It's not just theoretical. You want what you're working on to apply to a real problem."[124]

As he was preparing to enroll in Stanford's graduate school in the fall of 1995, Page participated in an orientation program that included a day in San Francisco. His guide was a gregarious second-year grad student named Sergey Brin. Page was naturally quiet, but Brin kept peppering him with opinions, and soon they were arguing about topics ranging from computers to urban zoning. They hit it off perfectly. "I remember thinking he was pretty obnoxious," Page admitted. "That's still the case. And maybe vice versa."[125] Yes, the feeling was mutual. "We both found each other obnoxious," Brin conceded. "But we say it a little bit jokingly. Obviously we spent a lot of time talking to each other, so there was something there. We had a kind of bantering thing going."[126]

Sergey Brin's parents were also academics, both mathematicians, but his childhood was very different from Page's. Brin was born in Moscow, where his father taught math at Moscow State University and his mother was a research engineer at the Soviet Oil and Gas Institute. Because they were Jewish, their careers were curtailed. "We were quite poor," Sergey told the journalist Ken Auletta. "My parents, both of them, went through periods of hardship." When his father applied to emigrate, he and his wife lost their jobs. Their exit visas came through in May 1979, when Sergey was five. With the help of the Hebrew Immigrant Aid Society, they settled in a working-class neighborhood near the University of Maryland, where his father got a job as a math professor and his mother became a researcher at the nearby NASA Goddard Space Flight Center.

Sergey went to a Montessori school, where independent thinking was nurtured. "It's not like somebody is telling you what to do," he said. "You have to plot your own path."[127] It was something he shared with Page. When asked later whether having parents who were professors was a key to their success, they both cited going to Montessori schools as a more important factor. "I think it was part of that training of not following rules and orders, and being self-motivated, questioning what's going on in the world and doing things a little bit differently," Page contended.[128]

Another thing Brin shared with Page was that his parents gave him a computer when he was very young, a Commodore 64 for his ninth birthday. "The ability to program your own computer was much more accessible than it is today," he recalled. "The computer came with a built-in BASIC interpreter,* and you could right away start writing your own programs." In middle school, Brin and a friend wrote programs that tried to simulate artificial intelligence by carrying on a text conversation with the user. "I don't think that kids starting out with computers today get as welcome of an entry to programming as I did."[129]

His rebellious attitude toward authority almost got him in trouble when his father took him back to Moscow for a visit when he was turning seventeen. Seeing a police car, he started throwing pebbles at it. The two officers got out of the car to confront Sergey, but his parents were able to defuse the situation. "My rebelliousness, I think, came out of being born in Moscow. I'd say this is something that followed me into adulthood."[130]

Among the books that inspired Brin were the memoirs of the physicist Richard Feynman, who touted the power that comes from joining art to science the way that Leonardo da Vinci did. "I remember he had an excerpt where he was explaining how he really wanted to be a Leonardo, an artist and a scientist," said Brin. "I found that pretty inspiring. I think that leads to having a fulfilling life."[131]

He was able to graduate from high school in three years and do the same at the University of Maryland, where he got an undergraduate degree in math and computer science. For a while, he and his fellow computer geeks liked to hang around on the Internet's bulletin boards and chat rooms, until he got bored with "ten-year-old boys trying to talk about sex." He then got involved in the text-based online games known as Multi-User Dungeons, writing one of his own that involved a mailman who delivered explosive packages. "I spent enough time MUDding to think it was cool," Brin recalled.[132] In the

---

*The one written by Bill Gates.

spring of 1993, his final year at Maryland, he downloaded the Mosaic browser, which Andreessen had just released, and became mesmerized by the Web.

Brin went to Stanford on a National Science Foundation scholarship, where he decided to focus on the study of data mining. (In a double whammy, to itself if not to them, MIT rejected him as well as Page.) There were eight comprehensive tests he needed to pass for his doctorate, and he aced seven of them soon after he arrived. "The one I thought I was best at, I didn't pass," he recalled. "I went to the prof and debated the answers. I wound up talking him into it. So I passed all eight."[133] That left him free to dabble in whatever courses he wanted and to indulge his quirky athletic interests in acrobatics, trapeze, sailing, gymnastics, and swimming. He could walk on his hands and, so he claimed, once considered running away and joining the circus. He was also an avid rollerblader, and was often seen zipping through the halls.

A few weeks after Page arrived at Stanford, he and Brin moved with the rest of the Computer Science Department into the new Gates Computer Science Building.* Annoyed by the uninspired numbering system for offices that the architect had provided, Brin devised a new system, which was adopted, that conveyed better the location of each room and the distance between them. "It was very intuitive, if I may say so," he said.[134] Page was assigned to a room with three other graduate students, and Brin made that his base as well. There were hanging plants with a computer-controlled watering system, a piano connected to a computer, an assortment of electronic toys, and sleeping pads for naps and all-nighters.

The inseparable duo became linked, in CamelCase fashion, as LarryAndSergey, and when engaged in argument or banter they were like two swords sharpening each other. Tamara Munzner, the only woman in the group, had a phrase for it: "goofy smart," she called

---

*Gates donated to computer buildings at Harvard, Stanford, MIT, and Carnegie Mellon. The one at Harvard, cofunded with Steve Ballmer, was named Maxwell Dworkin, after their mothers.

them, especially when they took to debating absurd concepts, such as whether it was possible to construct something the size of a building using only lima beans. "They were fun guys to share an office with," she said. "We all kept crazy hours. I remember once at three in the morning on a Saturday night, the office was full."[135] The duo was notable not only for their brilliance but for their boldness. "They didn't have this false respect for authority," according to Professor Rajeev Motwani, one of their advisors. "They were challenging me all the time. They had no compunction in saying to me, 'You're full of crap!' "[136]

Like many great partners in innovation, LarryAndSergey had complementary personalities. Page was not a social animal; he could make eye contact with a screen more easily than with a stranger. A chronic vocal cord problem, stemming from a viral infection, meant that he spoke in a whispery and raspy voice, and he had a disconcerting (although in many ways admirable) habit of simply not talking at times, which made his utterances, when they did occur, all the more memorable. He could be impressively detached but was sometimes intensely engaging. His smile was quick and real, his face expressive, and he listened with a focus that could be both flattering and unnerving. Intellectually rigorous, he could find logical flaws in the most mundane comments and effortlessly steer a shallow conversation into a deep discussion.

Brin, for his part, could be charmingly brash. He would barge into offices without knocking, blurt out ideas and requests, and engage on any subject. Page was more reflective and reserved. Whereas Brin was satisfied knowing that something worked, Page would ruminate about why it worked. The intense and talkative Brin dominated a room, but Page's quiet comments at the end of a discussion made people lean forward and listen. "I was probably a little bit more shy than Sergey, although he's shy in some ways," Page observed. "We had a great partnership, because I maybe thought more broadly and had different skills. I'm trained as a computer engineer. I'm more knowledgeable about the hardware. He has more of a mathematical background."[137]

What particularly amazed Page was how smart Brin was. "I mean, he was just unusually smart, even for somebody at the Computer Science Department." In addition, Brin's outgoing personality helped him bring people together. When Page arrived at Stanford, he was given a desk in an open room known as the bullpen with the other new graduate students. "Sergey was pretty social," Page said. "He would meet all the students and come hang out in the bullpen with us." Brin even had a knack for befriending the professors. "Sergey had this way of walking into professors' offices and hanging out with them, which was kind of unusual for a grad student. I think they tolerated it because he was so smart and knowledgeable. He could contribute on all sorts of random things."[138]

Page joined the Human-Computer Interaction Group, which explored ways to enhance the symbiosis between humans and machines. It was the field that had been pioneered by Licklider and Engelbart, and it had been the subject of his favorite course at Michigan. He became an adherent of the concept of user-centered design, which insisted that software and computer interfaces must be intuitive and that the user was always right. He had gone to Stanford knowing that he wanted as his advisor Terry Winograd, a joyful, Einstein-haired professor. Winograd had studied artificial intelligence but, after reflecting on the essence of human cognition, changed his focus, as Engelbart had, to how machines could augment and amplify (rather than replicate and replace) human thinking. "I shifted my view away from what would be thought of as artificial intelligence to the broader question, 'How do you want to interact with a computer?'" Winograd explained.[139]

The field of human-computer interactions and interface design, despite its noble heritage from Licklider, was still considered a rather soft discipline, looked down upon by hardnosed computer scientists as something usually taught by mere psychology professors, which Licklider and Judith Olson had once been. "For people studying Turing machines or whatever, dealing with human responses was considered very touchy-feely, almost like you're stuck in the humanities," according to Page. Winograd helped make the field more reputable. "Terry had a hard computer science background from his time working on artificial intelligence, but he was also interested in human-computer interaction, a field that nobody much was working on and

I think didn't get enough respect." One of Page's favorite courses was Film Craft in User Interface Design. "It showed how the language and techniques of film can actually be applied to computer interface designs," he said.[140]

Brin's academic focus was on data mining. With Professor Motwani, he started a group called Mining Data at Stanford, or MIDAS. Among the papers they produced (along with another graduate student, Craig Silverstein, who would become the first hire when they founded Google) were two on market basket analysis, a technique that assesses to what extent a consumer who buys items A and B is more or less likely also to buy items C and D.[141] From that Brin became interested in ways to analyze patterns from the data trove on the Web.

With Winograd's help, Page began casting around for a dissertation topic. He considered close to a dozen ideas, including one on how to design self-driving cars, as Google would later do. Eventually he homed in on studying how to assess the relative importance of different sites on the Web. His method came from growing up in an academic environment. One criterion that determines the value of a scholarly paper is how many other researchers cite it in their notes and bibliography. By the same theory, one way to determine the value of a Web page was to look at how many other Web pages linked to it.

There was a problem. The way that Tim Berners-Lee had designed the Web, much to the consternation of hypertext purists such as Ted Nelson, anyone could create a link to another page without getting permission, registering the link in a database, or having the link work in both directions. That permitted the Web to expand willy-nilly. But it also meant that there was no simple way of knowing the number of links pointing to a Web page or where those links might be coming from. You could look at a Web page and see all the links going out, but you couldn't see the number or the quality of the links pointing into it. "The Web was a poorer version of other collaboration systems I had seen because its hypertext had a flaw: it didn't have bidirectional links," said Page.[142]

So Page set about trying to figure out a way to gather a huge database of the links so that he could follow them in reverse and see which sites were linking to each page. One motivation was to foster collaboration. His scheme would allow folks to annotate another page. If Harry wrote a comment and linked it to Sally's website, then people looking at her website could go see his comment. "By reversing the links, making it possible to trace them backwards, it would allow people to comment on or annotate a site simply by linking to it," Page explained.[143]

Page's method for reversing links was based on an audacious idea that struck him in the middle of the night when he woke up from a dream. "I was thinking: What if we could download the whole Web, and just keep the links," he recalled. "I grabbed a pen and started writing. I spent the middle of that night scribbling out the details and convincing myself it would work."[144] His nocturnal burst of activity served as a lesson. "You have to be a little silly about the goals you are going to set," he later told a group of Israeli students. "There is a phrase I learned in college called, 'Having a healthy disregard for the impossible.' That is a really good phrase. You should try to do things that most people would not."[145]

Mapping the web was not a simple task. Even back then, in January 1996, there were 100,000 websites with a total of 10 million documents and close to a billion links between them, and it was growing exponentially each year. Early that summer, Page created a Web crawler that was designed to start on his home page and follow all of the links it encountered. As it darted like a spider through the Web, it would store the text of each hyperlink, the titles of the pages, and a record of where each link came from. He called the project BackRub.

Page told his advisor Winograd that, according to his rough estimate, his Web crawler would be able to accomplish the task in a few weeks. "Terry nodded knowingly, fully aware it would take much longer but wise enough to not tell me," Page recalled. "The optimism of youth is often underrated!"[146] The project was soon taking up almost half of Stanford's entire Internet bandwidth, and it caused at least one campuswide outage. But university officials were indulgent. "I am almost out of disk space," Page emailed Winograd on July 15, 1996, after he had collected 24 million URLs and more than 100 million links. "I have only about 15% of the pages but it seems very promising."[147]

Both the audacity and the complexity of Page's project appealed to the mathematical mind of Sergey Brin, who had been searching for a dissertation topic. He was thrilled to join forces with his friend: "This was the most exciting project, both because it tackled the Web, which represents human knowledge, and because I liked Larry."[148]

BackRub was still, at that point, intended to be a compilation of backlinks on the Web that would serve as the basis for a possible annotation system and citation analysis. "Amazingly, I had no thought of building a search engine," Page admitted. "The idea wasn't even on the radar." As the project evolved, he and Brin conjured up more sophisticated ways to assess the value of each page, based on the number and quality of links coming into it. That's when it dawned on the BackRub Boys that their index of pages ranked by importance could become the foundation for a high-quality search engine. Thus was Google born. "When a really great dream shows up," Page later said, "grab it!"[149]

At first the revised project was called PageRank, because it ranked each page captured in the BackRub index and, not incidentally, played to Page's wry humor and touch of vanity. "Yeah, I was referring to myself, unfortunately," he later sheepishly admitted. "I feel kind of bad about it."[150]

That page-ranking goal led to yet another layer of complexity. Instead of just tabulating the number of links that pointed to a page, Page and Brin realized that it would be even better if they could also assign a value to each of those incoming links. For example, an incoming link from the *New York Times* should count for more than a link from Justin Hall's dorm room at Swarthmore. That set up a recursive process with multiple feedback loops: each page was ranked by the number and quality of links coming into it, and the quality of these links was determined by the number and quality of links to the pages that originated them, and so on. "It's all recursive," Page explained. "It's all a big circle. But mathematics is great. You can solve this."[151]

This was the type of mathematical complexity that Brin could truly appreciate. "We actually developed a lot of math to solve that problem," he recalled. "We converted the entire web into a big equation with several hundred million variables, which are the page ranks of all the web pages."[152] In a paper they coauthored with their two academic advisors, they spelled out the complex math formulas based on how many incoming links a page had and the relative rank of each of these links. Then they put it in simple words for the layperson: "A page has a high rank if the sum of the ranks of its backlinks is high. This covers both the case when a page has many backlinks and when a page has a few highly ranked backlinks."[153]

The billion-dollar question was whether PageRank would actually produce better search results. So they did a comparison test. One example they used was searching *university*. In AltaVista and other engines, that would turn up a list of random pages that might happen to use that word in their title. "I remember asking them, 'Why are you giving people garbage?'" Page said. The answer he got was that the poor results were his fault, that he should refine his search query. "I had learned from my human-computer interaction course that blaming the user is not a good strategy, so I knew they fundamentally weren't doing the right thing. That insight, the user is never wrong, led to this idea that we could produce a search engine that was better."[154] With PageRank, the top results for a search on *university* were Stanford, Harvard, MIT, and the University of Michigan, which pleased them immensely. "Wow," Page recalled saying to himself. "It was pretty clear to me and the rest of the group that if you have a way of ranking things based not just on the page itself but based on what the world thought of that page, that would be a really valuable thing for search."[155]

Page and Brin proceeded to refine PageRank by adding more factors, such as the frequency, type size, and location of keywords on a Web page. Extra points were added if the keyword was in the URL or was capitalized or was in the title. They would look at each set of results, then tweak and refine the formula. They discovered that it was important to give a lot of weight to the anchor text, the words that were underlined as a hyperlink. For example, the words *Bill Clinton* were the anchor text for many links leading to whitehouse.gov, so that Web page went to the top when a user searched *Bill Clinton*, even though the whitehouse.gov site did not have Bill Clinton's name prominently on its home page. One competitor, by contrast, had "Bill Clinton Joke of the Day" as its number-one result when a user searched *Bill Clinton*.[156]

Partly because of the huge number of pages and links involved, Page and Brin named their search engine Google, playing off *googol*, the term for the number 1 followed by a hundred zeros. It was a suggestion made by one of their Stanford officemates, Sean Anderson, and when they typed in *Google* to see if the domain name was available, it was. So Page snapped it up. "I'm not sure that we realized that we had made a spelling error," Brin later said. "But googol was taken, anyway. There was this guy who'd already registered Googol.com, and I tried to buy it from him, but he was fond of it. So we went with Google."[157] It was a playful word, easy to remember, type, and turn into a verb.*

Page and Brin pushed to make Google better in two ways. First, they deployed far more bandwidth, processing power, and storage capacity to the task than any rival, revving up their Web crawler so that it was indexing a hundred pages per second. In addition, they were fanatic in studying user behavior so that they could constantly tweak their algorithms. If users clicked on the top result and then didn't return to the results list, it meant they had gotten what they wanted. But if they did a search and returned right away to revise their query, it meant that they were dissatisfied and the engineers should learn, by looking at the refined search query, what they had been seeking in the first place. Anytime users scrolled to the second or third page of the search results, it was a sign that they were unhappy with the order of results they received. As the journalist Steven Levy pointed out, this feedback loop helped Google learn that when users typed in *dogs* they also were looking for *puppies*, and when they typed in *boiling* they might also be referring to *hot water*, and eventually Google

---

*The Oxford English Dictionary added *google* as a verb in 2006.

also learned that when they typed in *hot dog* they were *not* looking for *boiling puppies*.[158]

One other person came up with a link-based scheme very similar to PageRank: a Chinese engineer named Yanhong (Robin) Li, who studied at SUNY Buffalo and then joined a division of Dow Jones based in New Jersey. In the spring of 1996, just as Page and Brin were creating PageRank, Li came up with an algorithm he dubbed RankDex that determined the value of search results by the number of inbound links to a page and the content of the text that anchored those links. He bought a self-help book on how to patent the idea, and then did so with the help of Dow Jones. But the company did not pursue the idea, so Li moved west to work for Infoseek and then back to China. There he cofounded Baidu, which became that country's largest search engine and one of Google's most powerful global competitors.

By early 1998 Page and Brin's database contained maps of close to 518 million hyperlinks, out of approximately 3 billion by then on the Web. Page was eager that Google not remain just an academic project but would also become a popular product. "It was like Nikola Tesla's problem," he said. "You make an invention you think is great, and so you want it to be used by many people as soon as possible."[159]

The desire to turn their dissertation topic into a business made Page and Brin reluctant to publish or give formal presentations on what they had done. But their academic advisors kept pushing them to publish something, so in the spring of 1998 they produced a twenty-page paper that managed to explain the academic theories behind PageRank and Google without opening their kimono so wide that it revealed too many secrets to competitors. Titled "The Anatomy of a Large-Scale Hypertextual Web Search Engine," it was delivered at a conference in Australia in April 1998.

"In this paper, we present Google, a prototype of a large-scale search engine which makes heavy use of the structure present in hypertext," they began.[160] By mapping more than a half billion of the Web's 3 billion links, they were able to calculate a PageRank for at least 25 million Web pages, which "corresponds well with people's subjective idea of importance." They detailed the "simple iterative algorithm" that produced PageRanks for every page. "Academic citation literature has been applied to the web, largely by counting citations or backlinks to a given page. This gives some approximation of a page's importance or quality. PageRank extends this idea by not counting links from all pages equally."

The paper included many technical details about ranking, crawling, indexing, and iterating the algorithms. There were also a few paragraphs about useful directions for future research. But by the end, it was clear this was not an academic exercise or purely scholarly pursuit. They were engaged in what would clearly become a commercial enterprise. "Google is designed to be a scalable search engine," they declared in conclusion. "The primary goal is to provide high quality search results."

This may have been a problem at universities where research was supposed to be pursued primarily for scholarly purposes, not commercial applications. But Stanford not only permitted students to work on commercial endeavors, it encouraged and facilitated it. There was even an office to assist with the patenting process and licensing arrangements. "We have an environment at Stanford that promotes entrepreneurship and risk-taking research," President John Hennessy declared. "People really understand here that sometimes the biggest way to deliver an effect to the world is not by writing a paper but by taking technology you believe in and making something of it."[161]

Page and Brin began by trying to license their software to other companies, and they met with the CEOs of Yahoo!, Excite, and AltaVista. They asked for a $1 million fee, which was not exorbitant since it would include the rights to their patents as well as the personal services of the two of them. "Those companies were worth hundreds of millions or more at the time," Page later said. "It wasn't that significant of an expense to them. But it was a lack of insight at the leadership level. A lot of them told us, 'Search is not that important.'"[162]

As a result, Page and Brin decided to start a company of their own. It helped that within a few miles of the campus there were successful entrepreneurs to act as angel investors, as well as eager venture capitalists just up Sand Hill Road to provide working capital. David Cheriton, one of their professors at Stanford, had founded an Ethernet product company with one such investor, Andy Bechtolsheim, which they had sold to Cisco Systems. In August 1998 Cheriton suggested to Page and Brin that they meet with Bechtolsheim, who had also cofounded Sun Microsystems. Late one night, Brin sent him an email. He got an instant reply, and early the next morning they all met on Cheriton's Palo Alto porch.

Even at that unholy hour for students, Page and Brin were able to give a compelling demo of their search engine, showing that they could download, index, and page-rank much of the Web on racks of minicomputers. It was a comfortable meeting at the height of the dotcom boom, and Bechtolsheim's questions were encouraging. Unlike the scores of pitches that came to him each week, this was not a PowerPoint presentation of some vaporware that didn't yet exist. He could actually type in queries, and answers popped up instantly that were far better than what AltaVista produced. Plus the two founders were whip smart and intense, the type of entrepreneurs he liked to bet on. Bechtolsheim appreciated that they were not throwing large amounts of money—or any money, for that matter—at marketing. They knew that Google was good enough to spread by word of mouth, so every penny they had went to components for the computers they were assembling themselves. "Other Web sites took a good chunk of venture funding and spent it on advertising," Bechtolsheim said. "This was the opposite approach. Build something of value and deliver a service compelling enough that people would just use it."[163]

Even though Brin and Page were averse to accepting advertising, Bechtolsheim knew that it would be simple—and not corrupting—to put clearly labeled display ads on the search results page. That meant there was an obvious revenue stream waiting to be tapped. "This is the single best idea I have heard in years," he told them. They talked about valuation for a minute, and Bechtolsheim said they were setting their price too low. "Well, I don't want to waste time," he concluded, since he had to get to work. "I'm sure it'll help you guys if I just write a check." He went to the car to get his checkbook and wrote one made out to Google Inc. for $100,000. "We don't have a bank account yet," Brin told him. "Deposit it when you get one," Bechtolsheim replied. Then he rode off in his Porsche.

Brin and Page went to Burger King to celebrate. "We thought we should get something that tasted really good, though it was really unhealthy," Page said. "And it was cheap. It seemed like the right combination of ways to celebrate the funding."[164]

Bechtolsheim's check made out to Google Inc. provided a spur to get themselves incorporated. "We had to quickly get a lawyer," Brin said.[165] Page recalled, "It was like, wow, maybe we really should start a company now."[166] Because of Bechtolsheim's reputation—and because of the impressive nature of Google's product—other funders came in, including Amazon's Jeff Bezos. "I just fell in love with Larry and Sergey," Bezos declared. "They had a vision. It was a customer-focused point of view."[167] The favorable buzz around Google grew so loud that, a few months later, it was able to pull off the rare feat of getting investments from both of the valley's rival top venture capital firms, Sequoia Capital and Kleiner Perkins.

Silicon Valley had one other ingredient, in addition to a helpful university and eager mentors and venture capitalists: a lot of garages, like the ones in which Hewlett and Packard designed their first products and Jobs and Wozniak assembled the first Apple I boards. When Page and Brin realized that it was time to put aside plans for dissertations and leave the Stanford nest, they found a garage—a two-car garage, which came with a hot tub and a couple of spare rooms inside the house—that they could rent for $1,700 a month at the Menlo Park house of a Stanford friend, Susan Wojcicki, who soon joined Google. In September 1998, one month after they met with Bechtolsheim, Page and Brin incorporated their company, opened a bank account, and cashed his check. On the wall of the garage they put up a whiteboard emblazoned "Google Worldwide Headquarters."

In addition to making all of the World Wide Web's information accessible, Google represented a climactic leap in the relationship between humans and machines—the "man-computer symbiosis" that Licklider had envisioned four decades earlier. Yahoo! had attempted a more

primitive version of this symbiosis by using both electronic searches and human-compiled directories. The approach that Page and Brin took might appear, at first glance, to be a way of removing human hands from this formula by having the searches performed by Web crawlers and computer algorithms only. But a deeper look reveals that their approach was in fact a melding of machine and human intelligence. Their algorithm relied on the billions of human judgments made by people when they created links from their own websites. It was an automated way to tap into the wisdom of humans—in other words, a higher form of human-computer symbiosis. "The process might seem completely automated," Brin explained, "but in terms of how much human input goes into the final product, there are millions of people who spend time designing their webpages, determining who to link to and how, and that human element goes into it."[168]

In his seminal 1945 essay "As We May Think," Vannevar Bush had set forth the challenge: "The summation of human experience is being expanded at a prodigious rate, and the means we use for threading through the consequent maze to the momentarily important item is the same as was used in the days of square-rigged ships." In the paper they submitted to Stanford just before they left to launch their company, Brin and Page made the same point: "The number of documents in the indices has been increasing by many orders of magnitude, but the user's ability to look at documents has not." Their words were less eloquent than Bush's, but they had succeeded in fulfilling his dream of a human-machine collaboration to deal with information overload. In doing so, Google became the culmination of a sixty-year process to create a world in which humans, computers, and networks were intimately linked. Anyone could share with people anywhere and, as the Victorian-era almanac promised, enquire within upon everything.

CHAPTER TWELVE

ADA FOREVER

LADY LOVELACE'S OBJECTION

Ada Lovelace would have been pleased. To the extent that we are permitted to surmise the thoughts of someone who's been dead for more than 150 years, we can imagine her writing a proud letter boasting about her intuition that calculating devices would someday become general-purpose computers, beautiful machines that can not only manipulate numbers but make music and process words and "combine together general symbols in successions of unlimited variety."

Machines such as these emerged in the 1950s, and during the subsequent thirty years there were two historic innovations that caused them to revolutionize how we live: microchips allowed computers to become small enough to be personal appliances, and packet-switched networks allowed them to be connected as nodes on a web. This merger of the personal computer and the Internet allowed digital creativity, content sharing, community formation, and social networking to blossom on a mass scale. It made real what Ada called "poetical science," in which creativity and technology were the warp and woof, like a tapestry from Jacquard's loom.

Ada might also be justified in boasting that she was correct, at least thus far, in her more controversial contention: that no computer, no matter how powerful, would ever truly be a "thinking" machine. A century after she died, Alan Turing dubbed this "Lady Lovelace's Objection" and tried to dismiss it by providing an operational definition of a thinking machine—that a person submitting questions could not distinguish the machine from a human—and predicting that a computer would pass this test within a few decades. But it's now been more than sixty years, and the machines that attempt to fool people on the test are at best engaging in lame conversation tricks rather than actual thinking. Certainly none has cleared Ada's higher bar of being able to "originate" any thoughts of its own.

Ever since Mary Shelley conceived her Frankenstein tale during a vacation with Ada's father, Lord Byron, the prospect that a man-made contraption might originate its own thoughts has unnerved generations. The Frankenstein motif became a staple of science fiction. A vivid example was Stanley Kubrick's 1968 movie, *2001: A Space Odyssey*, featuring the frighteningly intelligent computer HAL. With its calm voice, HAL exhibits attributes of a human: the ability to speak, reason, recognize faces, appreciate beauty, show emotion, and (of course) play chess. When HAL appears to malfunction, the human astronauts decide to shut it down. HAL becomes aware of the plan and kills all but one of them. After a lot of heroic struggle, the remaining astronaut gains access to HAL's cognitive circuits and disconnects them one by one. HAL regresses until, at the end, it intones "Daisy Bell"—an homage to the first computer-generated song, sung by an IBM 704 at Bell Labs in 1961.

Artificial intelligence enthusiasts have long been promising, or threatening, that machines like HAL would soon emerge and prove Ada wrong. Such was the premise of the 1956 conference at Dartmouth organized by John McCarthy and Marvin Minsky, where the field of artificial intelligence was launched. The conferees concluded that a breakthrough was about twenty years away. It wasn't. Decade after decade, new waves of experts have claimed that artificial intelligence was on the visible horizon, perhaps only twenty years away. Yet it has remained a mirage, always about twenty years away.

John von Neumann was working on the challenge of artificial intelligence shortly before he died in 1957. Having helped devise the architecture of modern digital computers, he realized that the architecture of the human brain is fundamentally different. Digital computers deal in precise units, whereas the brain, to the extent we understand it, is also partly an analog system, which deals with a continuum of possibilities. In other words, a human's mental process includes many signal pulses and analog waves from different nerves that flow together to produce not just binary yes-no data but also answers such as "maybe" and "probably" and infinite other nuances, including occasional bafflement. Von Neumann suggested that the future of intelligent computing might require abandoning the purely digital approach and creating "mixed procedures" that include a combination of digital and analog methods. "Logic will have to undergo a pseudomorphosis to neurology," he declared, which, roughly translated, meant that computers were going to have to become more like the human brain.[1]

In 1958 a Cornell professor, Frank Rosenblatt, attempted to do this by devising a mathematical approach for creating an artificial neural network like that of the brain, which he called a Perceptron. Using weighted statistical inputs, it could, in theory, process visual data. When the Navy, which was funding the work, unveiled the system, it drew the type of press hype that has accompanied many subsequent artificial intelligence claims. "The Navy revealed the embryo of an electronic computer today that it expects will be able to walk, talk, see, write, reproduce itself and be conscious of its existence," the *New York Times* reported. The *New Yorker* was equally enthusiastic: "The Perceptron, . . . as its name implies, is capable of what amounts to original thought. . . . It strikes us as the first serious rival to the human brain ever devised."[2]

That was almost sixty years ago. The Perceptron still does not exist.[3] Nevertheless, almost every year since then there have been breathless reports about some marvel on the horizon that would replicate and surpass the human brain, many of them using almost the exact same phrases as the 1958 stories about the Perceptron.

Discussion about artificial intelligence flared up a bit, at least in the popular press, after IBM's Deep Blue, a chess-playing machine, beat the world champion Garry Kasparov in 1997 and then Watson, its natural-language question-answering computer, won at *Jeopardy!* against champions Brad Rutter and Ken Jennings in 2011. "I think it awakened the entire artificial intelligence community," said IBM CEO Ginni Rometty.[4] But as she was the first to admit, these were not true breakthroughs of humanlike artificial intelligence. Deep Blue won its chess match by brute force; it could evaluate 200 million positions per second and match them against 700,000 past grandmaster games. Deep Blue's calculations were fundamentally different, most of us would agree, from what we mean by *real* thinking. "Deep Blue was only intelligent the way your programmable alarm clock is intelligent," Kasparov said. "Not that losing to a $10 million alarm clock made me feel any better."[5]

Likewise, Watson won at *Jeopardy!* by using megadoses of computing power: it had 200 million pages of information in its four terabytes of storage, of which the entire Wikipedia accounted for merely 0.2 percent. It could search the equivalent of a million books per second. It was also rather good at processing colloquial English. Still, no one who watched would bet on its passing the Turing Test. In fact, the IBM team leaders were afraid that the show's writers might try to turn the game into a Turing Test by composing questions designed to trick a machine, so they insisted that only old questions from unaired contests be used. Nevertheless, the machine tripped up in ways that showed it wasn't human. For example, one question was about the "anatomical oddity" of the former Olympic gymnast George Eyser. Watson answered, "What is a leg?" The correct answer was that Eyser was missing a leg. The problem was understanding *oddity*, explained David Ferrucci, who ran the Watson project at IBM. "The computer wouldn't know that a missing leg is odder than anything else."[6]

John Searle, the Berkeley philosophy professor who devised the "Chinese room" rebuttal to the Turing Test, scoffed at the notion that Watson represented even a glimmer of artificial intelligence. "Watson did not understand the questions, nor its answers, nor that some of its answers were right and some wrong, nor that it was playing a game, nor that it won—because it doesn't understand anything," Searle contended. "IBM's computer was not and could not have been designed to understand. Rather, it was designed to simulate understanding, to act as if it understood."[7]

Even the IBM folks agreed with that. They never held Watson out to be an "intelligent" machine. "Computers today are brilliant idiots," said the company's director of research, John E. Kelly III, after the Deep Blue and Watson victories. "They have tremendous capacities for storing information and performing numerical calculations—far superior to those of any human. Yet when it comes to another class of skills, the capacities for understanding, learning, adapting, and interacting, computers are woefully inferior to humans."[8]

Rather than demonstrating that machines are getting close to artificial intelligence, Deep Blue and Watson actually indicated the contrary. "These recent achievements have, ironically, underscored the limitations of computer science and artificial intelligence," argued Professor Tomaso Poggio, director of the Center for Brains, Minds, and Machines at MIT. "We do not yet understand how the brain gives rise to intelligence, nor do we know how to build machines that are as broadly intelligent as we are."[9]

Douglas Hofstadter, a professor at Indiana University, combined the arts and sciences in his unexpected 1979 best seller, *Gödel, Escher, Bach*. He believed that the only way to achieve meaningful artificial intelligence was to understand how human imagination worked. His approach was pretty much abandoned in the 1990s, when researchers found it more cost-effective to tackle complex tasks by throwing massive processing power at huge amounts of data, the way Deep Blue played chess.[10]

This approach produced a peculiarity: computers can do some of the toughest tasks in the world (assessing billions of possible chess positions, finding correlations in hundreds of Wikipedia-size information repositories), but they cannot perform some of the tasks that seem most simple to us mere humans. Ask Google a hard question like "What is the depth of the Red Sea?" and it will instantly respond,

"7,254 feet," something even your smartest friends don't know. Ask it an easy one like "Can a crocodile play basketball?" and it will have no clue, even though a toddler could tell you, after a bit of giggling.[11]

At Applied Minds near Los Angeles, you can get an exciting look at how a robot is being programmed to maneuver, but it soon becomes apparent that it still has trouble navigating an unfamiliar room, picking up a crayon, and writing its name. A visit to Nuance Communications near Boston shows the wondrous advances in speech-recognition technologies that underpin Siri and other systems, but it's also apparent to anyone using Siri that you still can't have a truly meaningful conversation with a computer, except in a fantasy movie. At the Computer Science and Artificial Intelligence Laboratory of MIT, interesting work is being done on getting computers to perceive objects visually, but even though the machine can discern pictures of a girl with a cup, a boy at a water fountain, and a cat lapping up cream, it cannot do the simple abstract thinking required to figure out that they are all engaged in the same activity: drinking. A visit to the New York City police command system in Manhattan reveals how computers scan thousands of feeds from surveillance cameras as part of a Domain Awareness System, but the system still cannot reliably identify your mother's face in a crowd.

All of these tasks have one thing in common: even a four-year-old can do them. "The main lesson of thirty-five years of AI research is that the hard problems are easy and the easy problems are hard," according to Steven Pinker, the Harvard cognitive scientist.[12] As the futurist Hans Moravec and others have noted, this paradox stems from the fact that the computational resources needed to recognize a visual or verbal pattern are huge.

Moravec's paradox reinforces von Neumann's observations from a half century ago about how the carbon-based chemistry of the human brain works differently from the silicon-based binary logic circuits of a computer. Wetware is different from hardware. The human brain not only combines analog and digital processes, it also is a distributed system, like the Internet, rather than a centralized one, like a computer. A computer's central processing unit can execute instructions much faster than a brain's neuron can fire. "Brains more than make up for this, however, because all the neurons and synapses are active simultaneously, whereas most current computers have only one or at most a few CPUs," according to Stuart Russell and Peter Norvig, authors of the foremost textbook on artificial intelligence.[13]

So why not make a computer that mimics the processes of the human brain? "Eventually we'll be able to sequence the human genome and replicate how nature did intelligence in a carbon-based system," Bill Gates speculates. "It's like reverse-engineering someone else's product in order to solve a challenge."[14] That won't be easy. It took scientists forty years to map the neurological activity of the one-millimeter-long roundworm, which has 302 neurons and 8,000 synapses.* The human brain has 86 billion neurons and up to 150 trillion synapses.[15]

At the end of 2013, the *New York Times* reported on "a development that is about to turn the digital world on its head" and "make possible a new generation of artificial intelligence systems that will perform some functions that humans do with ease: see, speak, listen, navigate, manipulate and control." The phrases were reminiscent of those used in its 1958 story on the Perceptron ("will be able to walk, talk, see, write, reproduce itself and be conscious of its existence"). Once again, the strategy was to replicate the way the human brain's neural networks operate. As the *Times* explained, "the new computing approach is based on the biological nervous system, specifically on how neurons react to stimuli and connect with other neurons to interpret information."[16] IBM and Qualcomm each disclosed plans to build "neuromorphic," or brainlike, computer processors, and a European research consortium called the Human Brain Project announced that it had built a neuromorphic microchip that incorporated "fifty million plastic synapses and 200,000 biologically realistic neuron models on a single 8-inch silicon wafer."[17]

Perhaps this latest round of reports does in fact mean that, in a

---

*A neuron is a nerve cell that transmits information using electrical or chemical signals. A synapse is a structure or pathway that carries a signal from a neuron to another neuron or cell.

few more decades, there will be machines that think like humans. "We are continually looking at the list of things machines cannot do—play chess, drive a car, translate language—and then checking them off the list when machines become capable of these things," said Tim Berners-Lee. "Someday we will get to the end of the list."[18]

These latest advances may even lead to the *singularity*, a term that von Neumann coined and the futurist Ray Kurzweil and the science fiction writer Vernor Vinge popularized, which is sometimes used to describe the moment when computers are not only smarter than humans but also can design themselves to be even supersmarter, and will thus no longer need us mortals. Vinge says this will occur by 2030.[19]

On the other hand, these latest stories might turn out to be like the similarly phrased ones from the 1950s, glimpses of a receding mirage. True artificial intelligence may take a few more generations or even a few more centuries. We can leave that debate to the futurists. Indeed, depending on your definition of consciousness, it may never happen. We can leave *that* debate to the philosophers and theologians. "Human ingenuity," wrote Leonardo da Vinci, whose Vitruvian Man became the ultimate symbol of the intersection of art and science, "will never devise any inventions more beautiful, nor more simple, nor more to the purpose than Nature does."

There is, however, yet another possibility, one that Ada Lovelace would like, which is based on the half century of computer development in the tradition of Vannevar Bush, J. C. R. Licklider, and Doug Engelbart.

HUMAN-COMPUTER SYMBIOSIS: "WATSON, COME HERE"

"The Analytical Engine has no pretensions whatever to *originate* anything," Ada Lovelace declared. "It can do whatever we know how to order it to perform." In her mind, machines would not replace humans but instead become their partners. What humans would bring to this relationship, she said, was originality and creativity.

This was the idea behind an alternative to the quest for pure artificial intelligence: pursuing instead the augmented intelligence that occurs when machines become partners with people. The strategy of combining computer and human capabilities, of creating a human-computer symbiosis, turned out to be more fruitful than the pursuit of machines that could think on their own.

Licklider helped chart that course back in 1960 in his paper "Man-Computer Symbiosis," which proclaimed: "Human brains and computing machines will be coupled together very tightly, and the resulting partnership will think as no human brain has ever thought and process data in a way not approached by the information-handling machines we know today."[20] His ideas built on the memex personal computer that Vannevar Bush had imagined in his 1945 essay, "As We May Think." Licklider also drew on his work designing the SAGE air defense system, which required an intimate collaboration between humans and machines.

The Bush-Licklider approach was given a friendly interface by Engelbart, who in 1968 demonstrated a networked computer system with an intuitive graphical display and a mouse. In a manifesto titled "Augmenting Human Intellect," he echoed Licklider. The goal, Engelbart wrote, should be to create "an integrated domain where hunches, cut-and-try, intangibles, and the human 'feel for a situation' usefully co-exist with . . . high-powered electronic aids." Richard Brautigan, in his poem "All Watched Over by Machines of Loving Grace," expressed that dream a bit more lyrically: "a cybernetic meadow / where mammals and computers / live together in mutually / programming harmony."

The teams that built Deep Blue and Watson have adopted this symbiosis approach rather than pursue the objective of the artificial intelligence purists. "The goal is not to replicate human brains," says John Kelly, the director of IBM Research. Echoing Licklider, he adds, "This isn't about replacing human thinking with machine thinking. Rather, in the era of cognitive systems, humans and machines will collaborate to produce better results, each bringing their own superior skills to the partnership."[21]

An example of the power of this human-computer symbiosis arose from a realization that struck Kasparov after he was beaten by Deep Blue. Even in a rule-defined game such as chess, he came to believe, "what computers are good at is where humans are weak, and vice versa." That gave him an idea for an experiment: "What if instead of human versus machine we played as partners?" When he and another grandmaster tried that, it created the symbiosis that Licklider had envisioned. "We could concentrate on strategic planning instead of spending so much time on calculations," Kasparov said. "Human creativity was even more paramount under these conditions."

A tournament along these lines was held in 2005. Players could work in teams with computers of their choice. Many grandmasters entered the fray, as did the most advanced computers. But neither the best grandmaster nor the most powerful computer won. Symbiosis did. "The teams of human plus machine dominated even the strongest computers," Kasparov noted. "Human strategic guidance combined with the tactical acuity of a computer was overwhelming." The final winner was not a grandmaster nor a state-of-the-art computer, nor even a combination of both, but two American amateurs who used three computers at the same time and knew how to manage the process of collaborating with their machines. "Their skill at manipulating and coaching their computers to look very deeply into positions effectively counteracted the superior chess understanding of their grandmaster opponents and the greater computational power of other participants," according to Kasparov.[22]

In other words, the future might belong to people who can best partner and collaborate with computers.

In a similar fashion, IBM decided that the best use of Watson, the *Jeopardy!*-playing computer, would be for it to collaborate with humans rather than try to top them. One project involved using the machine to work in partnership with doctors on cancer treatment plans. "The *Jeopardy!* challenge pitted man against machine," said IBM's Kelly. "With Watson and medicine, man and machine are taking on a challenge together—and going beyond what either could do on its own."[23] The Watson system was fed more than 2 million pages from medical journals and 600,000 pieces of clinical evidence, and could search up to 1.5 million patient records. When a doctor put in a patient's symptoms and vital information, the computer provided a list of recommendations ranked in order of its confidence.[24]

In order to be useful, the IBM team realized, the machine needed to interact with human doctors in a manner that made collaboration pleasant. David McQueeney, the vice president of software at IBM Research, described programming a pretense of humility into the machine: "Our early experience was with wary physicians who resisted by saying, 'I'm licensed to practice medicine, and I'm not going to have a computer tell me what to do.' So we reprogrammed our system to come across as humble and say, 'Here's the percentage likelihood that this is useful to you, and here you can look for yourself.'" Doctors were delighted, saying that it felt like a conversation with a knowledgeable colleague. "We aim to combine human talents, such as our intuition, with the strengths of a machine, such as its infinite breadth," said McQueeney. "That combination is magic, because each offers a piece that the other one doesn't have."[25]

That was one of the aspects of Watson that impressed Ginni Rometty, an engineer with a background in artificial intelligence who took over as CEO of IBM at the beginning of 2012. "I watched Watson interact in a collegial way with the doctors," she said. "It was the clearest testament of how machines can truly be partners with humans rather than try to replace them. I feel strongly about that."[26] She was so impressed that she decided to launch a new IBM division based on Watson. It was given a $1 billion investment and a new headquarters in the Silicon Alley area near Manhattan's Greenwich Village. Its mission was to commercialize "cognitive computing," meaning computing systems that can take data analysis to the next level by teaching themselves to complement the thinking skills of the human brain. Instead of giving the new division a technical name, Rometty simply called it Watson. It was in honor of Thomas Watson Sr., the IBM founder who ran the company for more than forty years, but it also evoked Sherlock Holmes's companion Dr. John ("Elementary, my dear") Watson and Alexander Graham Bell's assistant Thomas ("Come here, I want to see you") Watson. Thus the name helped to convey that Watson the computer should be seen as a collaborator and companion, not a threat like *2001*'s HAL.

Watson was a harbinger of a third wave of computing, one that blurred the line between augmented human intelligence and artificial intelligence. "The first generation of computers were machines that counted and tabulated," Rometty says, harking back to IBM's roots in Herman Hollerith's punch-card tabulators used for the 1890 census. "The second generation involved programmable machines that used the von Neumann architecture. You had to tell them what to do." Beginning with Ada Lovelace, people wrote algorithms that instructed these computers, step by step, how to perform tasks. "Because of the proliferation of data," Rometty adds, "there is no choice but to have a third generation, which are systems that are not programmed, they learn."[27]

But even as this occurs, the process could remain one of partnership and symbiosis with humans rather than one designed to relegate humans to the dustbin of history. Larry Norton, a breast cancer specialist at New York's Memorial Sloan-Kettering Cancer Center, was part of the team that worked with Watson. "Computer science is going to evolve rapidly, and medicine will evolve with it," he said. "This is coevolution. We'll help each other."[28]

This belief that machines and humans will get smarter together is a process that Doug Engelbart called "bootstrapping" and "coevolution."[29] It raises an interesting prospect: perhaps no matter how fast computers progress, artificial intelligence may never outstrip the intelligence of the human-machine partnership.

Let us assume, for example, that a machine someday exhibits all of the mental capabilities of a human: giving the outward appearance of recognizing patterns, perceiving emotions, appreciating beauty, creating art, having desires, forming moral values, and pursuing goals. Such a machine might be able to pass a Turing Test. It might even pass what we could call the Ada Test, which is that it could appear to "originate" its own thoughts that go beyond what we humans program it to do.

There would, however, be still another hurdle before we could say that artificial intelligence has triumphed over augmented intelligence. We can call it the Licklider Test. It would go beyond asking whether a machine could replicate all the components of human intelligence to ask whether the machine accomplishes these tasks better when whirring away completely on its own or when working in conjunction with humans. In other words, is it possible that humans and machines working in partnership will be indefinitely more powerful than an artificial intelligence machine working alone?

If so, then "man-computer symbiosis," as Licklider called it, will remain triumphant. Artificial intelligence need not be the holy grail of computing. The goal instead could be to find ways to optimize the collaboration between human and machine capabilities—to forge a partnership in which we let the machines do what they do best, and they let us do what we do best.

SOME LESSONS FROM THE JOURNEY

Like all historical narratives, the story of the innovations that created the digital age has many strands. So what lessons, in addition to the power of human-machine symbiosis just discussed, might be drawn from the tale?

First and foremost is that creativity is a collaborative process. Innovation comes from teams more often than from the lightbulb moments of lone geniuses. This was true of every era of creative ferment. The Scientific Revolution, the Enlightenment, and the Industrial Revolution all had their institutions for collaborative work and their networks for sharing ideas. But to an even greater extent, this has been true of the digital age. As brilliant as the many inventors of the Internet and computer were, they achieved most of their advances through teamwork. Like Robert Noyce, some of the best of them tended to resemble Congregational ministers rather than lonely prophets, madrigal singers rather than soloists.

Twitter, for example, was invented by a team of people who were collaborative but also quite contentious. When one of the cofounders, Jack Dorsey, started taking a lot of the credit in media interviews, another cofounder, Evan Williams, a serial entrepreneur who had previously created Blogger, told him to chill out, according to Nick Bilton of the *New York Times*. "But I invented Twitter," Dorsey said.

"No, you didn't invent Twitter," Williams replied. "I didn't invent Twitter either. Neither did Biz [Stone, another cofounder]. People don't invent things on the Internet. They simply expand on an idea that already exists."[30]

Therein lies another lesson: the digital age may seem revolutionary, but it was based on expanding the ideas handed down from previous generations. The collaboration was not merely among contemporaries, but also between generations. The best innovators were those who understood the trajectory of technological change and took the baton from innovators who preceded them. Steve Jobs built on the work of Alan Kay, who built on Doug Engelbart, who built on J. C. R. Licklider and Vannevar Bush. When Howard Aiken was devising his digital computer at Harvard, he was inspired by a fragment of Charles Babbage's Difference Engine that he found, and he made his crew members read Ada Lovelace's "Notes."

The most productive teams were those that brought together people with a wide array of specialties. Bell Labs was a classic example. In its long corridors in suburban New Jersey, there were theoretical physicists, experimentalists, material scientists, engineers, a few businessmen, and even some telephone-pole climbers with grease under their fingernails. Walter Brattain, an experimentalist, and John Bardeen, a theorist, shared a workspace, like a librettist and a composer sharing a piano bench, so they could perform a call-and-response all day about how to make what became the first transistor.

Even though the Internet provided a tool for virtual and distant collaborations, another lesson of digital-age innovation is that, now as in the past, physical proximity is beneficial. There is something special, as evidenced at Bell Labs, about meetings in the flesh, which cannot be replicated digitally. The founders of Intel created a sprawling, team-oriented open workspace where employees from Noyce on down all rubbed against one another. It was a model that became common in Silicon Valley. Predictions that digital tools would allow workers to telecommute were never fully realized. One of Marissa Mayer's first acts as CEO of Yahoo! was to discourage the practice of working from home, rightly pointing out that "people are more collaborative and innovative when they're together." When Steve

Jobs designed a new headquarters for Pixar, he obsessed over ways to structure the atrium, and even where to locate the bathrooms, so that serendipitous personal encounters would occur. Among his last creations was the plan for Apple's new signature headquarters, a circle with rings of open workspaces surrounding a central courtyard.

Throughout history the best leadership has come from teams that combined people with complementary styles. That was the case with the founding of the United States. The leaders included an icon of rectitude, George Washington; brilliant thinkers such as Thomas Jefferson and James Madison; men of vision and passion, including Samuel and John Adams; and a sage conciliator, Benjamin Franklin. Likewise, the founders of the ARPANET included visionaries such as Licklider, crisp decision-making engineers such as Larry Roberts, politically adroit people handlers such as Bob Taylor, and collaborative oarsmen such as Steve Crocker and Vint Cerf.

Another key to fielding a great team is pairing visionaries, who can generate ideas, with operating managers, who can execute them. Visions without execution are hallucinations.[31] Robert Noyce and Gordon Moore were both visionaries, which is why it was important that their first hire at Intel was Andy Grove, who knew how to impose crisp management procedures, force people to focus, and get things done.

Visionaries who lack such teams around them often go down in history as merely footnotes. There is a lingering historical debate over who most deserves to be dubbed the inventor of the electronic digital computer: John Atanasoff, a professor who worked almost alone at Iowa State, or the team led by John Mauchly and Presper Eckert at the University of Pennsylvania. In this book I give more credit to members of the latter group, partly because they were able to get their machine, ENIAC, up and running and solving problems. They did so with the help of dozens of engineers and mechanics plus a cadre of women who handled programming duties. Atanasoff's machine, by contrast, never fully worked, partly because there was no team to help him figure out how to make his punch-card burner operate. It ended up being consigned to a basement, then discarded when no one could remember exactly what it was.

Like the computer, the ARPANET and Internet were designed by collaborative teams. Decisions were made through a process, begun by a deferential graduate student, of sending around proposals as "Requests for Comments." That led to a weblike packet-switched network, with no central authority or hubs, in which power was fully distributed to every one of the nodes, each having the ability to create and share content and route around attempts to impose controls. A collaborative process thus produced a system designed to facilitate collaboration. The Internet was imprinted with the DNA of its creators.

The Internet facilitated collaboration not only within teams but also among crowds of people who didn't know each other. This is the advance that is closest to being revolutionary. Networks for collaboration have existed ever since the Persians and Assyrians invented postal systems. But never before has it been easy to solicit and collate contributions from thousands or millions of unknown collaborators. This led to innovative systems—Google page ranks, Wikipedia entries, the Firefox browser, the GNU/Linux software—based on the collective wisdom of crowds.

There were three ways that teams were put together in the digital age. The first was through government funding and coordination. That's how the groups that built the original computers (Colossus, ENIAC) and networks (ARPANET) were organized. This reflected the consensus, which was stronger back in the 1950s under President Eisenhower, that the government should undertake projects, such as the space program and interstate highway system, that benefited the common good. It often did so in collaboration with universities and private contractors as part of a government-academic-industrial triangle that Vannevar Bush and others fostered. Talented federal bureaucrats (not always an oxymoron), such as Licklider, Taylor, and Roberts, oversaw the programs and allocated public funds.

Private enterprise was another way that collaborative teams were formed. This happened at the research centers of big companies, such as Bell Labs and Xerox PARC, and at entrepreneurial new companies, such as Texas Instruments and Intel, Atari and Google, Microsoft and

Apple. A key driver was profits, both as a reward for the players and as a way to attract investors. That required a proprietary attitude to innovation that led to patents and intellectual property protections. Digital theorists and hackers often disparaged this approach, but a private enterprise system that financially rewarded invention was a component of a system that led to breathtaking innovation in transistors, chips, computers, phones, devices, and Web services.

Throughout history, there has been a third way, in addition to government and private enterprises, that collaborative creativity has been organized: through peers freely sharing ideas and making contributions as part of a voluntary common endeavor. Many of the advances that created the Internet and its services occurred in this fashion, which the Harvard scholar Yochai Benkler has labeled "commons-based peer production."[32] The Internet allowed this form of collaboration to be practiced on a much larger scale than before. The building of Wikipedia and the Web were good examples, along with the creation of free and open-source software such as Linux and GNU, OpenOffice and Firefox. As the technology journalist Steven Johnson has noted, "their open architecture allows others to build more easily on top of existing ideas, just as Berners-Lee built the Web on top of the Internet."[33] This commons-based production by peer networks was driven not by financial incentives but by other forms of reward and satisfaction.

The values of commons-based sharing and of private enterprise often conflict, most notably over the extent to which innovations should be patent-protected. The commons crowd had its roots in the hacker ethic that emanated from the MIT Tech Model Railroad Club and the Homebrew Computer Club. Steve Wozniak was an exemplar. He went to Homebrew meetings to show off the computer circuit he built, and he handed out freely the schematics so that others could use and improve it. But his neighborhood pal Steve Jobs, who began accompanying him to the meetings, convinced him that they should quit sharing the invention and instead build and sell it. Thus Apple was born, and for the subsequent forty years it has been at the forefront of aggressively patenting and profiting from its innovations. The instincts of both Steves were useful in creating the digital age. Innovation is most vibrant in the realms where open-source systems compete with proprietary ones.

Sometimes people advocate one of these modes of production over the others based on ideological sentiments. They prefer a greater government role, or exalt private enterprise, or romanticize peer sharing. In the 2012 election, President Barack Obama stirred up controversy by saying to people who owned businesses, "You didn't build that." His critics saw it as a denigration of the role of private enterprise. Obama's point was that any business benefits from government and peer-based community support: "If you were successful, somebody along the line gave you some help. There was a great teacher somewhere in your life. Somebody helped to create this unbelievable American system that we have that allowed you to thrive. Somebody invested in roads and bridges." It was not the most elegant way for him to dispel the fantasy that he was a closet socialist, but it did point to a lesson of modern economics that applies to digital-age innovation: that a combination of all of these ways of organizing production—governmental, market, and peer sharing—is stronger than favoring any one of them.

None of this is new. Babbage got most of his funding from the British government, which was generous in financing research that could strengthen its economy and empire. He adopted ideas from private industry, most notably the punch cards that had been developed by the textile firms for automated looms. He and his friends were founders of a handful of new peer-network clubs, including the British Association for the Advancement of Science, and though it may seem a stretch to view that august group as a fancy-dress forerunner to the Homebrew Computer Club, both existed to facilitate commons-based peer collaboration and the sharing of ideas.

The most successful endeavors in the digital age were those run by leaders who fostered collaboration while also providing a clear vision. Too often these are seen as conflicting traits: a leader is either very inclusive or a passionate visionary. But the best leaders could be both. Robert Noyce was a good example. He and Gordon Moore drove Intel forward based on a sharp vision of where semiconductor technology was heading, and they both were collegial and nonauthoritarian to a fault. Even Steve Jobs and Bill Gates, with all of their prickly intensity, knew how to build strong teams around them and inspire loyalty.

Brilliant individuals who could not collaborate tended to fail. Shockley Semiconductor disintegrated. Similarly, collaborative groups that lacked passionate and willful visionaries also failed. After inventing the transistor, Bell Labs went adrift. So did Apple after Jobs was ousted in 1985.

Most of the successful innovators and entrepreneurs in this book had one thing in common: they were product people. They cared about, and deeply understood, the engineering and design. They were not primarily marketers or salesmen or financial types; when such folks took over companies, it was often to the detriment of sustained innovation. "When the sales guys run the company, the product guys don't matter so much, and a lot of them just turn off," Jobs said. Larry Page felt the same: "The best leaders are those with the deepest understanding of the engineering and product design."[34]

Another lesson of the digital age is as old as Aristotle: "Man is a social animal." What else could explain CB and ham radios or their successors, such as WhatsApp and Twitter? Almost every digital tool, whether designed for it or not, was commandeered by humans for a social purpose: to create communities, facilitate communication, collaborate on projects, and enable social networking. Even the personal computer, which was originally embraced as a tool for individual creativity, inevitably led to the rise of modems, online services, and eventually Facebook, Flickr, and Foursquare.

Machines, by contrast, are not social animals. They don't join Facebook of their own volition nor seek companionship for its own sake. When Alan Turing asserted that machines would someday behave like humans, his critics countered that they would never be able to show affection or crave intimacy. To indulge Turing, perhaps we could program a machine to feign affection and pretend to seek intimacy, just as humans sometimes do. But Turing, more than almost anyone, would probably know the difference.

According to the second part of Aristotle's quote, the nonsocial nature of computers suggests that they are "either a beast or a god." Actually, they are neither. Despite all of the proclamations of artificial intelligence engineers and Internet sociologists, digital tools have no personalities, intentions, or desires. They are what we make of them.

ADA'S LASTING LESSON: POETICAL SCIENCE

That leads to a final lesson, one that takes us back to Ada Lovelace. As she pointed out, in our symbiosis with machines we humans have brought one crucial element to the partnership: creativity. The history of the digital age—from Bush to Licklider to Engelbart to Jobs, from SAGE to Google to Wikipedia to Watson—has reinforced this idea. And as long as we remain a creative species, this is likely to hold true. "The machines will be more rational and analytic," IBM's research director John Kelly says. "People will provide judgment, intuition, empathy, a moral compass, and human creativity."[35]

We humans can remain relevant in an era of cognitive computing because we are able to think different, something that an algorithm, almost by definition, can't master. We possess an imagination that, as Ada said, "brings together things, facts, ideas, conceptions in new, original, endless, ever-varying combinations." We discern patterns and appreciate their beauty. We weave information into narratives. We are storytelling as well as social animals.

Human creativity involves values, intentions, aesthetic judgments, emotions, personal consciousness, and a moral sense. These are what the arts and humanities teach us—and why those realms are as valuable a part of education as science, technology, engineering, and math. If we mortals are to uphold our end of the human-computer symbiosis, if we are to retain a role as the creative partners of our machines, we must continue to nurture the wellsprings of our imagination and originality and humanity. That is what we bring to the party.

At his product launches, Steve Jobs would conclude with a slide, projected on the screen behind him, of street signs showing the intersection of the Liberal Arts and Technology. At his last such appearance, for the iPad 2 in 2011, he stood in front of that image and declared, "It's in Apple's DNA that technology alone is not enough—

that it's technology married with liberal arts, married with the humanities, that yields us the result that makes our heart sing." That's what made him the most creative technology innovator of our era.

The converse to this paean to the humanities, however, is also true. People who love the arts and humanities should endeavor to appreciate the beauties of math and physics, just as Ada did. Otherwise, they will be left as bystanders at the intersection of arts and science, where most digital-age creativity will occur. They will surrender control of that territory to the engineers.

Many people who celebrate the arts and the humanities, who applaud vigorously the tributes to their importance in our schools, will proclaim without shame (and sometimes even joke) that they don't understand math or physics. They extoll the virtues of learning Latin, but they are clueless about how to write an algorithm or tell BASIC from C++, Python from Pascal. They consider people who don't know *Hamlet* from *Macbeth* to be Philistines, yet they might merrily admit that they don't know the difference between a gene and a chromosome, or a transistor and a capacitor, or an integral and a differential equation. These concepts may seem difficult. Yes, but so, too, is *Hamlet*. And like *Hamlet*, each of these concepts is beautiful. Like an elegant mathematical equation, they are expressions of the glories of the universe.

C. P. Snow was right about the need to respect both of "the two cultures," science and the humanities. But even more important today is understanding how they intersect. Those who helped lead the technology revolution were people in the tradition of Ada, who could combine science and the humanities. From her father came a poetic streak and from her mother a mathematical one, and it instilled in her a love for what she called "poetical science." Her father defended the Luddites who smashed mechanical looms, but Ada loved how punch cards instructed those looms to weave beautiful patterns, and she envisioned how this wondrous combination of art and technology could be manifest in computers.

The next phase of the Digital Revolution will bring even more new methods of marrying technology with the creative industries, such as media, fashion, music, entertainment, education, literature, and the arts. Much of the first round of innovation involved pouring old wine—books, newspapers, opinion pieces, journals, songs, television shows, movies—into new digital bottles. But new platforms, services, and social networks are increasingly enabling fresh opportunities for individual imagination and collaborative creativity. Role-playing games and interactive plays are merging with collaborative forms of storytelling and augmented realities. This interplay between technology and the arts will eventually result in completely new forms of expression and formats of media.

This innovation will come from people who are able to link beauty to engineering, humanity to technology, and poetry to processors. In other words, it will come from the spiritual heirs of Ada Lovelace, creators who can flourish where the arts intersect with the sciences and who have a rebellious sense of wonder that opens them to the beauty of both.

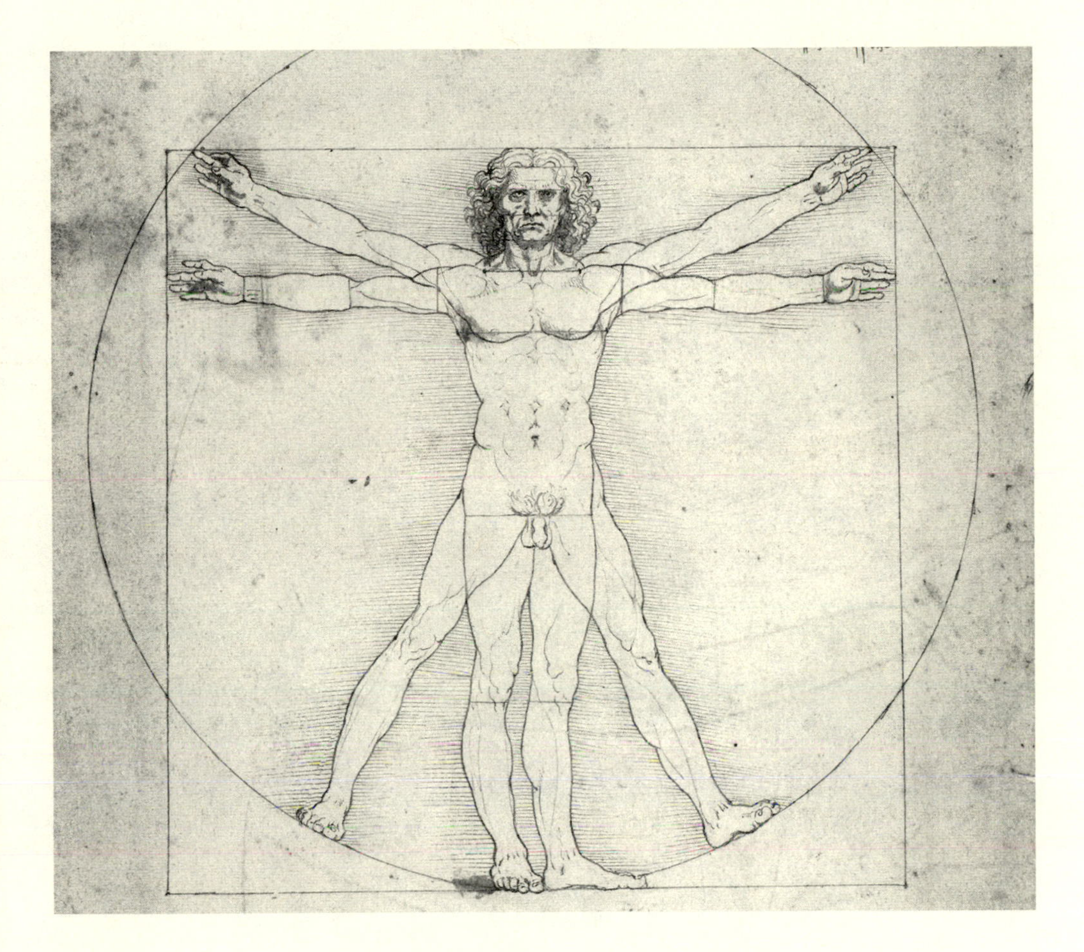

ACKNOWLEDGMENTS

I want to thank the people who gave me interviews and provided information, including Bob Albrecht, Al Alcorn, Marc Andreessen, Tim Berners-Lee, Stewart Brand, Dan Bricklin, Larry Brilliant, John Seeley Brown, Nolan Bushnell, Jean Case, Steve Case, Vint Cerf, Wes Clark, Steve Crocker, Lee Felsenstein, Bob Frankston, Bob Kahn, Alan Kay, Bill Gates, Al Gore, Andy Grove, Justin Hall, Bill Joy, Jim Kimsey, Leonard Kleinrock, Tracy Licklider, Liza Loop, David McQueeney, Gordon Moore, John Negroponte, Larry Page, Howard Rheingold, Larry Roberts, Arthur Rock, Virginia Rometty, Ben Rosen, Steve Russell, Eric Schmidt, Bob Taylor, Paul Terrell, Jimmy Wales, Evan Williams, and Steve Wozniak. I'm also grateful to people who gave useful advice along the way, including Ken Auletta, Larry Cohen, David Derbes, John Doerr, John Hollar, John Markoff, Lynda Resnick, Joe Zeff, and Michael Moritz.

Rahul Mehta at the University of Chicago and Danny Z. Wilson at Harvard read an early draft to fix any math or engineering mistakes; no doubt I snuck a few in when they weren't looking, so they shouldn't be blamed for any lapses. I'm particularly grateful to Strobe Talbott, who read and made extensive comments on a draft. He has done the same for each book I've written, going back to *The Wise Men* in 1986, and I've kept every set of his detailed notes as a testament to his wisdom and generosity.

I also tried something different for this book: crowdsourcing suggestions and corrections on many of the chapters. This isn't a new thing. Sending around papers for comments is one reason why the Royal Society was created in London in 1660 and why Benjamin

Franklin founded the American Philosophical Society. At *Time* magazine, we had a practice of sending story drafts to all bureaus for their "comments and corrections," which was very useful. In the past, I've sent parts of my drafts to dozens of people I knew. By using the Internet, I could solicit comments and corrections from thousands of people I didn't know.

This seemed fitting, because facilitating the collaborative process was one reason the Internet was created. One night when I was writing about that, I realized that I should try using the Internet for this original purpose. It would, I hoped, both improve my drafts and allow me to understand better how today's Internet-based tools (compared to Usenet and the old bulletin board systems) facilitate collaboration.

I experimented on many sites. The best, it turned out, was Medium, which was invented by Ev Williams, a character in this book. One excerpt was read by 18,200 people in its first week online. That's approximately 18,170 more draft readers than I've ever had in the past. Scores of readers posted comments, and hundreds sent me emails. This led to many changes and additions as well as an entirely new section (on Dan Bricklin and VisiCalc). I want to thank the hundreds of collaborators, some of whom I have now gotten to know, who helped me in this crowdsourcing process. (Speaking of which, I hope that someone will soon invent a cross between an enhanced eBook and a wiki so that new forms of multimedia history can emerge that are partly author-guided and partly crowdsourced.)

I also want to thank Alice Mayhew and Amanda Urban, who have been my editor and agent for thirty years, and the team at Simon & Schuster: Carolyn Reidy, Jonathan Karp, Jonathan Cox, Julia Prosser, Jackie Seow, Irene Kheradi, Judith Hoover, Ruth Lee-Mui, and Jonathan Evans. At the Aspen Institute, I am indebted to Pat Zindulka and Leah Bitounis, among many others. I'm also lucky to have three generations of my family willing to read and comment on a draft of this book: my father, Irwin (an electrical engineer); my brother, Lee (a computer consultant); and my daughter, Betsy (a tech writer, who first turned me on to Ada Lovelace). Most of all, I am grateful to my wife, Cathy, the wisest reader and most loving person I've ever known.

NOTES

INTRODUCTION

1. Henry Kissinger, background briefing for reporters, Jan. 15, 1974, from file in *Time* magazine archives.
2. Steven Shapin, *The Scientific Revolution* (University of Chicago Press, 1996), 1, 5.

CHAPTER ONE: ADA, COUNTESS OF LOVELACE

1. Lady Byron to Mary King, May 13, 1833. The Byron family letters, including those of Ada, are in the Bodleian Library, Oxford. Transcriptions of Ada's are in Betty Toole, *Ada, the Enchantress of Numbers: A Selection from the Letters* (Strawberry, 1992) and in Doris Langley Moore, *Ada, Countess of Lovelace* (John Murray, 1977). In addition to sources cited below, this section also draws on Joan Baum, *The Calculating Passion of Ada Byron* (Archon, 1986); William Gibson and Bruce Sterling, *The Difference Engine* (Bantam, 1991); Dorothy Stein, *Ada* (MIT Press, 1985); Doron Swade, *The Difference Engine* (Viking, 2001); Betty Toole, *Ada: Prophet of the Computer Age* (Strawberry, 1998); Benjamin Woolley, *The Bride of Science* (Macmillan, 1999); Jeremy Bernstein, *The Analytical Engine* (Morrow, 1963); James Gleick, *The Information* (Pantheon, 2011), chapter 4. Unless otherwise noted, quotes from Ada's letters rely on the Toole transcriptions.

   Writers about Ada Lovelace range from canonizers to debunkers. The most sympathetic books are those by Toole, Woolley, and Baum; the most scholarly and balanced is Stein's. For a debunking of Ada Lovelace, see Bruce Collier, "The Little Engines That Could've," PhD dissertation, Harvard, 1970, http://robroy.dyndns.info/collier/. He writes, "She was a manic depressive with the most amazing delusions about her talents. . . . Ada was as mad as a hatter, and contributed little more to the 'Notes' than trouble."
2. Lady Byron to Dr. William King, June 7, 1833.
3. Richard Holmes, *The Age of Wonder* (Pantheon, 2008), 450.
4. Laura Snyder, *The Philosophical Breakfast Club* (Broadway, 2011), 190.
5. Charles Babbage, *The Ninth Bridgewater Treatise* (1837), chapters 2 and 8, http://www.victorianweb.org/science/science_texts/bridgewater/intro.htm; Snyder, *The Philosophical Breakfast Club*, 192.
6. Toole, *Ada, the Enchantress of Numbers*, 51.
7. Sophia De Morgan, *Memoir of Augustus De Morgan* (Longmans, 1882), 9; Stein, *Ada*, 41.
8. Holmes, *The Age of Wonder*, xvi.
9. Ethel Mayne, *The Life and Letters of Anne Isabella, Lady Noel Byron* (Scribner's, 1929), 36; Malcolm Elwin, *Lord Byron's Wife* (Murray, 1974), 106.
10. Lord Byron to Lady Melbourne, Sept. 28, 1812, in John Murray, editor, *Lord Byron's Correspondence* (Scribner's, 1922), 88.
11. Stein, *Ada*, 14, from Thomas Moore's biography of Byron based on Byron's destroyed journals.
12. Woolley, *The Bride of Science*, 60.
13. Stein, *Ada*, 16; Woolley, *The Bride of Science*, 72.
14. Woolley, *The Bride of Science*, 92.
15. Woolley, *The Bride of Science*, 94.
16. John Galt, *The Life of Lord Byron* (Colburn and Bentley, 1830), 316.

17. Ada to Dr. William King, Mar. 9, 1834, Dr. King to Ada, Mar. 15, 1834; Stein, *Ada,* 42.
18. Ada to Dr. William King, Sept. 1, 1834; Stein, *Ada,* 46.
19. Woolley, *The Bride of Science,* 172.
20. Catherine Turney, *Byron's Daughter: A Biography of Elizabeth Medora Leigh* (Readers Union, 1975), 160.
21. Velma Huskey and Harry Huskey, "Lady Lovelace and Charles Babbage," *IEEE Annals of the History of Computing,* Oct.–Dec. 1980.
22. Ada to Charles Babbage, Nov. 1839.
23. Ada to Charles Babbage, July 30, 1843.
24. Ada to Lady Byron, Jan. 11, 1841.
25. Toole, *Ada, the Enchantress of Numbers,* 136.
26. Ada to Lady Byron, Feb. 6, 1841; Stein, *Ada,* 87.
27. Stein, *Ada,* 38.
28. Harry Wilmot Buxton and Anthony Hyman, *Memoir of the Life and Labours of the Late Charles Babbage* (ca. 1872; reprinted by Charles Babbage Institute/MIT Press, 1988), 46.
29. Martin Campbell Kelly and William Aspray, *Computer: A History of the Information Machine* (Westview, 2009), 6.
30. Swade, *The Difference Engine,* 42; Bernstein, *The Analytical Engine,* 46 and passim.
31. James Essinger, *Jacquard's Web* (Oxford , 2004), 23.
32. Ada to Charles Babbage, Feb. 16, 1840.
33. Ada to Charles Babbage, Jan. 12, 1841.
34. Charles Babbage, *Passages from the Life of a Philosopher* (Longman Green, 1864), 136.
35. Luigi Menabrea, with notes upon the memoir by the translator, Ada, Countess of Lovelace, "Sketch of the Analytical Engine, Invented by Charles Babbage," Oct. 1842, http://www.fourmilab.ch/babbage/sketch.html.
36. Babbage, *Passages from the Life of a Philosopher,* 136; John Füegi and Jo Francis, "Lovelace & Babbage and the Creation of the 1843 'Notes,'" *Annals of the History of Computing,* Oct. 2003.
37. All quotes from Menabrea and Lovelace's notes are from Menabrea, "Sketch of the Analytical Engine."
38. Charles Babbage to Ada, 1843, in Toole, *Ada, the Enchantress of Numbers,* 197.
39. Spoken in the film *Ada Byron Lovelace: To Dream Tomorrow,* directed and produced by John Füegi and Jo Francis (Flare Productions, 2003); also, Füegi and Francis, "Lovelace & Babbage."
40. Ada to Charles Babbage, July 5, 1843.
41. Ada to Charles Babbage, July 2, 1843.
42. Ada to Charles Babbage, Aug. 6, 1843; Woolley, *The Bride of Science,* 278; Stein, *Ada,* 114.
43. Ada to Lady Byron, Aug. 8, 1843.
44. Ada to Charles Babbage, Aug. 14, 1843.
45. Ada to Charles Babbage, Aug. 14, 1843.
46. Ada to Charles Babbage, Aug. 14, 1843.
47. Ada to Lady Lovelace, Aug. 15, 1843.
48. Stein, *Ada,* 120.
49. Ada to Lady Byron, Aug. 22, 1843.
50. Ada to Robert Noel, Aug. 9, 1843.

CHAPTER TWO: THE COMPUTER

1. Andrew Hodges, *Alan Turing: The Enigma* (Simon & Schuster, 1983; locations refer to the Kindle "Centenary Edition"), 439. In addition to the sources cited below, this section draws on Hodges's biography and his website, http://www.turing.org.uk/; the correspondence and documents in the Turing Archive, http://www.turingarchive.org/; David Leavitt, *The Man Who Knew Too Much* (Atlas Books, 2006); S. Barry Cooper and Jan van Leeuwen, *Alan Turing: His Work and Impact* (Elsevier, 2013); Sara Turing, *Alan M. Turing* (Cambridge, 1959; locations refer to the Kindle "Centenary Edition," with an afterword by John F. Turing, published in 2012); Simon Lavington, editor, *Alan Turing and His Contemporaries* (BCS, 2012).
2. John Turing in Sara Turing, *Alan M. Turing,* 146.
3. Hodges, *Alan Turing,* 590.
4. Sara Turing, *Alan M. Turing,* 56.
5. Hodges, *Alan Turing,* 1875.

LaDuke, *Pioneering Women in American Mathematics: The pre-1940 PhDs* (American Mathematical Society, 2009), 53; Beyer, *Grace Hopper*, 25 and 26.
4. Hopper oral history, Smithsonian, July 5, 1972.
5. Hopper oral history, Smithsonian, July 1968; Rosario Rausa, "In Profile, Grace Murray Hopper," *Naval History*, Fall 1992.
6. Hopper oral histories (she told the same story), Computer History Museum and Smithsonian, July 5, 1972.
7. The Staff of the Harvard Computation Library [Grace Hopper and Howard Aiken], *A Manual of Operation for the Automatic Sequence Controlled Calculator* (Harvard, 1946).
8. Grace Hopper oral history, Computer History Museum.
9. Beyer, *Grace Hopper*, 130.
10. Beyer, *Grace Hopper*, 135.
11. Richard Bloch oral history, Charles Babbage Institute, University of Minnesota.
12. Beyer, *Grace Hopper*, 53.
13. Grace Hopper and Richard Bloch panel discussion comments, Aug. 30, 1967, in Henry S. Tropp, "The 20th Anniversary Meeting of the Association for Computing Machinery," *IEEE Annals*, July 1987.
14. Beyer, *Grace Hopper*, 5.
15. Hopper oral history, Smithsonian, July 5, 1972.
16. Howard Aiken oral history, conducted by Henry Tropp and I. Bernard Cohen, Smithsonian Institution, Feb. 1973.
17. Grace Hopper and John Mauchly, "Influence of Programming Techniques on the Design of Computers," *Proceedings of the IRE*, Oct. 1953.
18. Harvard computer log, Sept. 9, 1947, http://www.history.navy.mil/photos/images/h96000/h96566k.jpg.
19. Grace Hopper oral history, Smithsonian, Nov. 1968.
20. *The Moore School Lectures*, Charles Babbage Institute, reprint (MIT Press, 1985).
21. Hopper oral history, Smithsonian, Nov. 1968.
22. In addition to the sources cited below, this section draws on Jean Jennings Bartik, *Pioneer Programmer* (Truman State, 2013; locations refer to the Kindle edition); Jean Bartik oral history, conducted by Gardner Hendrie, Computer History Museum, July 1, 2008; Jean Bartik oral history, conducted by Janet Abbate, IEEE Global History Network, Aug. 3, 2001; Steve Lohr, "Jean Bartik, Software Pioneer, Dies at 86," *New York Times*, Apr. 7, 2011; Jennifer Light, "When Computers Were Women," *Technology and Culture*, July 1999.
23. Jordynn Jack, *Science on the Home Front: American Women Scientists in World War II* (University of Illinois, 2009), 3.
24. Jennings Bartik, *Pioneer Programmer*, 1282.
25. W. Barkley Fritz, "The Women of ENIAC," *IEEE Annals of the History of Computing*, Fall 1996.
26. Fritz, "The Women of ENIAC."
27. Jennings Bartik, *Pioneer Programmer*, 1493. See also LeAnn Erickson, "Top Secret Rosies: The Female Computers of WWII" (Video, PBS, 2002); Bill Mauchly, ENIAC website, https://sites.google.com/a/opgate.com/eniac/; Thomas Petzinger Jr., "History of Software Begins with Work of Some Brainy Women," *Wall Street Journal*, Nov. 15, 1996. Kathy Kleiman helped bring recognition to the women programmers after first meeting them when researching her Harvard undergraduate thesis on women in computing in 1986, and she coproduced a twenty-minute documentary called *The Computers*, which premiered in 2014. See ENIAC Programmers Project website, http://eniacprogrammers.org/.
28. Kay McNulty Mauchly Antonelli, "The Kathleen McNulty Mauchly Antonelli Story," ENIAC website, https://sites.google.com/a/opgate.com/eniac/Home/kay-mcnulty-mauchly-antonelli.
29. Fritz, "The Women of ENIAC."
30. Jennings Bartik, *Pioneer Programmer*, 1480.
31. Autumn Stanley, *Mothers and Daughters of Invention* (Rutgers, 1995), 443.
32. Fritz, "The Women of ENIAC."
33. Oral history of Jean Jennings Bartik and Betty Snyder Holberton, conducted by Henry Tropp, Smithsonian, Apr. 27, 1973.
34. Jennings Bartik oral history, Computer History Museum.
35. Jennings Bartik oral history, Computer History Museum.

6. Alan Turing to Sara Turing, Feb. 16, 1930, Turing archive; Sara Turing, *Alan M. Turing*, 25.
7. Hodges, *Alan Turing*, 2144.
8. Hodges, *Alan Turing*, 2972.
9. Alan Turing, "On Computable Numbers," *Proceedings of the London Mathematical Society*, read on Nov. 12, 1936.
10. Alan Turing, "On Computable Numbers," 241.
11. Max Newman to Alonzo Church, May 31, 1936, in Hodges, *Alan Turing*, 3439; Alan Turing to Sara Turing, May 29, 1936, Turing Archive.
12. Alan Turing to Sara Turing, Feb. 11 and Feb. 22, 1937, Turing Archive; Alonzo Church, "Review of A. M. Turing's 'On computable numbers,'" *Journal of Symbolic Logic*, 1937.
13. This Shannon section draws on Jon Gertner, *The Idea Factory: Bell Labs and the Great Age of American Innovation* (Penguin, 2012; locations refer to the Kindle edition), chapter 7; M. Mitchell Waldrop, "Claude Shannon: Reluctant Father of the Digital Age," *MIT Technology Review*, July 2001; Graham Collins, "Claude E. Shannon: Founder of Information Theory," *Scientific American*, Oct. 2012; James Gleick, *The Information* (Pantheon, 2011), chapter 7.
14. Peter Galison, *Image and Logic* (University of Chicago, 1997), 781.
15. Claude Shannon, "A Symbolic Analysis of Relay and Switching Circuits," *Transactions of the American Institute of Electrical Engineers*, Dec. 1938. For a clear explanation, see Daniel Hillis, *The Pattern on the Stone* (Perseus, 1998), 2–10.
16. Paul Ceruzzi, *Reckoners: The Prehistory of the Digital Computer* (Greenwood, 1983), 79. See also Computer History Museum, "George Stibitz," http://www.computerhistory.org/revolution/birth-of-the-computer/4/85.
17. Howard Aiken oral history, conducted by Henry Tropp and I. Bernard Cohen, Smithsonian Institution, Feb. 1973.
18. Howard Aiken, "Proposed Automatic Calculating Machine," *IEEE Spectrum*, Aug. 1964; Cassie Ferguson, "Howard Aiken: Makin' a Computer Wonder," *Harvard Gazette*, Apr. 9, 1998.
19. I. Bernard Cohen, *Howard Aiken: Portrait of a Computer Pioneer* (MIT, 1999), 9.
20. Kurt Beyer, *Grace Hopper and the Invention of the Information Age* (MIT, 2009), 75.
21. Cohen, *Howard Aiken*, 115.
22. Cohen, *Howard Aiken*, 98 and passim.
23. Beyer, *Grace Hopper*, 80.
24. Ceruzzi, *Reckoners*, 65.
25. Horst Zuse (son), *The Life and Work of Konrad Zuse*, http://www.horst-zuse.homepage.t-online.de/Konrad_Zuse_index_english_html/biography.html.
26. Konrad Zuse archive, http://www.zib.de/zuse/home.php/Main/KonradZuse; Ceruzzi, *Reckoners*, 26.
27. Horst Zuse, *The Life and Work of Konrad Zuse*, part 4; Ceruzzi, *Reckoners*, 28.
28. The story of John Atanasoff and the controversy over the credit he deserves has led to some impassioned writings. A historical and legal battle pitted him against the creators of ENIAC, John Mauchly and Presper Eckert. The four main books about Atanasoff are all written by people who sought to take his side in this dispute. Alice Burks, *Who Invented the Computer?* (Prometheus, 2003; locations refer to the Kindle edition), is partly based on the documents of the legal battle. Alice Burks and Arthur Burks, *The First Electronic Computer: The Atanasoff Story* (University of Michigan, 1988) is an earlier, more technical book; Arthur Burks was an engineer on the ENIAC team who ended up being critical of Eckert and Mauchly. Clark Mollenhoff, *Atanasoff: Forgotten Father of the Computer* (Iowa State, 1988) was written by a Pulitzer Prize–winning reporter who was the Washington bureau chief of the *Des Moines Register* and after hearing of Atanasoff sought to resurrect him from being forgotten by history. Jane Smiley, *The Man Who Invented the Computer* (Doubleday, 2010) is by the acclaimed novelist who immersed herself in computer history and became an advocate for Atanasoff. For the personal background and involvement of Alice and Arthur Burks, see their "Memoir of the 1940s," *Michigan Quarterly Review*, Spring 1997, http://hdl.handle.net/2027/spo.act2080.0036.201. This section also draws on Allan Mackintosh, "Dr. Atanasoff's Computer," *Scientific American*, Aug. 1988; Jean Berry, "Clifford Edward Berry: His Role in Early Computers," *Annals of the History of Computing*, July 1986; William Broad, "Who Should Get the Glory for Inventing the Computer?" *New York Times*, Mar. 22, 1983.

29. John Atanasoff, "Advent of Electronic Digital Computing," *Annals of the History of Computing*, July 1984, 234.
30. Atanasoff, "Advent of Electronic Digital Computing," 238.
31. Atanasoff, "Advent of Electronic Digital Computing," 243.
32. Katherine Davis Fishman, *The Computer Establishment* (Harper and Row, 1981), 22.
33. Atanasoff testimony, *Honeywell v. Sperry Rand*, June 15, 1971, transcript p. 1700, in Burks, *Who Invented the Computer?*, 1144. The archives for the trial are at the University of Pennsylvania, http://www.archives.upenn.edu/faids/upd/eniactrial/upd8_10.html, and at the Charles Babbage Institute of the University of Minnesota, http://discover.lib.umn.edu/cgi/f/findaid/findaid-idx?c=umfa;cc=umfa;rgn=main;view=text;didno=cbi00001.
34. Atanasoff testimony, transcript p. 1703.
35. Atanasoff, "Advent of Electronic Digital Computing," 244.
36. John Atanasoff, "Computing Machine for the Solution of Large Systems of Linear Algebraic Equations," 1940, available online from Iowa State, http://jva.cs.iastate.edu/img/Computing%20machine.pdf. For detailed analysis, see Burks and Burks, *The First Electronic Computer*, 7 and passim.
37. Robert Stewart, "The End of the ABC," *Annals of the History of Computing*, July 1984; Mollenhoff, *Atanasoff*, 73.
38. This section draws on John Mauchly oral history, conducted by Henry Tropp, Jan. 10, 1973, Smithsonian Institution; John Mauchly oral history, conducted by Nancy Stern, May 6, 1977, American Institute of Physics (AIP); Scott McCartney, *ENIAC* (Walker, 1999); Herman Goldstine, *The Computer from Pascal to von Neumann* (Princeton, 1972; locations refer to Kindle edition); Kathleen Mauchly, "John Mauchly's Early Years," *Annals of the History of Computing*, Apr. 1984; David Ritchie, *The Computer Pioneers* (Simon & Schuster, 1986); Bill Mauchly and others, "The ENIAC" website, http://the-eniac.com/first/; Howard Rheingold, *Tools for Thought* (MIT, 2000); Joel Shurkin, *Engines of the Mind: A History of the Computer* (Washington Square Press, 1984).
39. John Costello, "The Twig Is Bent: The Early Life of John Mauchly," *IEEE Annals of the History of Computing*, 1996.
40. Mauchly oral history, AIP.
41. Costello, "The Twig Is Bent."
42. McCartney, *ENIAC*, 82.
43. Kay McNulty Mauchly Antonelli, "The Kathleen McNulty Mauchly Antonelli Story," Mar. 26, 2004, ENIAC website, https://sites.google.com/a/opgate.com/eniac/Home/kay-mcnulty-mauchly-antonelli; McCartney, *ENIAC*, 32.
44. Ritchie, *The Computer Pioneers*, 129; Rheingold, *Tools for Thought*, 80.
45. McCartney, *ENIAC*, 34.
46. Kathleen Mauchly, "John Mauchly's Early Years."
47. McCartney, *ENIAC*, 36.
48. Kathleen Mauchly, "John Mauchly's Early Years."
49. John Mauchly to H. Helm Clayton, Nov. 15, 1940.
50. John Mauchly to John de Wire, Dec. 4, 1940; Kathleen Mauchly, "John Mauchly's Early Years."
51. Mauchly to Atanasoff, Jan. 19, 1941; Atanasoff to Mauchly, Jan. 23, 1941; Mauchly oral history, Smithsonian; Burks, *Who Invented the Computer?*, 668.
52. The battle over what happened was fought out in the *Annals of the History of Computing*, with multiple articles, comments, and bitter letters. This section and that on the legal battle, below, derive from them. They include Arthur Burks and Alice Burks, "The ENIAC: First General-Purpose Electronic Computer," with comments by John Atanasoff, J. Presper Eckert, Kathleen R. Mauchly, and Konrad Zuse, and a response by Burks and Burks, *Annals of the History of Computing*, Oct. 1981, 310–99 (more than eighty pages of this issue were devoted to the assertions and rebuttals, prompting some discomfort on the part of the editors); Kathleen Mauchly, "John Mauchly's Early Years," *Annals of the History of Computing*, Apr. 1984; John Mauchly, "Mauchly: Unpublished Remarks," with an afterword by Arthur Burks and Alice Burks, *Annals of the History of Computing*, July 1982; Arthur Burks, "Who Invented the General Purpose Computer?" talk at the University of Michigan, Apr. 2, 1974; James McNulty, letter to the editor, *Datamation*, June 1980.
53. Lura Meeks Atanasoff testimony, *Sperry v. Honeywell*; Burks, *Who Invented the Computer?*, 1445.

54. Mollenhoff, *Atanasoff*, 114.
55. Mauchly oral history, Smithsonian; John Mauchly, "Fireside Chat," Nov. 13, 1973, *Annal the History of Computing*, July 1982.
56. Ritchie, *The Computer Pioneers*, 142.
57. Mauchly oral history, Smithsonian.
58. John Mauchly testimony, *Sperry v. Honeywell*; Burks, *Who Invented the Computer?*, 429.
59. John Mauchly to John Atanasoff, Sept. 30, 1941, *Sperry v. Honeywell* trial records.
60. Atanasoff to Mauchly, Oct. 7, 1941, *Sperry v. Honeywell* trial records.
61. In addition to the sources cited below, this section draws from Peter Eckstein, "Prespe Eckert," *Annals of the History of Computing*, Spring 1996; J. Presper Eckert oral history, conducted by Nancy Stern, Oct. 28, 1977, Charles Babbage Institute, University of Minnesota Nancy Stern, *From ENIAC to UNIVAC* (Digital Press, 1981); J. Presper Eckert, "Thought on the History of Computing," *Computer*, Dec. 1976; J. Presper Eckert, "The ENIAC," John Mauchly, "The ENIAC," and Arthur W. Burks, "From ENIAC to the Stored Program Computer," all in Nicholas Metropolis et al., editors, *A History of Computing in the Twentieth Century* (Academic Press, 1980); Alexander Randall, "A Lost Interview with Presper Eckert," *Computerworld*, Feb. 4, 2006.
62. Eckert oral history, Charles Babbage Institute.
63. Eckstein, "Presper Eckert."
64. Ritchie, *The Computer Pioneers*, 148.
65. Eckert oral history, Charles Babbage Institute.
66. John W. Mauchly, "The Use of High Speed Vacuum Tube Devices for Calculating," 1942, in Brian Randell, editor, *The Origins of Digital Computers: Selected Papers* (Springer-Verlag, 1973), 329. See also John G. Brainerd, "Genesis of the ENIAC," *Technology and Culture*, July 1976, 482.
67. Mauchly oral history, Smithsonian; Goldstine, *The Computer from Pascal to von Neumann*, 3169; McCartney, *ENIAC*, 61.
68. Burks, *Who Invented the Computer?*, 71.
69. McCartney, *ENIAC*, 89.
70. Eckert oral history, Charles Babbage Institute.
71. Eckert oral history, Charles Babbage Institute.
72. Eckert oral history, Charles Babbage Institute; Randall, "A Lost Interview with Presper Eckert."
73. Hodges, *Alan Turing*, 3628.
74. In addition to the Hodges biography, *Alan Turing*, this section draws on B. Jack Copeland, *Colossus: The Secrets of Bletchley Park's Codebreaking Computers* (Oxford, 2006); I. J. Good, "Early Work on Computers at Bletchley," *Annals of the History of Computing*, July 1979; Tommy Flowers, "The Design of Colossus," *Annals of the History of Computing*, July 1983; Simon Lavington, editor, *Alan Turing and His Contemporaries* (BCS, 2012); Sinclair McKay, *The Secret Life of Bletchley Park: The History of the Wartime Codebreaking Centre by the Men and Women Who Were There* (Aurum Press, 2010); and my visit to Bletchley Park and the scholars, tour guides, displays, and material available there.
75. Randall, "A Lost Interview with Presper Eckert."
76. The archives for the *Honeywell v. Sperry Rand* trial. See also Charles E. McTiernan, "The ENIAC Patent," *Annals of the History of Computing*, Apr. 1998.
77. Judge Earl Richard Larson decision, *Honeywell v. Sperry Rand*.
78. Randall, "A Lost Interview with Presper Eckert."

CHAPTER THREE: PROGRAMMING

1. Alan Turing, "Intelligent Machinery," National Physical Laboratory report, July 1948, available at http://www.AlanTuring.net/intelligent_machinery.
2. In addition to the sources cited below, this section draws from Kurt Beyer, *Grace Hopper and the Invention of the Information Age* (MIT, 2009), and the following trove of Grace Hopper oral histories: Smithsonian (five sessions), July 1968, Nov. 1968, Jan. 7, 1969, Feb. 4, 1969, July 5, 1972; the Computer History Museum, Dec. 1980; Grace Hopper interview, Sept. 1982, Women in Federal Government oral history project, Radcliffe Institute, Harvard.
3. Kurt Beyer mistakenly calls her the first to get a math doctorate from Yale. Charlotte Barnum was the first in 1895, and there were ten before Hopper. See Judy Green and Jeanne

36. Jennings Bartik, *Pioneer Programmer*, 557.
37. Eckert and Mauchly, "Progress Report on ENIAC," Dec. 31, 1943, in Nancy Stern, *From ENIAC to UNIVAC* (Digital Press, 1981).
38. John Mauchly, "Amending the ENIAC Story," letter to the editor of *Datamation*, Oct. 1979.
39. Presper Eckert, "Disclosure of a Magnetic Calculating Machine," Jan. 29, 1944, declassified trial exhibit, in Don Knuth archives, Computer History Museum; Mark Priestley, *A Science of Operations* (Springer, 2011), 127; Stern, *From ENIAC to UNIVAC*, 28.
40. In addition to specific notes below, this section draws on William Aspray, *John von Neumann and the Origins of Modern Computing* (MIT, 1990); Nancy Stern, "John von Neumann's Influence on Electronic Digital Computing, 1944–1946," *IEEE Annals of the History of Computing*, Oct.–Dec. 1980; Stanislaw Ulam, "John von Neumann," *Bulletin of the American Mathematical Society*, Feb. 1958; George Dyson, *Turing's Cathedral* (Random House, 2012; locations refer to Kindle edition); Herman Goldstine, *The Computer from Pascal to von Neumann* (Princeton, 1972; locations refer to Kindle edition).
41. Dyson, *Turing's Cathedral*, 41.
42. Nicholas Vonneumann, "John von Neumann as Seen by His Brother" (Privately printed, 1987), 22, excerpted as "John von Neumann: Formative Years," *IEEE Annals*, Fall 1989.
43. Dyson, *Turing's Cathedral*, 45.
44. Goldstine, *The Computer from Pascal to von Neumann*, 3550.
45. Dyson, *Turing's Cathedral*, 1305.
46. Dyson, *Turing's Cathedral*, 1395.
47. Hopper oral history, Smithsonian, Jan. 7, 1969.
48. Bloch oral history, Feb. 22, 1984, Charles Babbage Institute.
49. Robert Slater, *Portraits in Silicon* (MIT Press, 1987), 88; Beyer, *Grace Hopper and the Invention of the Information Age*, 9.
50. Goldstine, *The Computer from Pascal to von Neumann*, 3634.
51. Goldstine, *The Computer from Pascal to von Neumann*, 3840.
52. Goldstine, *The Computer from Pascal to von Neumann*, 199; Goldstine to Gillon, Sept. 2, 1944; Beyer, *Grace Hopper and the Invention of the Information Age*, 120. See also John Mauchly, "Amending the ENIAC Story," letter to the editor of *Datamation*, Oct. 1979; Arthur W. Burks, "From ENIAC to the Stored Program Computer," in Nicholas Metropolis et al., editors, *A History of Computing in the Twentieth Century* (Academic Press, 1980).
53. Jean Jennings Bartik and Betty Snyder Holberton oral history, Smithsonian, Apr. 27, 1973.
54. McCartney, *ENIAC*, 116.
55. Jean Jennings Bartik and Betty Snyder Holberton oral history, Smithsonian, Apr. 27, 1973.
56. Dyson, *Turing's Cathedral*, 53.
57. Burks, *Who Invented the Computer?*, 161; Norman Macrae, *John von Neumann* (American Mathematical Society, 1992), 281.
58. Ritchie, *The Computer Pioneers*, 178.
59. Presper Eckert oral history, conducted by Nancy Stern, Charles Babbage Institute, Oct. 28, 1977; Dyson, *Turing's Cathedral*, 1952.
60. John von Neumann, "First Draft of a Report on the EDVAC," U.S. Army Ordnance Department and the University of Pennsylvania, June 30, 1945. The report is available at http://www.virtualtravelog.net/wp/wp-content/media/2003-08-TheFirstDraft.pdf.
61. Dyson, *Turing's Cathedral*, 1957. See also Aspray, *John von Neumann and the Origins of Modern Computing*.
62. Eckert oral history, Charles Babbage Institute. See also McCartney, *ENIAC*, 125, quoting Eckert: "We were clearly suckered by John von Neumann, who succeeded in some circles at getting my ideas called the 'von Neumann architecture.'"
63. Jennings Bartik, *Pioneer Programmer*, 518.
64. Charles Duhigg and Steve Lohr, "The Patent, Used as a Sword," *New York Times*, Oct. 7, 2012.
65. McCartney, *ENIAC*, 103.
66. C. Dianne Martin, "ENIAC: The Press Conference That Shook the World," *IEEE Technology and Society*, Dec. 1995.
67. Jennings Bartik, *Pioneer Programmer*, 1878.
68. Fritz, "The Women of ENIAC."
69. Jennings Bartik, *Pioneer Programmer*, 1939.
70. Jean Jennings Bartik and Betty Snyder Holberton oral history, Smithsonian, Apr. 27, 1973.

71. Jennings Bartik, *Pioneer Programmer*, 672, 1964, 1995, 1959.
72. T. R Kennedy, "Electronic Computer Flashes Answers," *New York Times*, Feb. 15, 1946.
73. McCartney, *ENIAC*, 107.
74. Jennings Bartik, *Pioneer Programmer*, 2026, 2007.
75. Jean Jennings Bartik oral history, Computer History Museum.
76. McCartney, *ENIAC*, 132.
77. Steven Henn, "The Night a Computer Predicted the Next President," NPR, Oct. 31, 2012; Alex Bochannek, "Have You Got a Prediction for Us, UNIVAC?" Computer History Museum, http://www.computerhistory.org/atchm/have-you-got-a-prediction-for-us-univac/. Some reports say that CBS did not air the Eisenhower prediction because preelection polls had predicted that Stevenson would win. This is not true; polls had predicted an Eisenhower win.
78. Hopper oral history, Computer History Museum, Dec. 1980.
79. Beyer, *Grace Hopper*, 277.
80. Von Neumann to Stanley Frankel, Oct. 29, 1946; Joel Shurkin, *Engines of the Mind* (Washington Square Press, 1984), 204; Dyson, *Turing's Cathedral*, 1980; Stern, "John von Neumann's Influence on Electronic Digital Computing."
81. Eckert oral history, Charles Babbage Institute.
82. Goldstine, *The Computer from Pascal to von Neumann*, 5077.
83. Crispin Rope, "ENIAC as a Stored-Program Computer: A New Look at the Old Records," *IEEE Annals of the History of Computing*, Oct. 2007; Dyson, *Turing's Cathedral*, 4429.
84. Fritz, "The Women of ENIAC."
85. Maurice Wilkes, "How Babbage's Dream Came True," *Nature*, Oct. 1975.
86. Hodges, *Alan Turing*, 10622.
87. Dyson, *Turing's Cathedral*, 2024. See also Goldstine, *The Computer from Pascal to von Neumann*, 5376.
88. Dyson, *Turing's Cathedral*, 6092.
89. Hodges, *Alan Turing*, 6972.
90. Alan Turing, "Lecture to the London Mathematical Society," Feb. 20, 1947, available at http://www.turingarchive.org/; Hodges, *Alan Turing*, 9687.
91. Dyson, *Turing's Cathedral*, 5921.
92. Geoffrey Jefferson, "The Mind of Mechanical Man," Lister Oration, June 9, 1949, Turing Archive, http://www.turingarchive.org/browse.php/B/44.
93. Hodges, *Alan Turing*, 10983.
94. For an online version, see http://loebner.net/Prizef/TuringArticle.html.
95. John Searle, "Minds, Brains and Programs," *Behavioral and Brain Sciences*, 1980. See also "The Chinese Room Argument," *The Stanford Encyclopedia of Philosophy*, http://plato.stanford.edu/entries/chinese-room/.
96. Hodges, *Alan Turing*, 11305; Max Newman, "Alan Turing, An Appreciation," the *Manchester Guardian*, June 11, 1954.
97. M. H. A. Newman, Alan M. Turing, Sir Geoffrey Jefferson, and R. B. Braithwaite, "Can Automatic Calculating Machines Be Said to Think?" 1952 BBC broadcast, reprinted in Stuart Shieber, editor, *The Turing Test: Verbal Behavior as the Hallmark of Intelligence* (MIT, 2004); Hodges, *Alan Turing*, 12120.
98. Hodges, *Alan Turing*, 12069.
99. Hodges, *Alan Turing*, 12404. For discussions of Turing's suicide and character, see Robin Gandy, unpublished obituary of Alan Turing for the *Times*, and other items in the Turing Archives, http://www.turingarchive.org/. His mother, Sara, liked to believe that Turing's suicide was actually an accident caused when he was using cyanide to gold-plate a spoon. She sent to his archive a spoon she found in his lab with her note, "This is the spoon which I found in Alan Turing's laboratory. It is similar to the one which he gold-plated himself. It seems quite probable he was intending to gold-plate this one using cyanide of potassium of his own manufacture." Exhibit AMT/A/12, Turing Archive, http://www.turingarchive.org/browse.php/A/12.

CHAPTER FOUR: THE TRANSISTOR

1. Jon Gertner, *The Idea Factory: Bell Labs and the Great Age of American Innovation* (Penguin, 2012; locations refer to the Kindle edition). In addition to specific citations below, sources for this section include Joel Shurkin, *Broken Genius: The Rise and Fall of William*

Shockley (Macmillan, 2006; locations refer to the Kindle edition); Lillian Hoddeson and Vicki Daitch, *True Genius: The Life and Science of John Bardeen* (National Academies, 2002); Michael Riordan and Lillian Hoddeson, *Crystal Fire: The Invention of the Transistor and the Birth of the Information Age* (Norton, 1998); William Shockley, "The Invention of the Transistor—An Example of Creative-Failure Methodology," National Bureau of Standards Special Publication, May 1974, 47–89; William Shockley, "The Path to the Conception of the Junction Transistor," *IEEE Transactions of Electron Device*, July 1976; David Pines, "John Bardeen," *Proceedings of the American Philosophical Society*, Sept. 2009; "Special Issue: John Bardeen," *Physics Today*, Apr. 1992, with remembrances by seven of his colleagues; John Bardeen, "Semiconductor Research Leading to the Point Contact Transistor," Nobel Prize lecture, Dec. 11, 1956; John Bardeen, "Walter Houser Brattain: A Biographical Memoir," National Academy of Sciences, 1994; *Transistorized!*, PBS, transcripts and interviews, 1999, http://www.pbs.org/transistor/index.html; William Shockley oral history, American Institute of Physics (AIP), Sept. 10, 1974; Oral History of Shockley Semiconductor, Computer History Museum, Feb. 27, 2006; John Bardeen oral history, AIP, May 12, 1977; Walter Brattain oral history, AIP, Jan. 1964.

2. Gertner, *The Idea Factory*, 2255.
3. Shurkin, *Broken Genius*, 2547.
4. John Pierce, "Mervin Joe Kelly: 1894–1971," National Academy of Sciences, Biographical Memoirs, 1975, http://www.nasonline.org/publications/biographical-memoirs/memoir-pdfs/kelly-mervin.pdf; Gertner, *The Idea Factory*, 2267.
5. Shurkin, *Broken Genius*, 178.
6. Shurkin, *Broken Genius*, 231.
7. Shurkin, *Broken Genius*, 929; Lillian Hoddeson, "The Discovery of the Point-Contact Transistor," *Historical Studies in the Physical Sciences* 12, no. 1 (1981): 76.
8. John Pierce interview, *Transistorized!*, PBS, 1999.
9. Shurkin, *Broken Genius*, 935; Shockley, "The Path to the Conception of the Junction Transistor."
10. Gertner, *The Idea Factory*, 1022.
11. Gertner, *The Idea Factory*, 1266.
12. Gertner, *The Idea Factory*, 1336.
13. Brattain oral history, AIP.
14. Pines, "John Bardeen."
15. Bardeen, "Walter Houser Brattain."
16. Brattain oral history, AIP.
17. Riordan and Hoddeson, *Crystal Fire*, 126.
18. Shockley, "The Path to the Conception of the Junction Transistor"; Michael Riordan, "The Lost History of the Transistor," *IEEE Spectrum*, May 2004.
19. Riordan and Hoddeson, *Crystal Fire*, 121.
20. Brattain oral history, AIP.
21. Riordan and Hoddeson, *Crystal Fire*, 131.
22. Bardeen, "Semiconductor Research Leading to the Point Contact Transistor," Nobel Prize lecture.
23. Brattain oral history, AIP.
24. Brattain oral history, AIP.
25. Shurkin, *Broken Genius*, 1876.
26. Riordan and Hoddeson, *Crystal Fire*, 4, 137.
27. Riordan and Hoddeson, *Crystal Fire*, 139.
28. Shurkin, *Broken Genius*, 1934.
29. Shockley, "The Path to the Conception of the Junction Transistor."
30. Brattain oral history, AIP.
31. Riordan and Hoddeson, *Crystal Fire*, 148.
32. Shockley, "The Path to the Conception of the Junction Transistor."
33. Shockley, "The Path to the Conception of the Junction Transistor."
34. Shockley, "The Invention of the Transistor"; Gertner, *The Idea Factory*, 1717.
35. Brattain interview, "Naming the Transistor," PBS, 1999; Pierce interview, PBS, 1999.
36. Mervin Kelly, "The First Five Years of the Transistor," *Bell Telephone* magazine, Summer 1953.
37. Nick Holonyak oral history, AIP, Mar. 23, 2005.

38. Riordan and Hoddeson, *Crystal Fire*, 207; Mark Burgess, "Early Semiconductor History of Texas Instruments," https://sites.google.com/site/transistorhistory/Home/us-semiconductor-manufacturers/ti.
39. Gordon Teal talk, "Announcing the Transistor," Texas Instruments strategic planning conference, Mar. 17, 1980.
40. Riordan and Hoddeson, *Crystal Fire*, 211; Regency TR1 manual, http://www.regencytr1.com/images/Owners%20Manual%20-%20TR-1G.pdf.
41. T. R. Reid, *The Chip* (Simon & Schuster, 1984; locations refer to the Kindle edition), 2347.
42. Regency trivia page, http://www.regencytr1.com/TRivia_CORNER.html.
43. Brattain oral history, AIP.
44. John Bardeen to Mervin Kelly, May 25, 1951; Ronald Kessler, "Absent at the Creation," *Washington Post* magazine, Apr. 6, 1997; Pines, "John Bardeen."
45. Gertner, *The Idea Factory*, 3059; Shurkin, *Broken Genius*, 2579.
46. Riordan and Hoddeson, *Crystal Fire*, 231 and passim.
47. Arnold Thackray and Minor Myers, *Arnold O. Beckman: One Hundred Years of Excellence*, vol. 1 (Chemical Heritage Foundation, 2000), 6.
48. Walter Isaacson, *Steve Jobs* (Simon & Schuster, 2011), 9.
49. Sources for the passages on Silicon Valley include Leslie Berlin's *The Man Behind the Microchip: Robert Noyce and the Invention of Silicon Valley* (Oxford, 2005; locations refer to the Kindle edition), 1332 and passim. Berlin is the project historian for the Silicon Valley Archives at Stanford and is writing a book on the rise of Silicon Valley. Also: Rebecca Lowen, *Creating the Cold War University: The Transformation of Stanford* (University of California, 1997); Michael Malone, *The Intel Trinity* (HarperBusiness, 2014), *Infinite Loop* (Doubleday, 1999), *The Big Score: The Billion Dollar Story of Silicon Valley* (Doubleday, 1985), *The Valley of Heart's Delight: A Silicon Valley Notebook, 1963–2001* (Wiley, 2002), *Bill and Dave* (Portfolio, 2007); Christophe Lécuyer, *Making Silicon Valley* (MIT, 2007); C. Stewart Gillmore, *Fred Terman at Stanford: Building a Discipline, a University, and Silicon Valley* (Stanford, 2004); Margaret Pugh O'Mara, *Cities of Knowledge: Cold War Science and the Search for the Next Silicon Valley* (Princeton, 2005); Thomas Heinrich, "Cold War Armory: Military Contracting in Silicon Valley," *Enterprise & Society*, June 1, 2002; Steve Blank, "The Secret History of Silicon Valley," http://steveblank.com/secret-history/.
50. Berlin, *The Man Behind the Microchip*, 1246; Reid, *The Chip*, 1239. In addition to these two sources and those cited below, the section draws on my interviews with Gordon Moore and Andy Grove; Shurkin, *Broken Genius*; Michael Malone, *The Intel Trinity* (Harpers, 2014); Tom Wolfe, "The Tinkerings of Robert Noyce," *Esquire*, Dec. 1983; Bo Lojek, *History of Semiconductor Engineering* (Springer, 2007); notebooks and items in the Computer History Museum; Robert Noyce oral history, conducted by Michael F. Wolff, IEEE History Center, Sept. 19, 1975; Gordon Moore oral history, conducted by Michael F. Wolff, IEEE History Center, Sept. 19, 1975; Gordon Moore oral history, conducted by Daniel Morrow, Computerworld Honors Program, Mar. 28, 2000; Gordon Moore and Jay Last oral history, conducted by David Brock and Christophe Lécuyer, Chemical Heritage Foundation, Jan. 20, 2006; Gordon Moore oral history, conducted by Craig Addison, SEMI, Jan. 25, 2008; Gordon Moore interview, conducted by Jill Wolfson and Teo Cervantes, *San Jose Mercury News*, Jan. 26, 1997; Gordon Moore, "Intel: Memories and the Microprocessor," *Daedalus*, Spring 1966.
51. Shurkin, *Broken Genius*, 2980, from Fred Warshorfsky, *The Chip War* (Scribner's Sons, 1989).
52. Berlin, *The Man Behind the Microchip*, 276.
53. Berlin, *The Man Behind the Microchip*, 432, 434.
54. Wolfe, "The Tinkerings of Robert Noyce."
55. Robert Noyce interview, "Silicon Valley," PBS, 2013; Malone, *The Big Score*, 74.
56. Berlin, *The Man Behind the Microchip*, 552; Malone, *Intel Trinity*, 81.
57. Leslie Berlin writes that the transistors did not arrive until 1950, after Noyce graduated: "[Bell's research head] Buckley did not have any devices to spare, but he did send Gale copies of several technical monographs that Bell Labs had written on the transistor. These monographs formed the basis of Noyce's initial exposure to the device. No textbooks addressed transistors, and (although prevailing mythology claims otherwise) Bell Labs did not ship Gale a transistor until after Noyce graduated" (*The Man Behind the Microchip*, 650). Berlin cites as her source for this a March 1984 letter written by Professor Gale to a friend; Berlin writes in an endnote, "Gale mentions an 'attached original shipping invoice [for the transistors, sent from Bardeen to Gale] dated March 6, 1950' (now lost)." Berlin's reporting

conflicts with Noyce's recollections. Noyce's quote that "Grant Gale got hold of one of the first point contact transistors . . . during my junior year" is from Noyce's September 1975 IEEE History Center oral history, cited above. Tom Wolfe's *Esquire* profile of Noyce, based on his visits with Noyce, reports, "By the fall of 1948 Gale had obtained two of the first transistors ever made, and he presented the first academic instruction in solid-state electronics available anywhere in the world, for the benefit of the eighteen students [including Noyce] majoring in physics at Grinnell College" ("The Tinkerings of Robert Noyce"). Reid, *The Chip*, 1226, based on his 1982 interviews with Robert Noyce, writes, "Gale had been a classmate of John Bardeen in the engineering school at the University of Wisconsin, and thus he was able to obtain one of the first transistors and demonstrate it to his students. It was not a lecture the student was to forget. 'It hit me like the atom bomb,' Noyce recalled forty years later." Bardeen and other engineers at Bell Labs did send out many transistor samples to academic institutions that requested them beginning in July 1948.

58. Reid, *The Chip*, 1266; Berlin, *The Man Behind the Microchip*, 1411.
59. Gordon Moore interview, "Silicon Valley," PBS, 2013.
60. Author's interview with Gordon Moore.
61. Riordan and Hoddeson, *Crystal Fire,* 239.
62. Berlin, *The Man Behind the Microchip*, 1469.
63. Jay Last interview, "Silicon Valley," PBS, 2013.
64. Malone, *Intel Trinity*, 107.
65. Jay Last interview, "Silicon Valley," PBS, 2013; Berlin, *The Man Behind the Microchip*, 1649; Riordan and Hoddeson, *Crystal Fire*, 246.
66. Berlin, *The Man Behind the Microchip*, 1641.
67. Shurkin, *Broken Genius*, 3118.
68. Author's interview with Gordon Moore.
69. Arnold Beckman oral history, conducted by Jeffrey L. Sturchio and Arnold Thackray, Chemical Heritage Foundation, July 23, 1985.
70. Gordon Moore and Jay Last interviews, "Silicon Valley," PBS, 2013.
71. Regis McKenna and Michael Malone interviews, "Silicon Valley," PBS, 2013.
72. Berlin, *The Man Behind the Microchip*, 1852; author's interview with Arthur Rock.
73. Author's interview with Arthur Rock.
74. Arthur Rock interview, "Silicon Valley," PBS, 2013; author's interview and papers provided to me by Arthur Rock.
75. "Multifarious Sherman Fairchild," *Fortune*, May 1960; "Yankee Tinkerer" (cover story on Sherman Fairchild), *Time*, July 25, 1960.

CHAPTER FIVE: THE MICROCHIP

1. In addition to the sources cited below, this section draws from Jack Kilby, "Turning Potentials into Realities," Nobel Prize lecture, Dec. 8, 2000; Jack Kilby, "Invention of the Integrated Circuit," *IEEE Transactions on Electron Devices*, July 1976; T. R. Reid, *The Chip* (Simon & Schuster, 1984; locations refer to the Kindle edition).
2. Jack Kilby, biographical essay, Nobel Prize organization, 2000.
3. Reid, *The Chip*, 954.
4. Reid, *The Chip*, 921.
5. Reid, *The Chip*, 1138.
6. Berlin, *The Man Behind the Microchip*, 2386. The Fairchild notebooks are being preserved and are on display at the Computer History Museum in Mountain View, California.
7. Berlin, *The Man Behind the Microchip*, 2515.
8. Robert Noyce oral history, IEEE.
9. Reid, *The Chip*, 1336; Robert Noyce oral history, IEEE.
10. Robert Noyce journal entry, Jan. 23, 1959, Computer History Museum, Mountain View, California. For a picture of the page, see http://www.computerhistory.org/atchm/the-relics-of-st-bob/.
11. J. S. Kilby, "Capacitor for Miniature Electronic Circuits or the Like," patent application US 3434015 A, Feb. 6, 1959; Reid, *The Chip*, 1464.
12. R. N. Noyce, "Semiconductor Device-and-Lead Structure," patent application US 2981877 A, July 30, 1959; Reid, *The Chip*, 1440.
13. Reid, *The Chip*, 1611 and passim.
14. *Noyce v. Kilby*, U.S. Court of Customs and Patent Appeals, Nov. 6, 1969.

15. Reid, *The Chip*, 1648.
16. Jack Kilby oral history, conducted by Arthur L. Norberg, Charles Babbage Institute, University of Minnesota, June 21, 1984.
17. Craig Matsumoto, "The Quiet Jack Kilby," Valley Wonk column, *Heavy Reading*, June 23, 2005.
18. Reid, *The Chip*, 3755, 3775; Jack Kilby, Nobel Prize lecture, Dec. 8, 2000.
19. Paul Ceruzzi, *A History of Modern Computing* (MIT Press, 1998), 187.
20. Ceruzzi, *A History of Modern Computing*, chapter 6.
21. Reid, *The Chip*, 2363, 2443.
22. Robert Noyce, "Microelectronics," *Scientific American*, Sept. 1977.
23. Gordon Moore, "Cramming More Components onto Integrated Circuits," *Electronics*, Apr. 1965.
24. Berlin, *The Man Behind the Microchip*, 3177.
25. Gordon Moore interview, "American Experience: Silicon Valley," PBS, 2013.
26. Author's interview with Gordon Moore.
27. Berlin, *The Man Behind the Microchip*, 3529.
28. Author's interview with Arthur Rock.
29. John Wilson, *The New Venturers* (Addison-Wesley, 1985), chapter 2.
30. Author's interview with Arthur Rock; David Kaplan, *The Silicon Boys* (Morrow, 1999), 165 and passim.
31. Author's interview with Arthur Rock.
32. Author's interview with Arthur Rock.
33. Malone, *Intel Trinity*, 4, 8.
34. Berlin, *The Man Behind the Microchip*, 4393.
35. Andrew Grove, *Swimming Across* (Grand Central, 2001), 2. This section is also based on author's interviews and conversations with Grove over the years and on Joshua Ramo, "Man of the Year: A Survivor's Tale," *Time*, Dec. 29, 1997; Richard Tedlow, *Andy Grove* (Portfolio, 2006).
36. Tedlow, *Andy Grove*, 92.
37. Tedlow, *Andy Grove*, 96.
38. Berlin, *The Man Behind the Microchip*, 129.
39. Andrew Grove interview, "American Experience: Silicon Valley," PBS, 2013.
40. Tedlow, *Andy Grove*, 74; Andy Grove oral history conducted by Arnold Thackray and David C. Brock, July 14 and Sept. 1, 2004, Chemical Heritage Foundation.
41. Author's interview with Arthur Rock.
42. Michael Malone interview, "American Experience: Silicon Valley," PBS, 2013.
43. Berlin, *The Man Behind the Microchip*, 4400.
44. Ann Bowers interview, "American Experience: Silicon Valley," PBS, 2013.
45. Ted Hoff interview, "American Experience: Silicon Valley," PBS, 2013.
46. Wolfe, "The Tinkerings of Robert Noyce."
47. Malone, *Intel Trinity*, 115.
48. Author's interview with Gordon Moore.
49. Malone, *Intel Trinity*, 130.
50. Ann Bowers interview, "American Experience"; author's interview with Ann Bowers.
51. Reid, *The Chip*, 140; Malone, *Holy Trinity*, 148.
52. Ted Hoff interview, "American Experience: Silicon Valley," PBS, 2013.
53. Berlin, *The Man Behind the Microchip*, 4329.
54. Berlin, *The Man Behind the Microchip*, 4720.
55. Don Hoefler, "Silicon Valley USA," *Electronic News*, Jan. 11, 1971.

CHAPTER SIX: VIDEO GAMES

1. Steven Levy, *Hackers* (Anchor/Doubleday, 1984; locations refer to the twenty-fifth anniversary reissue, O'Reilly, 2010), 28. In this classic and influential book, which begins with a detailed account of MIT's Tech Model Railroad Club, Levy describes a "hacker ethic" that includes the following: "Access to computers—and anything else which might teach you about the way the world works—should be unlimited and total. Always yield to the Hands-On Imperative!" In addition to Levy's book and specific sources cited below, sources for this chapter include author's interviews with Steve Russell and Stewart Brand; Steve Russell oral history, conducted by Al Kossow, Aug. 9, 2008, Computer History Museum; J. Martin

Graetz, "The Origin of Spacewar," *Creative Computing*, Aug. 1981; Stewart Brand, "Spacewar," *Rolling Stone*, Dec. 7, 1972.

2. Levy, *Hackers*, 7.
3. "Definition of Hackers," website of the Tech Model Railroad Club, http://tmrc.mit.edu/hackers-ref.html.
4. Brand, "Spacewar."
5. Graetz, "The Origin of Spacewar."
6. Steve Russell oral history, Computer History Museum; Graetz, "The Origin of Spacewar."
7. Author's interview with Steve Russell.
8. Graetz, "The Origin of Spacewar."
9. Brand, "Spacewar."
10. Author's interview with Steve Russell.
11. Sources for this section include author's interviews with Nolan Bushnell, Al Alcorn, Steve Jobs (for previous book), and Steve Wozniak; Tristan Donovan, *Replay: The Story of Video Games* (Yellow Ant, 2010; locations refer to the Kindle edition); Steven Kent, *The Ultimate History of Video Games: From Pong to Pokemon* (Three Rivers, 2001); Scott Cohen, *Zap! The Rise and Fall of Atari* (McGraw-Hill, 1984); Henry Lowood, "Videogames in Computer Space: The Complex History of Pong," *IEEE Annals*, July 2009; John Markoff, *What the Dormouse Said* (Viking, 2005, locations refer to the Kindle edition); Al Alcorn interview, *Retro Gaming Roundup*, May 2011; Al Alcorn interview, conducted by Cam Shea, *IGN*, Mar. 10, 2008.
12. Kent, *The Ultimate History of Video Games*, 12.
13. Author's interview with Nolan Bushnell.
14. Nolan Bushnell talk to young entrepreneurs, Los Angeles, May 17, 2013 (author's notes).
15. Donovan, *Replay*, 429.
16. Donovan, *Replay*, 439.
17. Eddie Adlum, quoted in Kent, *The Ultimate History of Video Games*, 42.
18. Kent, *The Ultimate History of Video Games*, 45.
19. Author's interview with Nolan Bushnell.
20. Author's interview with Nolan Bushnell.
21. Author's interview with Al Alcorn.
22. Donovan, *Replay*, 520.
23. Author's interviews with Nolan Bushnell and Al Alcorn. This tale is told in much the same way in other sources, often with a few embellishments.
24. Author's interview with Nolan Bushnell.
25. Nolan Bushnell talk to young entrepreneurs, Los Angeles, May 17, 2013.
26. Author's interview with Nolan Bushnell.
27. Donovan, *Replay*, 664.
28. Author's interview with Nolan Bushnell.

CHAPTER SEVEN: THE INTERNET

1. Sources for Vannevar Bush include Vannevar Bush, *Pieces of the Action* (Morrow, 1970); Pascal Zachary, *Endless Frontier: Vannevar Bush, Engineer of the American Century* (MIT, 1999); "Yankee Scientist," *Time* cover story, Apr. 3, 1944; Jerome Weisner, "Vannevar Bush: A Biographical Memoir," National Academy of Sciences, 1979; James Nyce and Paul Kahn, editors, *From Memex to Hypertext: Vannevar Bush and the Mind's Machine* (Academic Press, 1992); Jennet Conant, *Tuxedo Park* (Simon & Schuster, 2002); Vannevar Bush oral history, American Institute of Physics, 1964.
2. Weisner, "Vannevar Bush."
3. Zachary, *Endless Frontier*, 23.
4. *Time*, Apr. 3, 1944.
5. *Time*, Apr. 3, 1944.
6. Bush, *Pieces of the Action*, 41.
7. Weisner, "Vannevar Bush."
8. Vannevar Bush, *Science, the Endless Frontier* (National Science Foundation, July 1945), vii.
9. Bush, *Science*, 10.
10. Bush, *Pieces of the Action*, 65.
11. Joseph V. Kennedy, "The Sources and Uses of U.S. Science Funding," *The New Atlantis*, Summer 2012.

12. Mitchell Waldrop, *The Dream Machine: J. C. R. Licklider and the Revolution That Made Computing Personal* (Penguin, 2001), 470. Other sources for this section include author's interviews with Tracy Licklider (son), Larry Roberts, and Bob Taylor; Katie Hafner and Matthew Lyon, *Where Wizards Stay Up Late: The Origins of the Internet* (Simon & Schuster, 1998); J. C. R. Licklider oral history, conducted by William Aspray and Arthur Norberg, Oct. 28, 1988, Charles Babbage Institute, University of Minnesota; J. C. R. Licklider interview, conducted by James Pelkey, "A History of Computer Communications," June 28, 1988 (Pelkey's material is only online, http://www.historyofcomputercommunications.info/index.html); Robert M. Fano, *Joseph Carl Robnett Licklider 1915–1990, a Biographical Memoir* (National Academies Press, 1998).
13. Licklider oral history, Charles Babbage Institute.
14. Norbert Wiener, "A Scientist's Dilemma in a Materialistic World" (1957), in *Collected Works*, vol. 4 (MIT, 1984), 709.
15. Author's interview with Tracy Licklider.
16. Author's interview with Tracy Licklider.
17. Waldrop, *The Dream Machine*, 237.
18. Bob Taylor, "In Memoriam: J. C. R. Licklider," Aug. 7, 1990, Digital Equipment Corporation publication.
19. J. C. R. Licklider interview, conducted by John A. N. Lee and Robert Rosin, "The Project MAC Interviews," *IEEE Annals of the History of Computing*, Apr. 1992.
20. Author's interview with Bob Taylor.
21. Licklider oral history, Charles Babbage Institute.
22. J. C. R. Licklider, "Man-Computer Symbiosis," *IRE Transactions on Human Factors in Electronics*, Mar. 1960, http://groups.csail.mit.edu/medg/people/psz/Licklider.html.
23. David Walden and Raymond Nickerson, editors, *A Culture of Innovation: Insider Accounts of Computing and Life at BBN* (privately printed at the Harvard bookstore, 2011), see http://walden-family.com/bbn/.
24. Licklider oral history, Charles Babbage Institute.
25. J. C. R. Licklider, *Libraries of the Future* (MIT, 1965), 53.
26. Licklider, *Libraries of the Future*, 4.
27. Sherman Adams, *Firsthand Report* (Harper, 1961), 415; Hafner and Lyon, *Where Wizards Stay Up Late*, 17.
28. James Killian interview, "War and Peace," WGBH, Apr. 18, 1986; James Killian, *Sputnik, Scientists, and Eisenhower* (MIT, 1982), 20.
29. Fred Turner, *From Counterculture to Cyberculture* (University of Chicago, 2006), 108.
30. Licklider oral history, Charles Babbage Institute.
31. Licklider interview, conducted by James Pelkey; see also James Pelkey, "Entrepreneurial Capitalism and Innovation," http://www.historyofcomputercommunications.info/Book/2/2.1-IntergalacticNetwork_1962-1964.html#_ftn1.
32. J. C. R. Licklider, "Memorandum for Members and Affiliates of the Intergalactic Computer Network," ARPA, Apr. 23, 1963. See also J. C. R. Licklider and Welden Clark, "Online Man-Computer Communications," *Proceedings of AIEE-IRE*, Spring 1962.
33. Author's interview with Bob Taylor.
34. Author's interview with Larry Roberts.
35. Bob Taylor oral history, Computer History Museum, 2008; author's interview with Bob Taylor.
36. Michael Hiltzik, *Dealers of Lightning* (Harper, 1999; locations refer to the Kindle edition), 536. 530.
37. Author's interview with Bob Taylor.
38. Author's interview with Bob Taylor.
39. Robert Taylor oral history, Computer History Museum; author's interview with Bob Taylor; Hafner and Lyon, *Where Wizards Stay Up Late*, 86.
40. Hafner and Lyon, *Where Wizards Stay Up Late*, 591, has the fullest description of this meeting. See also Hiltzik, *Dealers of Lightning*, 1120; Kleinrock oral history, "How the Web Was Won," *Vanity Fair*, July 2008.
41. Charles Herzfeld interview with Andreu Veà, "The Unknown History of the Internet," 2010, http://www.computer.org/comphistory/pubs/2010-11-vea.pdf.
42. Author's interview with Bob Taylor.
43. Author's interview with Larry Roberts.

44. Author's interview with Larry Roberts.
45. As with the tale of Herzfeld funding ARPANET after a twenty-minute meeting, this story of Taylor recruiting Roberts down to Washington has been oft-told. This version comes from author's interviews with Taylor and Roberts; Hafner and Lyon, *Where Wizards Stay Up Late*, 667; Stephen Segaller, *Nerds 2.0.1* (TV Books, 1998), 47; Bob Taylor oral history, Computer History Museum; Larry Roberts, "The Arpanet and Computer Networks," *Proceedings of the ACM Conference on the History of Personal Workstations*, Jan. 9, 1986.
46. Author's interview with Bob Taylor.
47. Author's interview with Bob Taylor.
48. Author's interview with Larry Roberts.
49. Larry Roberts oral history, Charles Babbage Institute.
50. Author's interview with Bob Taylor.
51. Janet Abbate, *Inventing the Internet* (MIT, 1999), 1012; Larry Roberts oral history, Charles Babbage Institute.
52. Wes Clark oral history, conducted by Judy O'Neill, May 3, 1990, Charles Babbage Institute.
53. There are differing versions of this story, including some that say it was a taxi ride. Bob Taylor insists it was in a car he had rented. Author's interviews with Bob Taylor and Larry Roberts; Robert Taylor oral history, conducted by Paul McJones, Oct. 2008, Computer History Museum; Hafner and Lyon, *Where Wizards Stay Up Late*, 1054; Segaller, *Nerds*, 62.
54. Author's interview with Vint Cerf.
55. Paul Baran, "On Distributed Computer Networks," *IEEE Transactions on Communications Systems*, Mar. 1964. This section on Baran draws on John Naughton, *A Brief History of the Future* (Overlook, 2000), chapter 6; Abbate, *Inventing the Internet*, 314 and passim; Hafner and Lyon, *Where Wizards Stay Up Late*, 723, 1119.
56. Paul Baran interview, in James Pelkey, "Entrepreneurial Capitalism and Innovation," http://www.historyofcomputercommunications.info/Book/2/2.4-Paul%20Baran-59-65.html#_ftn9.
57. Paul Baran oral history, "How the Web Was Won," *Vanity Fair*, July 2008; interview with Paul Baran, by Stewart Brand, *Wired*, Mar. 2001; Paul Baran oral history, conducted by David Hochfelder, Oct. 24, 1999, IEEE History Center; Clayton Christensen, *The Innovator's Dilemma* (Harper, 1997).
58. Donald Davies, "A Historical Study of the Beginnings of Packet Switching," *Computer Journal*, British Computer Society, 2001; Abbate, *Inventing the Internet*, 558; author's interview with Larry Roberts; Trevor Harris, "Who Is the Father of the Internet? The Case for Donald Davies," http://www.academia.edu.
59. Author's interview with Leonard Kleinrock; Leonard Kleinrock oral history, conducted by John Vardalas, IEEE History Center, Feb. 21, 2004.
60. Author's interview with Leonard Kleinrock.
61. Kleinrock oral history, IEEE.
62. Segaller, *Nerds*, 34.
63. Author's interviews with Kleinrock, Roberts; see also Hafner and Lyon, *Where Wizards Stay Up Late*, 1009; Segaller, *Nerds*, 53.
64. Leonard Kleinrock, "Information Flow in Large Communications Nets," proposal for a PhD thesis, MIT, May 31, 1961. See also Leonard Kleinrock, *Communication Nets: Stochastic Message Flow and Design* (McGraw-Hill, 1964).
65. Leonard Kleinrock personal website, http://www.lk.cs.ucla.edu/index.html.
66. Leonard Kleinrock, "Memoirs of the Sixties," in Peter Salus, *The ARPANET Sourcebook* (Peer-to-Peer, 2008), 96.
67. Leonard Kleinrock interview, *Computing Now*, IEEE Computer Society, 1996. Kleinrock is quoted in Peter Salus, *Casting the Net* (Addison-Wesley, 1995), 52: "I was the first to discuss the performance gains to be had by packet switching."
68. Author's interview with Taylor.
69. Author's interview with Kleinrock.
70. Donald Davies, "A Historical Study of the Beginnings of Packet Switching," *Computer Journal*, British Computer Society, 2001.
71. Alex McKenzie, "Comments on Dr. Leonard Kleinrock's Claim to Be 'the Father of Modern Data Networking,'" Aug. 16, 2009, http://alexmckenzie.weebly.com/comments-on-kleinrocks-claims.html.
72. Katie Hafner, "A Paternity Dispute Divides Net Pioneers," *New York Times*, Nov. 8, 2001;

Les Earnest, "Birthing the Internet," *New York Times,* Nov. 22, 2001. Earnest minimizes the distinction between a "store and forward" system and a "packet switch" one.

73. Leonard Kleinrock, "Principles and Lessons in Packet Communications," *Proceedings of the IEEE*, Nov. 1978.
74. Kleinrock oral history, Charles Babbage Institute, Apr. 3, 1990.
75. Leonard Kleinrock, "On Resource Sharing in a Distributed Communication Environment," *IEEE Communications Magazine*, May 2002. One loyalist did support Kleinrock's claims: his longtime friend, casino mate, and colleague Larry Roberts. "If you read Len's 1964 book, it's clear that he's breaking files into message units," Roberts told me in 2014. However, like Kleinrock, Roberts had previously given primary credit for packet switching to Baran. Roberts wrote in 1978, "The first published description of what we now call packet switching was an 11-volume analysis, On Distributed Communications, prepared by Paul Baran of the RAND Corporation in August 1964." See Lawrence Roberts, "The Evolution of Packet Switching," *Proceedings of the IEEE*, Nov. 1978.
76. Paul Baran oral history, "How the Web Was Won," *Vanity Fair*, July 2008.
77. Paul Baran interview, by Stewart Brand, *Wired*, Mar. 2001.
78. Paul Baran, "Introduction to Distributed Communications Networks," RAND, 1964, http://www.rand.org/pubs/research_memoranda/RM3420/RM3420-chapter1.html.
79. Segaller, *Nerds*, 70.
80. Author's interview with Bob Taylor. I was an editor of *Time* and remember the dispute.
81. Mitchell Waldrop, *The Dream Machine* (Viking, 2001), 279.
82. Stephen Lukasik, "Why the ARPANET Was Built," *IEEE Annals of the History of Computing*, Mar. 2011; Stephen Lukasik oral history, conducted by Judy O'Neill, Charles Babbage Institute, Oct. 17, 1991.
83. Charles Herzfeld, "On ARPANET and Computers," undated, http://inventors.about.com/library/inventors/bl_Charles_Herzfeld.htm.
84. "A Brief History of the Internet," Internet Society, Oct. 15, 2012, http://www.internetsociety.org/internet/what-internet/history-internet/brief-history-internet.
85. "NSFNET: A Partnership for High-Speed Networking: Final Report," 1995, http://www.merit.edu/documents/pdf/nsfnet/nsfnet_report.pdf.
86. Author's interview with Steve Crocker.
87. Author's interview with Leonard Kleinrock.
88. Author's interview with Robert Taylor.
89. Author's interview with Vint Cerf; Radia Joy Perlman, "Network Layer Protocols with Byzantine Robustness," PhD dissertation, MIT, 1988, http://dspace.mit.edu/handle/1721.1/14403.
90. Abbate, *Inventing the Internet*, 180.
91. Author's interview with Taylor.
92. Larry Roberts interview, conducted by James Pelkey, http://www.historyofcomputercommunications.info/Book/2/2.9-BoltBeranekNewman-WinningBid-68%20.html#_ftn26.
93. Hafner and Lyon, *Where Wizards Stay Up Late*, 1506 and passim.
94. Pelkey, "A History of Computer Communications," http://www.historyofcomputercommunications.info/index.html, 2.9; Hafner and Lyon, *Where Wizards Stay Up Late*, 1528.
95. The tale of Steve Crocker's RFCs has been told in many variations. This account comes from my interviews with Steve Crocker, Vint Cerf, Leonard Kleinrock; Hafner and Lyon, *Where Wizards Stay Up Late*, 2192 and passim; Abbate, *Inventing the Internet*, 1330 and passim; Stephen Crocker oral history, conducted by Judy E. O'Neill, Oct. 24, 1991, Charles Babbage Institute, University of Minnesota; Stephen Crocker, "How the Internet Got Its Rules," *New York Times*, Apr. 6, 2009; Cade Metz, "Meet the Man Who Invented the Instructions for the Internet," *Wired*, May 18, 2012; Steve Crocker, "The Origins of RFCs," in "The Request for Comments Guide," RFC 1000, Aug. 1987, http://www.rfc-editor.org/rfc/rfc1000.txt; Steve Crocker, "The First Pebble: Publication of RFC 1," RFC 2555, Apr. 7, 1999.
96. Author's interview with Steve Crocker.
97. Crocker, "How the Internet Got Its Rules."
98. Stephen Crocker, "Host Software," RFC 1, Apr. 7, 1969, http://tools.ietf.org/html/rfc1.
99. Crocker, "How the Internet Got Its Rules."
100. Vint Cerf, "The Great Conversation," RFC 2555, Apr. 7, 1999, http://www.rfc-editor.org/rfc/rfc2555.txt.

101. "The IMP Log: October 1969 to April 1970," Kleinrock Center for Internet Studies, UCLA, http://internethistory.ucla.edu/the-imp-log-october-1969-to-april-1970/; Segaller, *Nerds*, 92; Hafner and Lyon, *Where Wizards Stay Up Late*, 2336.
102. Vint Cerf oral history, conducted by Daniel Morrow, Nov. 21, 2001, Computerworld Honors Program; Hafner and Lyon, *Where Wizards Stay Up Late*, 2070 and passim; Abbate, *Inventing the Internet*, 127 and passim.
103. Cerf oral history, Computerworld.
104. Robert Kahn oral history, conducted by Michael Geselowitz, Feb. 17, 2004, IEEE History Center.
105. Vint Cerf oral history, conducted by Judy O'Neill, Apr. 24, 1990, Charles Babbage Institute; Vint Cerf, "How the Internet Came to Be," Nov. 1993, http://www.netvalley.com/archives/mirrors/cerf-how-inet.html.
106. Robert Kahn oral history, conducted by David Allison, Apr. 20, 1995, Computerworld Honors Program.
107. "The Poems," RFC 1121, Sept. 1989.
108. Author's interview with Vint Cerf.
109. Hafner and Lyon, *Where Wizards Stay Up Late*, 1163.
110. David D. Clark, "A Cloudy Crystal Ball," MIT Laboratory for Computer Science, July 1992, http://groups.csail.mit.edu/ana/People/DDC/future_ietf_92.pdf.
111. J. C. R. Licklider and Robert Taylor, "The Computer as a Communication Device," *Science and Technology*, Apr. 1968.

CHAPTER EIGHT: THE PERSONAL COMPUTER

1. Vannevar Bush, "As We May Think," *Atlantic*, July 1945.
2. Dave Ahl, who was at the meeting, said, "It fell to Ken Olsen to make a decision. I'll never forget his fateful words, "I can't see any reason that anyone would want a computer of his own." John Anderson, "Dave Tells Ahl," *Creative Computing*, Nov. 1984. For Olsen's defense, see http://www.snopes.com/quotes/kenolsen.asp, but this piece does not address Ahl's assertion that he made the statement when discussing with his staff whether a personal version of the PDP-8 should be developed.
3. In 1995, Stewart Brand wrote an essay for *Time*, which I had assigned, called "We Owe it All to the Hippies." It stressed the role of the counterculture in the birth of the personal computer. This chapter also draws on five well-reported and insightful books about how the counterculture helped to shape the personal computer revolution: Steven Levy, *Hackers* (Anchor/Doubleday, 1984; locations refer to the twenty-fifth anniversary reissue, O'Reilly, 2010); Paul Freiberger and Michael Swaine, *Fire in the Valley* (Osborne, 1984); John Markoff, *What the Dormouse Said* (Viking, 2005, locations refer to the Kindle edition); Fred Turner, *From Counterculture to Cyberculture* (University of Chicago, 2006); Theodore Roszak, *From Satori to Silicon Valley* (Don't Call It Frisco Press, 1986).
4. Liza Loop post on my crowdsourced draft on Medium and email to me, 2013.
5. Lee Felsenstein post on my crowdsourced draft on Medium, 2013. See also, "More Than Just Digital Quilting," *Economist*, Dec. 3, 2011; Victoria Sherrow, *Huskings, Quiltings, and Barn Raisings: Work-Play Parties in Early America* (Walker, 1992).
6. Posters and programs for the acid tests, in Phil Lesh, "The Acid Test Chronicles," http://www.postertrip.com/public/5586.cfm; Tom Wolfe, *The Electric Kool-Aid Acid Test* (Farrar, Straus and Giroux, 1987), 251 and passim.
7. Turner, *From Counterculture to Cyberculture*, 29, from Lewis Mumford, *Myth of the Machine* (Harcourt, Brace, 1967), 3.
8. Markoff, *What the Dormouse Said*, 165.
9. Charles Reich, *The Greening of America* (Random House, 1970), 5.
10. Author's interview with Ken Goffman, aka R. U. Sirius; Mark Dery, *Escape Velocity: Cyberculture at the End of the Century* (Grove, 1966), 22; Timothy Leary, *Cyberpunks CyberFreedom* (Ronin, 2008), 170.
11. First published in limited distribution by the Communication Company, San Francisco, 1967.
12. Brand's story was in a March 1995 special issue of *Time* on "Cyberspace," which was a sequel to a February 8, 1993, *Time* cover by Phil Elmer-Dewitt called "Cyberpunks" that also explored the countercultural influences surrounding the computer, online services such as The WELL, and the Internet.

13. This section is based on author's interviews with Stewart Brand; Stewart Brand, "'Whole Earth' Origin," 1976, http://sb.longnow.org/SB_homepage/WholeEarth_buton.html; Turner, *From Counterculture to Cyberculture*; Markoff, *What the Dormouse Said.* Turner's book is focused on Brand.
14. Author's interview with Stewart Brand; Stewart Brand public comments on early draft of this chapter posted on Medium.com.
15. Stewart Brand, "Spacewar: Fanatic Life and Symbolic Death among the Computer Bums," *Rolling Stone*, Dec. 7, 1972.
16. Stewart Brand comments on my crowdsourced draft on Medium; Stewart Brand interviews and emails with the author, 2013; poster and programs for the Trips Festival, http://www.postertrip.com/public/5577.cfm and http://www.lysergia.com/MerryPranksters/MerryPranksters_post.htm; Wolfe, *Electric Kool-Aid Test*, 259.
17. Turner, *From Counterculture to Cyberculture*, 67.
18. Author's interview with Stewart Brand; Brand, "'Whole Earth' Origin."
19. Brand, "'Whole Earth' Origin"; author's interview with Stewart Brand.
20. *Whole Earth Catalog*, Fall 1968, http://www.wholeearth.com/.
21. Author's interview with Lee Felsenstein.
22. The best account of Engelbart is Thierry Bardini, *Bootstrapping: Douglas Engelbart, Coevolution, and the Origins of Personal Computing* (Stanford, 2000). This section also draws on Douglas Engelbart oral history (four sessions), conducted by Judy Adams and Henry Lowood, Stanford, http://www-sul.stanford.edu/depts/hasrg/histsci/ssvoral/engelbart/start1.html; Douglas Engelbart oral history, conducted by Jon Eklund, the Smithsonian Institution, May 4, 1994; Christina Engelbart, "A Lifetime Pursuit," a biographical sketch written in 1986 by his daughter, http://www.dougengelbart.org/history/engelbart.html#10a; "Tribute to Doug Engelbart," a series of reminiscences by colleagues and friends, http://tribute2doug.wordpress.com/; Douglas Engelbart interviews, in Valerie Landau and Eileen Clegg, *The Engelbart Hypothesis: Dialogs with Douglas Engelbart* (Next Press, 2009) and http://engelbartbookdialogues.wordpress.com/; The Doug Engelbart Archives (includes many videos and interviews), http://dougengelbart.org/library/engelbart-archives.html; Susan Barnes, "Douglas Carl Engelbart: Developing the Underlying Concepts for Contemporary Computing," *IEEE Annals of the History of Computing*, July 1997; Markoff, *What the Dormouse Said*, 417; Turner, *From Counterculture to Cyberculture*, 110; Bardini, *Bootstrapping*, 138.
23. Douglas Engelbart oral history, Stanford, interview 1, Dec. 19, 1986.
24. The *Life* excerpt, Sept. 10, 1945, was heavily illustrated with drawings of the proposed memex. (The issue also had aerial photographs of Hiroshima after the dropping of the atom bomb.)
25. Douglas Engelbart oral history, Smithsonian, 1994.
26. Douglas Engelbart oral history, Stanford, interview 1, Dec. 19, 1986.
27. Landau and Clegg, *The Engelbart Hypothesis.*
28. Douglas Engelbart oral history, Stanford, interview 1, Dec. 19, 1986.
29. The quote is from Nilo Lindgren, "Toward the Decentralized Intellectual Workshop," *Innovation*, Sept. 1971, quoted in Howard Rheingold, *Tools for Thought* (MIT, 2000), 178. See also Steven Levy, *Insanely Great* (Viking, 1994), 36.
30. Douglas Engelbart oral history, Stanford, interview 3, Mar. 4, 1987.
31. Douglas Engelbart, "Augmenting Human Intellect," prepared for the director of Information Sciences, Air Force Office of Scientific Research, Oct. 1962.
32. Douglas Engelbart to Vannevar Bush, May 24, 1962, MIT/Brown Vannevar Bush Symposium, archives, http://www.dougengelbart.org/events/vannevar-bush-symposium.html.
33. Douglas Engelbart oral history, Stanford, interview 2, Jan. 14, 1987.
34. Author's interview with Bob Taylor.
35. Douglas Engelbart oral history, Stanford, interview 3, Mar. 4, 1987.
36. Landau and Clegg, "Engelbart on the Mouse and Keyset," in *The Engelbart Hypothesis*; William English, Douglas Engelbart, and Melvyn Berman, "Display Selection Techniques for Text Manipulation," *IEEE Transactions on Human-Factors in Electronics*, Mar. 1967.
37. Douglas Engelbart oral history, Stanford, interview 3, Mar. 4, 1987.
38. Landau and Clegg, "Mother of All Demos," in *The Engelbart Hypothesis.*
39. The video of the "Mother of All Demos" can be viewed at http://sloan.stanford.edu/MouseSite/1968Demo.html#complete. This section also draws from Landau and Clegg, "Mother of All Demos," in *The Engelbart Hypothesis.*

40. Rheingold, *Tools for Thought*, 190.
41. Author's interview with Stewart Brand; video of the Mother of All Demos.
42. Markoff, *What the Dormouse Said*, 2734. John Markoff found the reports of the Les Earnest demonstration in the Stanford microfilm archives. Markoff's book provides a good analysis of the distinction between augmented intellect and artificial intelligence.
43. Markoff, *What the Dormouse Said*, 2838.
44. Author's interview with Alan Kay. Kay read sections of this book and made comments and corrections. This section also draws on Alan Kay, "The Early History of Smalltalk," *ACM SIGPLAN Notices*, Mar. 1993; Michael Hiltzik, *Dealers of Lightning* (Harper, 1999; locations refer to the Kindle edition), chapter 6.
45. Author's interview with Alan Kay; Landau and Clegg, "Reflections by Fellow Pioneers," in *The Engelbart Hypothesis*; Alan Kay talk, thirtieth-anniversary panel on the Mother of All Demos, Internet archive, https://archive.org/details/XD1902_1EngelbartsUnfinishedRev30AnnSes2. See also Paul Spinrad, "The Prophet of Menlo Park," http://coe.berkeley.edu/news-center/publications/forefront/archive/copy_of_forefront-fall-2008/features/the-prophet-of-menlo-park-douglas-engelbart-carries-on-his-vision. After reading an early draft of this section, Kay clarified some of what he had said in earlier talks and interviews, and I modified a few of his quotes based on his suggestions.
46. Cathy Lazere, "Alan C. Kay: A Clear Romantic Vision," 1994, http://www.cs.nyu.edu/courses/fall04/G22.2110-001/kaymini.pdf.
47. Author's interview with Alan Kay. See also Alan Kay, "The Center of Why," Kyoto Prize lecture, Nov. 11, 2004.
48. Author's interview with Alan Kay; Ivan Sutherland, "Sketchpad," PhD dissertation, MIT, 1963; Howard Rheingold, "Inventing the Future with Alan Kay," The WELL, http://www.well.com/user/hlr/texts/Alan%20Kay.
49. Hiltzik, *Dealers of Lightning*, 1895; author's email exchange with Alan Kay.
50. Alan Kay talk, thirtieth-anniversary panel on the Mother of All Demos; Kay, "The Early History of Smalltalk."
51. Kay, "The Early History of Smalltalk."
52. Kay, "The Early History of Smalltalk." (Includes all quotes in preceding paragraphs.)
53. John McCarthy, "The Home Information Terminal—A 1970 View," June 1, 2000, http://www-formal.stanford.edu/jmc/hoter2.pdf.
54. Markoff, *What the Dormouse Said*, 4535.
55. Markoff, *What the Dormouse Said*, 2381.
56. In addition to citations below and Hiltzik's *Dealers of Lightning* and Kay's "The Early History of Smalltalk" cited above, this section draws on Douglas Smith and Robert Alexander, *Fumbling the Future: How Xerox Invented, Then Ignored, the First Personal Computer* (Morrow, 1988) and author's interviews with Alan Kay, Bob Taylor, and John Seeley Brown.
57. Charles P. Thacker, "Personal Distributed Computing: The Alto and Ethernet Hardware," ACM Conference on History of Personal Workstations, 1986. See also Butler W. Lampson, "Personal Distributed Computing: The Alto and Ethernet Software," ACM Conference on History of Personal Workstations, 1986. Both papers, with the same title, can be accessed at http://research.microsoft.com/en-us/um/people/blampson/38-AltoSoftware/Abstract.html.
58. Linda Hill, Greg Brandeau, Emily Truelove, and Kent Linebeck, *Collective Genius: The Art and Practice of Leading Innovation* (Harvard Business Review Press, 2014); Hiltzik, *Dealers of Lightning*, 2764; author's interview with Bob Taylor.
59. Author's interview with Bob Taylor.
60. Hiltzik, *Dealers of Lightning*, 1973, 2405.
61. Stewart Brand, "Spacewar," *Rolling Stone*, Dec. 7, 1972.
62. Alan Kay, "Microelectronics and the Personal Computer," *Scientific American*, Sept. 1977.
63. Alan Kay, "A Personal Computer for Children of All Ages," in *Proceedings of the ACM Annual Conference*, 1972. His typescript is at http://www.mprove.de/diplom/gui/Kay72a.pdf.
64. Kay, "The Early History of Smalltalk"; author's interview with Alan Kay.
65. Hiltzik, *Dealers of Lightning*, 3069.
66. Kay, "The Early History of Smalltalk"; Hiltzik, *Dealers of Lightning*, 3102.
67. Kay, "The Early History of Smalltalk"; author's interview with Alan Kay.
68. Kay, "The Early History of Smalltalk" (see section IV, "The First Real Smalltalk"); author's interviews with Alan Kay and Bob Taylor; Hiltzik, *Dealers of Lightning*, 3128; Markoff,

What the Dormouse Said, 3940; Butler Lampson, "Why Alto?" Xerox interoffice memo, Dec. 19, 1972, http://www.digibarn.com/friends/butler-lampson/.

69. Author's interview with Bob Taylor; Thacker, "Personal Distributed Computing."
70. Engelbart Oral History, Stanford, interview 4, Apr. 1, 1987.
71. Author's interview with Bob Taylor.
72. Alan Kay interview, conducted by Kate Kane, *Perspectives on Business Innovation*, May 2002.
73. Bob Taylor discussion, University of Texas, Sept. 17, 2009, conducted by John Markoff, http://transcriptvids.com/v/jvbGAPJSDJI.html.
74. Author's interview with Bob Taylor; Hiltzik, *Dealers of Lightning*, 4834.
75. Fred Moore's tale is detailed in Levy's *Hackers* and Markoff's *What the Dormouse Said.*
76. Author's interview with Lee Felsenstein.
77. Video of the Whole Earth Demise Party, http://mediaburn.org/video/aspects-of-demise-the-whole-earth-demise-party-2/; Levy, *Hackers*, 197; author's interview with Stewart Brand; Stewart Brand, "Demise Party, etc.," http://www.wholeearth.com/issue/1180/article/321/history.-.demise.party.etc.
78. Markoff, *What the Dormouse Said*, 3335.
79. In addition to the sources just cited, see Thomas Albright and Charles Moore, "The Last Twelve Hours of the Whole Earth," *Rolling Stone*, July 8, 1971; Barry Lopez, "Whole Earth's Suicide Party," *Washington Post*, June 14, 1971.
80. Author's interview with Bob Albrecht; Albrecht's notes provided to me.
81. Archive of the People's Computer Company and its related newsletters, http://www.digibarn.com/collections/newsletters/peoples-computer/.
82. Author's interview with Bob Albrecht.
83. Author's interview with Lee Felsenstein. This section is also based on a seventeen-chapter unpublished memoir Felsenstein wrote, which he provided to me; Felsenstein's articles "Tom Swift Lives!" and "Convivial Design" in *People's Computer Company*; his article "My Path through the Free Speech Movement and Beyond," February 22, 2005, which he provided to me; the autobiographical essays he has posted at http://www.leefelsenstein.com/; Freiberger and Swaine, *Fire in the Valley*, 99–102; Levy, *Hackers*, 153 and passim; Markoff, *What the Dormouse Said*, 4375 and passim.
84. Author's interview with Lee Felsenstein.
85. Author's interview with Felsenstein; Lee Felsenstein, "Philadelphia 1945–1963," http://www.leefelsenstein.com/?page_id=16; oral history of Lee Felsenstein, by Kip Crosby, May 7, 2008, Computer History Museum.
86. Felsenstein, "My Path through the Free Speech Movement and Beyond."
87. Author's interview with Lee Felsenstein.
88. Felsenstein, "My Path through the Free Speech Movement and Beyond."
89. Author's interview with Lee Felsenstein; Felsenstein unpublished memoir.
90. Felsenstein's unpublished memoir, provided to me, has an entire chapter on the police radio incident.
91. Felsenstein, "My Path through the Free Speech Movement and Beyond."
92. Lee Felsenstein, "Explorations in the Underground," http://www.leefelsenstein.com/?page_id=50.
93. Author's interview with Lee Felsenstein.
94. Author's interview with Lee Felsenstein; Felsenstein's unpublished memoir.
95. Author's interview with Lee Felsenstein.
96. Levy, *Hackers*, 160.
97. Ken Colstad and Efrem Lipkin, "Community Memory: A Public Information Network," *ACM SIGCAS Computers and Society*, Dec. 1975. For an archive of the Resource One Newsletter, see http://www.well.com/~szpak/cm/index.html.
98. Doug Schuler, "Community Networks: Building a New Participatory Medium," *Communications of the ACM,* Jan. 1994. See also Community Memory flyer, on The WELL, http://www.well.com/~szpak/cm/cmflyer.html: "We have a powerful tool—a genie—at our disposal."
99. R. U. Sirius and St. Jude, *How to Mutate and Take Over the World* (Ballantine, 1996); Betsy Isaacson, "St. Jude," undergraduate thesis, Harvard University, 2012.
100. Lee Felsenstein, "Resource One/Community Memory," http://www.leefelsenstein.com/?page_id=44.
101. Author's interview with Lee Felsenstein; Felsenstein, "Resource One/Community Memory."

102. Ivan Illich, *Tools for Conviviality* (Harper, 1973), 17.
103. Author's interview with Lee Felsenstein.
104. Lee Felsenstein, "The Maker Movement—Looks Like Revolution to Me," speech at Bay Area Maker Faire, May 18, 2013. See also Evgeny Morozov, "Making It," *New Yorker*, Jan. 13, 2014.
105. Lee Felsenstein, "Tom Swift Terminal, or a Convivial Cybernetic Device," http://www.leefelsenstein.com/wp-content/uploads/2013/01/TST_scan_150.pdf; Lee Felsenstein, "Social Media Technology," http://www.leefelsenstein.com/?page_id=125.
106. Homebrew Computer Club newsletter #1, DigiBarn Computer Museum, http://www.digibarn.com/collections/newsletters/homebrew/V1_01/; Levy, *Hackers*, 167.
107. Lee Felsenstein's comments on my crowdsourced first draft, Medium.com, Dec. 20, 2013. There is no evidence that any of Eisenhower's personal pilots ever had a sex change.
108. This section draws from Ed Roberts interview, conducted by Art Salsberg, *Modern Electronics*, Oct. 1984; Ed Roberts interview, conducted by David Greelish, *Historically Brewed* magazine, 1995; Levy, *Hackers*, 186 and passim; Forrest M. Mims III, "The Altair Story: Early Days at MITS," *Creative Computing*, Nov. 1984; Freiberger and Swaine, *Fire in the Valley*, 35 and passim.
109. Levy, *Hackers*, 186.
110. Mims, "The Altair Story."
111. Levy, *Hackers*, 187.
112. Levy, *Hackers*, 187.
113. Les Solomon, "Solomon's Memory," Atari Archives, http://www.atariarchives.org/deli/solomons_memory.php; Levy, *Hackers*, 189 and passim; Mims, "The Altair Story."
114. H. Edward Roberts and William Yates, "Altair 8800 Minicomputer," *Popular Electronics*, Jan. 1975.
115. Author's interview with Bill Gates.
116. Michael Riordan and Lillian Hoddeson, "Crystal Fire," *IEEE SCS News*, Spring 2007, adapted from *Crystal Fire* (Norton, 1977).
117. Author's interviews with Lee Felsenstein, Steve Wozniak, Steve Jobs, and Bob Albrecht. This section also draws from the accounts of the Homebrew Computer Club origins in Wozniak, *iWoz* (Norton, 2006); Markoff, *What the Dormouse Said*, 4493 and passim; Levy, *Hackers*, 201 and passim; Freiberger and Swaine, *Fire in the Valley*, 109 and passim; Steve Wozniak, "Homebrew and How the Apple Came to Be," http://www.atariarchives.org/deli/homebrew_and_how_the_apple.php; the Homebrew archives exhibit at the Computer History Museum; the Homebrew newsletter archives, http://www.digibarn.com/collections/newsletters/homebrew/; Bob Lash, "Memoir of a Homebrew Computer Club Member," http://www.bambi.net/bob/homebrew.html.
118. Steve Dompier, "Music of a Sort," *Peoples Computer Company*, May 1975. See also Freiberger and Swaine, *Fire in the Valley*, 129; Levy, *Hackers*, 204. For Dompier's code, see http://kevindriscoll.org/projects/ccswg2012/fool_on_a_hill.html.
119. Bill Gates, "Software Contest Winners Announced," *Computer Notes*, July 1975.

CHAPTER NINE: SOFTWARE

1. Author's interview with Bill Gates; Paul Allen, *Idea Man* (Portfolio, 2011, locations refer to the Kindle edition), 129. This section also draws from a formal interview in 2013 and other conversations I had with Bill Gates; the time I spent with him, his father, and colleagues for a *Time* cover story I wrote, "In Search of the Real Bill Gates," *Time*, Jan. 13, 1997; emails from Bill Gates Sr.; Stephen Manes and Paul Andrews, *Gates* (Doubleday, 1993, locations refer to Kindle edition); James Wallace and Jim Erickson, *Hard Drive* (Wiley, 1992); Bill Gates oral history, conducted by Mark Dickison, Henry Ford Innovation Series, June 30, 2009; Bill Gates interview, conducted by David Allison, Smithsonian Institution, Apr. 1995; other nonpublic oral histories provided by Bill Gates.
2. Wallace and Erickson, *Hard Drive*, 38.
3. Allen, *Idea Man*, 1069.
4. Author's interview with Bill Gates. See also Bill Gates oral history, Ford Innovation Series.
5. Isaacson, "In Search of the Real Bill Gates."
6. Isaacson, "In Search of the Real Bill Gates."
7. Author's interview with Bill Gates Sr.
8. Manes and Andrews, *Gates*, 715.

9. Author's interview with Bill Gates Sr. The law says: "A Scout is trustworthy, loyal, helpful, friendly, courteous, kind, obedient, cheerful, thrifty, brave, clean, and reverent."
10. Manes and Andrews, *Gates*, 583, 659.
11. Author's interview with Bill Gates Sr.
12. Wallace and Erickson, *Hard Drive*, 21.
13. Author's interview with Bill Gates.
14. Allen, *Idea Man*, 502.
15. Wallace and Erickson, *Hard Drive*, 25.
16. Allen, *Idea Man*, 511.
17. Wallace and Erickson, *Hard Drive*, 26.
18. Allen, *Idea Man*, 751.
19. Author's interview with Bill Gates; Isaacson, "In Search of the Real Bill Gates."
20. Author's interview with Bill Gates. (Also in other oral histories.)
21. Manes and Andrews, *Gates*, 924.
22. Author's interviews with Bill Gates and Bill Gates Sr.
23. Author's interview with Steve Russell.
24. Wallace and Erickson, *Hard Drive*, 31.
25. Author's interview with Bill Gates.
26. Allen, *Idea Man*, 616; author's interviews with Steve Russell and Bill Gates.
27. Author's interview with Bill Gates.
28. Paul Freiberger and Michael Swaine, *Fire in the Valley* (Osborne, 1984), 21; author's interview with Bill Gates; Wallace and Erickson, *Hard Drive*, 35.
29. Allen, *Idea Man*, 719.
30. Wallace and Erickson, *Hard Drive*, 42.
31. Author's interview with Bill Gates; Isaacson, "In Search of the Real Bill Gates."
32. Author's interview with Bill Gates; Bill Gates oral history with Larry Cohen and Brent Schlender, provided to me by Bill Gates.
33. Wallace and Erickson, *Hard Drive*, 43.
34. Author's interviews with Bill Gates.
35. Allen, *Idea Man*, 811.
36. Wallace and Erickson, *Hard Drive*, 43.
37. Author's interview with Bill Gates; Allen, *Idea Man*, 101.
38. Author's interview with Bill Gates; Allen, *Idea Man*, 849.
39. Allen, *Idea Man*, 860.
40. Wallace and Erickson, *Hard Drive*, 45; Manes and Andrews, *Gates*, 458.
41. Manes and Andrews, *Gates*, 1445; Allen, *Idea Man*, 917; author's interview with Bill Gates.
42. Allen, *Idea Man*, 942.
43. Author's interview with Bill Gates.
44. Allen, *Idea Man*, 969.
45. Wallace and Erickson, *Hard Drive*, 55. An earlier version of this section was published in the *Harvard Gazette*, and the current version reflects comments and corrections made by Gates and others on it.
46. Author's interview with Bill Gates.
47. Nicholas Josefowitz, "College Friends Remember Bill Gates," *Harvard Crimson*, June 4, 2002.
48. Manes and Andrews, *Gates*, 1564.
49. "Bill Gates to Sign Off at Microsoft," AFP, June 28, 2008.
50. William H. Gates and Christos P. Papadimitriou, "Bounds for Sorting by Prefix Reversal," *Discrete Mathematics*, 1979; Harry Lewis, "Reinventing the Classroom," *Harvard Magazine*, Sept. 2012; David Kestenbaum, "Before Microsoft, Gates Solved a Pancake Problem," NPR, July 4, 2008.
51. Allen, *Idea Man*, 62.
52. Author's interview with Bill Gates.
53. Allen, *Idea Man*, 1058.
54. Author's interview with Bill Gates.
55. Bill Gates and Paul Allen to Ed Roberts, Jan. 2, 1975; Manes and Andrews, *Gates*, 1810.
56. Allen, *Idea Man*, 160.
57. Allen, *Idea Man*, 1103.
58. Manes and Andrews, *Gates*, 1874.
59. Author's interview with Bill Gates; Allen, *Idea Man*, 1117.

60. Wallace and Erickson, *Hard Drive*, 76.
61. Allen, *Idea Man*, 1163.
62. Allen, *Idea Man*, 1204.
63. Allen, *Idea Man*, 1223; Wallace and Erickson, *Hard Drive*, 81.
64. Author's interview with Bill Gates.
65. Remarks of Bill Gates, *Harvard Gazette*, June 7, 2007.
66. Author's interview with Bill Gates.
67. The section on Gates in Albuquerque draws on Allen, *Idea Man*, 1214 and passim; Manes and Andrews, *Gates*, 2011 and passim; Wallace and Erickson, *Hard Drive*, 85 and passim.
68. Bill Gates oral history, Henry Ford Innovation Series.
69. Allen, *Idea Man*, 1513.
70. Author's interview with Bill Gates.
71. Allen, *Idea Man*, 1465; Manes and Andrews, *Gates*, 2975; Wallace and Erickson, *Hard Drive*, 130.
72. Author's interview with Bill Gates.
73. Allen, *Idea Man*, 1376.
74. Fred Moore, "It's a Hobby," Homebrew Computer Club newsletter, June 7, 1975.
75. John Markoff, *What the Dormouse Said* (Viking, 2005; locations refer to the Kindle edition), 4633; Steven Levy, *Hackers* (Anchor/Doubleday, 1984; locations refer to the twenty-fifth anniversary reissue, O'Reilly, 2010), 231.
76. Author's interview with Lee Felsenstein; Lee Felsenstein oral history, by Kip Crosby, Computer History Museum, May 7, 2008.
77. Homebrew Computer Club newsletter, Feb. 3, 1976, http://www.digibarn.com/collections/newsletters/homebrew/V2_01/gatesletter.html.
78. Author's interview with Bill Gates.
79. Harold Singer, "Open Letter to Ed Roberts," Micro-8 Computer User Group newsletter, Mar. 28, 1976.
80. Author's interview with Lee Felsenstein.
81. Bill Gates interview, *Playboy*, July 1994.
82. This section draws from my *Steve Jobs* (Simon & Schuster, 2011), which was based on interviews with Steve Jobs, Steve Wozniak, Nolan Bushnell, Al Alcorn, and others. The Jobs biography includes a bibliography and source notes. For this book, I reinterviewed Bushnell, Alcorn, and Wozniak. This section also draws on Steve Wozniak, *iWoz* (Norton, 1984); Steve Wozniak, "Homebrew and How the Apple Came to Be," http://www.atariarchives.org/deli/homebrew_and_how_the_apple.php.
83. When I posted an early draft of parts of this book for crowdsourced comments and corrections on Medium, Dan Bricklin offered useful suggestions. We got into an exchange about the creation of VisiCalc, and I subsequently added this section to the book. It is partly based on email exchanges with Bricklin and Bob Frankston and on chapter 12, "VisiCalc," in Dan Bricklin, *Bricklin on Technology* (Wiley, 2009).
84. Email from Dan Bricklin to the author; Dan Bricklin, "The Idea," http://www.bricklin.com/history/saiidea.htm.
85. Peter Ruell, "A Vision of Computing's Future," *Harvard Gazette*, Mar. 22, 2012.
86. Bob Frankston, "Implementing VisiCalc," unpublished, Apr. 6, 2002.
87. Frankston, "Implementing VisiCalc."
88. Author's interview with Steve Jobs.
89. IBM corporate history, "The Birth of the IBM PC," http://www-03.ibm.com/ibm/history/exhibits/pc25/pc25_birth.html.
90. Manes and Andrews, *Gates*, 3629.
91. Manes and Andrews, *Gates*, 3642; Steve Ballmer interview, "Triumph of the Nerds," part II, PBS, June 1996. See also James Chposky and Ted Leonsis, *Blue Magic* (Facts on File, 1988), chapter 9.
92. Bill Gates and Paul Allen interview, by Brent Schlender, *Fortune*, Oct. 2, 1995.
93. Steve Ballmer interview, "Triumph of the Nerds," part II, PBS, June 1996.
94. Jack Sams interview, "Triumph of the Nerds," part II, PBS, June 1996. See also Steve Hamm and Jay Greene, "The Man Who Could Have Been Bill Gates," *Business Week*, Oct. 24, 2004.
95. Tim Paterson and Paul Allen interviews, "Triumph of the Nerds," part II, PBS, June 1996.
96. Steve Ballmer and Paul Allen interviews, "Triumph of the Nerds," part II, PBS, June 1996; Manes and Andrews, *Gates*, 3798.

97. Bill Gates and Paul Allen interview, by Brent Schlender, *Fortune*, Oct. 2, 1995; Manes and Andrews, *Gates*, 3868.
98. Manes and Andrews, *Gates*, 3886, 3892.
99. Author's interview with Bill Gates.
100. Bill Gates and Paul Allen interview, by Brent Schlender, *Fortune*, Oct. 2, 1995.
101. Author's interview with Bill Gates.
102. Author's interview with Bill Gates.
103. Bill Gates and Paul Allen interview, by Brent Schlender, *Fortune*, Oct. 2, 1995.
104. Bill Gates interview by David Rubenstein, Harvard, Sept. 21, 2013, author's notes.
105. Bill Gates and Paul Allen interview, by Brent Schlender, *Fortune*, Oct. 2, 1995.
106. Bill Gates interview, conducted by David Bunnell, *PC* magazine, Feb. 1, 1982.
107. Isaacson, *Steve Jobs*, 135.
108. Isaacson, *Steve Jobs*, 94.
109. Author's interview with Steve Jobs.
110. Steve Jobs presentation, Jan. 1984, https://www.youtube.com/watch?v=2B-XwPjn9YY.
111. Isaacson, *Steve Jobs*, 173.
112. Author's interview with Andy Hertzfeld.
113. Author's interviews with Steve Jobs and Bill Gates.
114. Andy Hertzfeld, *Revolution in the Valley* (O'Reilly Media, 2005), 191. See also Andy Hertzfeld, http://www.folklore.org/StoryView.py?story=A_Rich_Neighbor_Named_Xerox.txt.
115. Author's interviews with Steve Jobs and Bill Gates.
116. Author's interview with Steve Jobs.
117. In addition to the sources cited below, this section is based on my interview with Richard Stallman; Richard Stallman, essays and philosophy, on http://www.gnu.org/gnu/gnu.html; Sam Williams, with revisions by Richard M. Stallman, *Free as in Freedom (2.0): Richard Stallman and the Free Software Revolution* (Free Software Foundation, 2010). An earlier edition of the Williams book was published by O'Reilly Media in 2002. As that edition was being completed, Stallman and Williams "parted on less than cordial terms" based on Stallman's objections and requests for corrections. Version 2.0 incorporated Stallman's objections and a significant rewriting of some segments of the book. These are described by Stallman in his foreword and Williams in his preface to version 2.0, which Stallman later called "my semi-autobiography." For comparison, the original text of the book can be found at http://oreilly.com/openbook/freedom/.
118. Author's interview with Richard Stallman. See also K. C. Jones, "A Rare Glimpse into Richard Stallman's World," *InformationWeek*, Jan. 6, 2006; Richard Stallman interview, in Michael Gross, "Richard Stallman: High School Misfit, Symbol of Free Software, MacArthur-Certified Genius," 1999, www.mgross.com/interviews/stallman1.html; Williams, *Free as in Freedom*, 26 and passim.
119. Richard Stallman, "The GNU Operating System and the Free Software Movement," in Chris DiBona and Sam Ockman, editors, *Open Sources: Voices from the Open Source Revolution* (O'Reilly, 1999).
120. Author's interview with Richard Stallman.
121. Richard Stallman, "The GNU Project," http://www.gnu.org/gnu/thegnuproject.html.
122. Williams, *Free as in Freedom*, 75.
123. Richard Stallman, "The GNU Manifesto," http://www.gnu.org/gnu/manifesto.html.
124. Richard Stallman, "What Is Free Software?" and "Why Open Source Misses the Point of Free Software," https://www.gnu.org/philosophy/.
125. Richard Stallman, "The GNU System," https://www.gnu.org/philosophy/.
126. Interview with Richard Stallman, conducted by David Betz and Jon Edwards, *BYTE*, July 1986.
127. "Linus Torvalds," Linux Information Project, http://www.linfo.org/linus.html.
128. Linus Torvalds with David Diamond, *Just for Fun* (HarperCollins, 2001), 4.
129. Torvalds and Diamond, *Just for Fun*, 74, 4, 17; Michael Learmonth, "Giving It All Away," *San Jose Metro*, May 8, 1997.
130. Torvalds and Diamond, *Just for Fun*, 52, 55, 64, 78, 72.
131. Linus Torvalds pronouncing "Linux": http://upload.wikimedia.org/wikipedia/commons/0/03/Linus-linux.ogg.
132. Learmonth, "Giving It All Away."
133. Torvalds and Diamond, *Just for Fun*, 58.

134. Linus Torvalds, "Free Minix-like Kernel Sources for 386-AT," posting to Newsgroups: comp.os.minix, Oct. 5, 1991, http://www.cs.cmu.edu/~awb/linux.history.html.
135. Torvalds and Diamond, *Just for Fun*, 87, 93, 97, 119.
136. Gary Rivlin, "Leader of the Free World," *Wired*, November 2003.
137. Yochai Benkler, *The Penguin and the Leviathan: How Cooperation Triumphs over Self-Interest* (Crown, 2011); Yochai Benkler, "Coase's Penguin, or, Linux and the Nature of the Firm," *Yale Law Journal* (2002), http://soc.ics.uci.edu/Resources/bibs.php?793.
138. Eric Raymond, *The Cathedral and the Bazaar* (O'Reilly Media, 1999), 30.
139. Alexis de Tocqueville, *Democracy in America* (originally published 1835–40; Packard edition), Kindle location 3041.
140. Torvalds and Diamond, *Just for Fun*, 122, 167, 120, 121.
141. Richard Stallman interview, *Reddit*, July 29, 2010, http://www.redditblog.com/2010/07/rms-ama.html.
142. Richard Stallman, "What's in a Name?" https://www.gnu.org/gnu/why-gnu-linux.html.
143. Torvalds and Diamond, *Just for Fun*, 164.
144. Linus Torvalds blog post, "Black and White," Nov. 2, 2008, http://torvalds-family.blogspot.com/2008/11/black-and-white.html.
145. Torvalds and Diamond, *Just for Fun*, 163.
146. Raymond, *The Cathedral and the Bazaar*, 1.

CHAPTER TEN: ONLINE

1. Lawrence Landweber email to the author, Feb. 5, 2014.
2. Ray Tomlinson, "The First Network Email," http://openmap.bbn.com/~tomlinso/ray/firstemailframe.html.
3. Larry Brilliant email to the author, Feb. 14, 2014.
4. Larry Brilliant interview, *Wired*, Dec. 20, 2007.
5. Larry Brilliant interview, *Wired*, Dec. 20, 2007.
6. Katie Hafner, *The Well* (Carroll & Graf, 2001), 10.
7. Hafner, *The Well*, 30; Turner, *From Counterculture to Cyberculture*, 145.
8. Howard Rheingold, *The Virtual Community* (Perseus, 1993), 9.
9. Tom Mandel, "Confessions of a Cyberholic," *Time*, Mar. 1, 1995. At that point, Mandel knew he was dying, and he asked his editors at *Time*—Phil Elmer-DeWitt, Dick Duncan, and myself—if he could write a farewell reflection about the online world.
10. Tom Mandel, posting on The WELL, http://www.well.com/~cynsa/tom/tom13.html. See also "To Our Readers" [signed by the publisher Elizabeth Long but written by Phil Elmer-DeWitt], *Time*, Apr. 17, 1995.
11. This section draws from interviews with Steve Case, Jim Kimsey, and Jean Case; Julius Duscha, "For Computers, a Marrying Sam," *New York Times*, Dec. 25, 1977; Michael Banks, *On the Way to the Web* (APress, 2008, locations refer to the Kindle edition); Kara Swisher, *AOL.com* (Random House, 1998); Alec Klein, *Stealing Time* (Simon & Schuster, 2003). Steve Case, a longtime friend and colleague, provided comments and corrections on an early draft.
12. Klein, *Stealing Time*, 11.
13. Banks, *On the Way to the Web*, 792, 743.
14. Banks, *On the Way to the Web*, 602, 1467.
15. Author's interview with Steve Case; Banks, *On the Way to the Web*, 1503; Swisher, *AOL.com*, 27.
16. Steve Case talk, JP Morgan Technology Conference, San Francisco, May 1, 2001.
17. Nina Munk, *Fools Rush In* (Collins, 2004), 73.
18. Author's interview with Steve Case.
19. Swisher, *AOL.com*, 25.
20. Steve Case speech, Stanford, May 25, 2010.
21. Steve Case speech, Stanford, May 25, 2010.
22. Author's interview with Steve Case.
23. Steve Case speech, Stanford, May 25, 2010.
24. Swisher, *AOL.com*, 27.
25. Author's interview with Steve Case.
26. Author's interview with Steve Case; Case email to author and comments on first draft published on Medium. Accounts differ on whether or not von Meister was eager to hire Steve Case or whether Dan Case pushed him to do it. Swisher, *AOL.com*, 28, says it was the

former. Banks, *On the Way to the Web*, 1507, says it was the latter. There are probably elements of truth in both versions.

27. Author's interview with Jim Kimsey.
28. Swisher, *AOL.com*, 53.
29. Swisher, *AOL.com*, 48.
30. Author's interviews with Steve Case, Steve Wozniak.
31. Steve Case speech, Stanford, May 25, 2010.
32. Author's interview with Steve Case.
33. Author's interview with Steve Case.
34. Steve Case oral history, conducted by Walter Isaacson, 2013, the Riptide Project, Harvard, http://www.niemanlab.org/riptide/person/steve-case/. I participated in this oral history project on the digital disruption of journalism, which was curated by John Huey, Paul Sagan, and Martin Nisenholtz.
35. Steve Case oral history, "How the Web Was Won," *Vanity Fair*, July 2008.
36. Author's interview with Jim Kimsey.
37. Steve Case speech, Stanford, May 25, 2010.
38. Dave Fischer post, newsgroup: alt.folklore.computers, Jan. 25, 1994, https://groups.google.com/forum/#!original/alt.folklore.computers/wF4CpYbWuuA/jS6ZOyJd10sJ.
39. Wendy Grossman, *Net. Wars* (NYU, 1977), 33.
40. Author's interview with Al Gore.
41. Al Gore interview with Wolf Blitzer, "Late Edition," CNN, Mar. 9, 1999, http://www.cnn.com/ALLPOLITICS/stories/1999/03/09/president.2000/transcript.gore/.
42. Robert Kahn and Vinton Cerf, "Al Gore and the Internet," an email to Declan McCullaugh and others, Sept. 28, 2000, http://www.politechbot.com/p-01394.html.
43. Newt Gingrich, speech to the American Political Science Association, Sept. 1, 2000.

CHAPTER ELEVEN: THE WEB

1. Tim Berners-Lee, *Weaving the Web* (HarperCollins, 1999), 4. See also Mark Fischetti, "The Mind Behind the Web," *Scientific American*, Mar. 12, 2009.
2. Author's interview with Tim Berners-Lee.
3. Author's interview with Tim Berners-Lee.
4. Author's interview with Tim Berners-Lee.
5. Author's interview with Tim Berners-Lee.
6. Tim Berners-Lee interview, Academy of Achievement, June 22, 2007.
7. Author's interview with Tim Berners-Lee.
8. Author's interview with Tim Berners-Lee.
9. *Enquire Within Upon Everything* (1894), http://www.gutenberg.org/files/10766/10766-h/10766-h.htm.
10. Berners-Lee, *Weaving the Web*, 1.
11. Author's interview with Tim Berners-Lee.
12. Tim Berners-Lee interview, Academy of Achievement, June 22, 2007.
13. Berners-Lee, *Weaving the Web*, 10.
14. Berners-Lee, *Weaving the Web*, 4.
15. Berners-Lee, *Weaving the Web*, 14.
16. Author's interview with Tim Berners-Lee.
17. Tim Berners-Lee interview, Academy of Achievement, June 22, 2007.
18. Berners-Lee, *Weaving the Web*, 15.
19. John Naish, "The NS Profile: Tim Berners-Lee," *New Statesman*, Aug. 15, 2011.
20. Berners-Lee, *Weaving the Web*, 16, 18.
21. Berners-Lee, *Weaving the Web*, 61.
22. Tim Berners-Lee, "Information Management: A Proposal," CERN, Mar. 1989, http://www.w3.org/History/1989/proposal.html.
23. James Gillies and Robert Cailliau, *How the Web Was Born* (Oxford, 2000), 180.
24. Berners-Lee, *Weaving the Web*, 26.
25. Gillies and Cailliau, *How the Web Was Born*, 198.
26. Gillies and Cailliau, *How the Web Was Born*, 190.
27. Robert Cailliau interview, "How the Web Was Won," *Vanity Fair*, July 2008.
28. Gillies and Cailliau, *How the Web Was Born*, 234.

29. Tim Smith and François Flückiger, "Licensing the Web," CERN, http://home.web.cern.ch/topics/birth-web/licensing-web.
30. Tim Berners-Lee, "The World Wide Web and the 'Web of Life,'" 1998, http://www.w3.org/People/Berners-Lee/UU.html.
31. Tim Berners-Lee, posting to the Newsgroup alt.hypertext, Aug. 6, 1991, http://www.w3.org/People/Berners-Lee/1991/08/art-6484.txt.
32. Nick Bilton, "As the Web Turns 25, Its Creator Talks about Its Future," *New York Times*, Mar. 11, 2014.
33. Gillies and Cailliau, *How the Web Was Born*, 203. See also Matthew Lasar, "Before Netscape," *Ars Technica*, Oct. 11, 2011.
34. Berners-Lee, *Weaving the Web*, 56.
35. Gillies and Cailliau, *How the Web Was Born*, 217.
36. Author's interview with Marc Andreessen.
37. Author's interview with Marc Andreessen.
38. Robert Reid, *Architects of the Web* (Wiley, 1997), 7.
39. Gillies and Cailliau, *How the Web Was Born*, 239; alt.hypertext Newsgroup, Friday, Jan. 29, 1993, 12:22:43 GMT, http://www.jmc.sjsu.edu/faculty/rcraig/mosaic.txt.
40. Author's interview with Marc Andreessen.
41. Gillies and Cailliau, *How the Web Was Born*, 240.
42. Author's interview with Marc Andreessen.
43. Berners-Lee, *Weaving the Web*, 70; author's interview with Tim Berners-Lee.
44. Author's interview with Marc Andreessen.
45. Author's interview with Tim Berners-Lee.
46. Berners-Lee, *Weaving the Web*, 70.
47. Berners-Lee, *Weaving the Web*, 65.
48. Ted Nelson, "Computer Paradigm," http://xanadu.com.au/ted/TN/WRITINGS/TCOMPARADIGM/tedCompOneLiners.html.
49. Jaron Lanier interview, by Eric Allen Bean, Nieman Journalism Lab, May 22, 2013.
50. John Huey, Martin Nisenholtz, and Paul Sagan, "Riptide," Harvard Kennedy School, http://www.niemanlab.org/riptide/.
51. Author's interview with Marc Andreessen.
52. Author's interview with Tim Berners-Lee.
53. Author's interview with Marc Andreessen.
54. John Markoff, "A Free and Simple Computer Link," *New York Times*, Dec. 8, 1993.
55. This section is primarily based on my interviews with Justin Hall and his own postings at http://www.links.net/.
56. Justin Hall, "Justin's Links," http://www.links.net/vita/web/story.html.
57. Author's interviews with Justin Hall, Joan Hall.
58. Author's interview with Howard Rheingold; Howard Rheingold, *The Virtual Community* (Perseus, 1993).
59. Author's interviews with Justin Hall, Howard Rheingold; Gary Wolf, *Wired—A Romance* (Random House, 2003), 110.
60. Scott Rosenberg, *Say Everything* (Crown, 2009), 24.
61. Rosenberg, *Say Everything*, 44.
62. Justin Hall, "Exposing Myself," posted by Howard Rheingold, http://www.well.com/~hlr/jam/justin/justinexposing.html.
63. Author's interview with Arianna Huffington.
64. Clive Thompson, *Smarter Than You Think* (Penguin, 2013), 68.
65. Hall, "Exposing Myself."
66. Author's interview with Ev Williams. This section also draws from the Ev Williams interview in Jessica Livingston, *Founders at Work* (Apress, 2007), 2701 and passim; Nick Bilton, *Hatching Twitter* (Portfolio, 2013), 9 and passim; Rosenberg, *Say Everything*, 104 and passim; Rebecca Mead, "You've Got Blog," *New Yorker*, Nov. 13, 2000.
67. Dave Winer, "Scripting News in XML," Dec. 15, 1997, http://scripting.com/davenet/1997/12/15/scriptingNewsInXML.html.
68. Livingston, *Founders at Work*, 2094.
69. Livingston, *Founders at Work*, 2109, 2123, 2218.
70. Meg Hourihan, "A Sad Kind of Day," http://web.archive.org/web/20010917033719/http://

www.megnut.com/archive.asp?which=2001_02_01_archive.inc; Rosenberg, *Say Everything*, 122.

71. Ev Williams, "And Then There Was One," Jan. 31, 2001, http://web.archive.org/web/2001 1214143830/http://www.evhead.com/longer/2200706_essays.asp.
72. Livingston, *Founders at Work*, 2252.
73. Livingston, *Founders at Work*, 2252.
74. Williams, "And Then There Was One."
75. Dan Bricklin, "How the Blogger Deal Happened," blog posting, Apr. 15, 2001, http://dan bricklin.com/log/blogger.htm; Dan Bricklin, *Bricklin on Technology* (Wiley, 2009), 206.
76. Livingston, *Founders at Work*, 2289, 2302.
77. Author's interview with Ev Williams.
78. Author's interview with Ev Williams.
79. Author's interview with Ev Williams.
80. Andrew Lih, *The Wikipedia Revolution* (Hyperion, 2009), 1111. See also Ward Cunningham and Bo Leuf, *The Wiki Way: Quick Collaboration on the Web* (Addison-Wesley, 2001); Ward Cunningham, "HyperCard Stacks," http://c2.com/~ward/HyperCard/; Ward Cunningham, keynote speech, Wikimania, Aug. 1, 2005.
81. Ward Cunningham, "Invitation to the Pattern List," May 1, 1995, http://c2.com/cgi/wiki? InvitationToThePatternsList.
82. Ward Cunningham, correspondence on the etymology of *wiki*, http://c2.com/doc/etymol ogy.html.
83. Tim Berners-Lee interview, Riptide Project, Schornstein Center, Harvard, 2013.
84. Kelly Kazek, "Wikipedia Founder, Huntsville Native Jimmy Wales, Finds Fame Really Cool," *News Courier* (Athens, AL), Aug. 12, 2006.
85. Author's interview with Jimmy Wales.
86. Author's interview with Jimmy Wales; Lih, *The Wikipedia Revolution*, 585.
87. Marshall Poe, "The Hive," *Atlantic*, Sept. 2006.
88. Jimmy Wales interview, conducted by Brian Lamb, C-SPAN, Sept. 25, 2005.
89. Author's interview with Jimmy Wales; Eric Raymond, "The Cathedral and the Bazaar," first presented in 1997, reprinted in *The Cathedral and the Bazaar* (O'Reilly Media, 1999).
90. Richard Stallman, "The Free Universal Encyclopedia and Learning Resource" (1999), http:// www.gnu.org/encyclopedia/free-encyclopedia.html.
91. Larry Sanger, "The Early History of Nupedia and Wikipedia," Slashdot, http://beta.slash dot.org/story/56499; and O'Reilly Commons, http://commons.oreilly.com/wiki/index.php /Open_Sources_2.0/Beyond_Open_Source:_Collaboration_and_Community/The_Early _History_of_Nupedia_and_Wikipedia:_A_Memoir.
92. Larry Sanger, "Become an Editor or Peer Reviewer!" Nupedia, http://archive.is/IWDNq.
93. Author's interview with Jimmy Wales; Lih, *The Wikipedia Revolution*, 960.
94. Author's interview with Jimmy Wales.
95. Larry Sanger, "Origins of Wikipedia," Sanger user page, http://en.wikipedia.org/wiki /User:Larry_Sanger/Origins_of_Wikipedia; Lih, *The Wikipedia Revolution*, 1049.
96. Ben Kovitz, "The Conversation at the Taco Stand," Kovitz user page, http://en.wikipedia .org/wiki/User:BenKovitz.
97. Jimmy Wales, "Re: Sanger's Memoirs" thread, Apr. 2005, http://lists.wikimedia.org/piper mail/wikipedia-l/2005-April/021463.html.
98. Jimmy Wales and Larry Sanger, "Re: Sanger's Memoirs" thread, Apr. 2005, http://lists .wikimedia.org/pipermail/wikipedia-l/2005-April/021460.html, http://lists.wikimedia.org /pipermail/wikipedia-l/2005-April/021469.html, and subsequent. See also Larry Sanger, "My Role in Wikipedia," http://larrysanger.org/roleinwp.html; "User:Larry Sanger/Origins of Wikipedia," http://en.wikipedia.org/wiki/User:Larry_Sanger/Origins_of_Wikipedia; "History of Wikipedia" and its talk page, http://en.wikipedia.org/wiki/History_of_Wiki pedia, along with Jimmy Wales edit changes to the article, http://en.wikipedia.org/w/index .php?title=Jimmy_Wales&diff=next&oldid=29849184; Talk: Bomis, revisions made by Jimmy Wales, http://en.wikipedia.org/w/index.php?diff=11139857.
99. Kovitz, "The Conversation at the Taco Stand."
100. Larry Sanger, "Let's Make a Wiki," Nupedia message thread, Jan. 10, 2001, http://archive.is /yovNt.
101. Lih, *The Wikipedia Revolution*, 1422.
102. Clay Shirky, "Wikipedia—An Unplanned Miracle," *Guardian*, Jan. 14, 2011; see also Clay

Shirky, *Here Comes Everybody: The Power of Organizing without Organizations* (Penguin, 2008) and *Cognitive Surplus: Creativity and Generosity in a Connected Age* (Penguin, 2010).

103. Author's interview with Jimmy Wales.
104. Larry Sanger, "Why Wikipedia Must Jettison Its Anti-Elitism," Dec. 31, 2004, www.LarrySanger.org.
105. Wikipedia press release, Jan. 15, 2002, http://en.wikipedia.org/wiki/Wikipedia:Press_releases/January_2002.
106. Author's interview with Jimmy Wales.
107. Shirky, "Wikipedia—An Unplanned Miracle."
108. Yochai Benkler, "Coase's Penguin, or, Linux and the Nature of the Firm," *Yale Law Journal* (2002), http://soc.ics.uci.edu/Resources/bibs.php?793; Yochai Benkler, *The Penguin and the Leviathan: How Cooperation Triumphs over Self-Interest* (Crown, 2011).
109. Daniel Pink, "The Buck Stops Here," *Wired*, Mar. 2005; Tim Adams, "For Your Information," *Guardian*, June 30, 2007; Lord Emsworth user page, http://en.wikipedia.org/wiki/User:Lord_Emsworth; Peter Steiner, *New Yorker* cartoon, July 5, 1993, at http://en.wikipedia.org/wiki/On_the_Internet,_nobody_knows_you're_a_dog.
110. Jonathan Zittrain, *The Future of the Internet and How to Stop It* (Yale, 2008), 147.
111. Author's interview with Jimmy Wales.
112. Author's interview with Jimmy Wales.
113. John Battelle, *The Search* (Portfolio, 2005; locations refer to the Kindle edition), 894.
114. Battelle, *The Search*, 945; author's visit with Srinija Srinivasan.
115. In addition to the sources cited below, this section is based on my interview and conversations with Larry Page; Larry Page commencement address at the University of Michigan, May 2, 2009; Larry Page and Sergey Brin interviews, Academy of Achievement, Oct. 28, 2000; "The Lost Google Tapes," interviews by John Ince with Sergey Brin, Larry Page, and others, Jan. 2000, http://www.podtech.net/home/?s=Lost+Google+Tapes; John Ince, "Google Flashback—My 2000 Interviews," *Huffington Post*, Feb. 6, 2012; Ken Auletta, *Googled* (Penguin, 2009); Battelle, *The Search*; Richard Brandt, *The Google Guys* (Penguin, 2011); Steven Levy, *In the Plex* (Simon & Schuster, 2011); Randall Stross, *Planet Google* (Free Press, 2008); David Vise, *The Google Story* (Delacorte, 2005); Douglas Edwards, *I'm Feeling Lucky: The Confessions of Google Employee Number 59* (Mariner, 2012); Brenna McBride, "The Ultimate Search," *College Park* magazine, Spring 2000; Mark Malseed, "The Story of Sergey Brin," *Moment* magazine, Feb. 2007.
116. Author's interview with Larry Page.
117. Larry Page interview, Academy of Achievement.
118. Larry Page interview, by Andy Serwer, *Fortune*, May 1, 2008.
119. Author's interview with Larry Page.
120. Author's interview with Larry Page.
121. Author's interview with Larry Page.
122. Larry Page, Michigan commencement address.
123. Author's interview with Larry Page.
124. Author's interview with Larry Page.
125. Author's interview with Larry Page.
126. Battelle, *The Search*, 1031.
127. Auletta, *Googled*, 28.
128. Interview with Larry Page and Sergey Brin, conducted by Barbara Walters, *ABC News*, Dec. 8, 2004.
129. Sergey Brin talk, Breakthrough Learning conference, Google headquarters, Nov. 12, 2009.
130. Malseed, "The Story of Sergey Brin."
131. Sergey Brin interview, Academy of Achievement.
132. McBride, "The Ultimate Search."
133. Auletta, *Googled*, 31.
134. Auletta, *Googled*, 32.
135. Vise, *The Google Story*, 33.
136. Auletta, *Googled*, 39.
137. Author's interview with Larry Page.
138. Author's interview with Larry Page.
139. Terry Winograd interview, conducted by Bill Moggridge, http://www.designinginteractions.com/interviews/TerryWinograd.

140. Author's interview with Larry Page.
141. Craig Silverstein, Sergey Brin, Rajeev Motwani, and Jeff Ullman, "Scalable Techniques for Mining Causal Structures," *Data Mining and Knowledge Discovery*, July 2000.
142. Author's interview with Larry Page.
143. Author's interview with Larry Page.
144. Larry Page, Michigan commencement address.
145. Vise, *The Google Story*, 10.
146. Larry Page, Michigan commencement address.
147. Battelle, *The Search*, 1183.
148. Battelle, *The Search*, 1114.
149. Larry Page, Michigan commencement address.
150. Author's interview with Larry Page.
151. Levy, *In the Plex*, 415, citing Page's remarks at the 2001 PC Forum, held in Scottsdale, Arizona.
152. Sergey Brin interview, conducted by John Ince, "The Lost Google Tapes," part 2.
153. Sergey Brin, Rajeev Motwani, Larry Page, Terry Winograd, "What Can You Do with a Web in Your Pocket?" *Bulletin of the IEEE Computer Society Technical Committee on Data Engineering*, 1998.
154. Author's interview with Larry Page.
155. Levy, *In the Plex*, 358.
156. Levy, *In the Plex*, 430.
157. Sergey Brin interview, conducted by John Ince, "The Lost Google Tapes," part 2, http://www.podtech.net/home/1728/podventurezone-lost-google-tapes-part-2-sergey-brin.
158. Levy, *In the Plex*, 947.
159. Author's interview with Larry Page.
160. Sergey Brin and Larry Page, "The Anatomy of a Large-Scale Hypertextual Web Search Engine," seventh International World-Wide Web Conference, Apr. 1998, Brisbane, Australia.
161. Vise, *The Google Story*, 30.
162. Author's interview with Larry Page.
163. David Cheriton, Mike Moritz, and Sergey Brin interviews, conducted by John Ince, "The Lost Google Tapes"; Vise, *The Google Story*, 47; Levy, *In the Plex*, 547.
164. Vise, *The Google Story*, 47; Battelle, *The Search*, 86.
165. Sergey Brin interview, conducted by John Ince, "The Lost Google Tapes."
166. Larry Page interview, conducted by John Ince, "The Lost Google Tapes."
167. Auletta, *Googled*, 44.
168. Sergey Brin interview, conducted by John Ince, "The Lost Google Tapes," part 2.

CHAPTER TWELVE: ADA FOREVER

1. Dyson, *Turing's Cathedral*, 6321; John von Neumann, *The Computer and the Brain* (Yale, 1958), 80.
2. Gary Marcus, "Hyping Artificial Intelligence, Yet Again," *New Yorker*, Jan. 1, 2014, citing "New Navy Device Learns by Doing" (UPI wire story), *New York Times*, July 8, 1958; "Rival," *New Yorker*, Dec. 6, 1958.
3. Marvin Minsky and Seymour Papert, the original gurus of artificial intelligence, challenged some of Rosenblatt's premises, after which the excitement surrounding the Perceptron faded and the entire field entered a decline known as the "AI winter." See Danny Wilson, "Tantalizingly Close to a Mechanized Mind: The Perceptrons Controversy and the Pursuit of Artificial Intelligence," undergraduate thesis, Harvard, December 2012; Frank Rosenblatt, "The Perceptron: A Probabilistic Model for Information Storage and Organization in the Brain," *Psychological Review*, Fall 1958; Marvin Minsky and Seymour Papert, *Perceptrons* (MIT, 1969).
4. Author's interview with Ginni Rometty.
5. Garry Kasparov, "The Chess Master and the Computer," *New York Review of Books*, Feb. 11, 2010; Clive Thompson, *Smarter Than You Think* (Penguin, 2013), 3.
6. "Watson on Jeopardy," IBM's Smarter Planet website, Feb. 14, 2011, http://asmarterplanet.com/blog/2011/02/watson-on-jeopardy-day-one-man-vs-machine-for-global-bragging-rights.html.
7. John Searle, "Watson Doesn't Know It Won on Jeopardy," *Wall Street Journal*, Feb. 23, 2011.
8. John E. Kelly III and Steve Hamm, *Smart Machines* (Columbia, 2013), 4. Steve Hamm is a

technology journalist now working as a writer and communications strategist at IBM. I have attributed the opinions in the book to Kelly, who is the director of IBM research.

9. Larry Hardesty, "Artificial-Intelligence Research Revives Its Old Ambitions," *MIT News*, Sept. 9, 2013.
10. James Somers, "The Man Who Would Teach Computers to Think," *Atlantic*, Nov. 2013.
11. Gary Marcus, "Why Can't My Computer Understand Me," *New Yorker*, Aug. 16, 2013.
12. Steven Pinker, *The Language Instinct* (Harper, 1994), 191.
13. Stuart Russell and Peter Norvig, *Artificial Intelligence: A Modern Approach* (Prentice Hall, 1995), 566.
14. Author's interview with Bill Gates.
15. Nicholas Wade, "In Tiny Worm, Unlocking Secrets of the Brain," *New York Times*, June 20, 2011; "The Connectome of a Decision-Making Neural Network," *Science*, July 27, 2012; The Dana Foundation, https://www.dana.org/News/Details.aspx?id=43512.
16. John Markoff, "Brainlike Computers, Learning from Experience," *New York Times*, Dec. 28, 2013. Markoff, who has long done thoughtful reporting on this field, is writing a book that explores the implications of machines that can replace human labor.
17. "Neuromorphic Computing Platform," the Human Brain Project, https://www.human brainproject.eu/neuromorphic-computing-platform1; Bennie Mols, "Brainy Computer Chip Ditches Digital for Analog," *Communications of the ACM*, Feb. 27, 2014; Klint Finley, "Computer Chips That Work Like a Brain Are Coming—Just Not Yet," *Wired*, Dec. 31, 2013. Beau Cronin of O'Reilly Media has proposed a drinking game: "take a shot every time you find a news article or blog post that describes a new AI system as working or thinking 'like the brain'" (http://radar.oreilly.com/2014/05/it-works-like-the-brain-so.html), and he maintains a pinboard of stories making such claims (https://pinboard.in/u:beaucronin /t:like-the-brain/#).
18. Author's interview with Tim Berners-Lee.
19. Vernor Vinge, "The Coming Technological Singularity," *Whole Earth Review*, Winter 1993. See also Ray Kurzweil, "Accelerating Intelligence," http://www.kurzweilai.net/.
20. J. C. R. Licklider, "Man-Computer Symbiosis," *IRE Transactions on Human Factors in Electronics*, Mar. 1960.
21. Kelly and Hamm, *Smart Machines*, 7.
22. Kasparov, "The Chess Master and the Computer."
23. Kelly and Hamm, *Smart Machines*, 2.
24. "Why Cognitive Systems?" IBM Research website, http://www.research.ibm.com/cogni tive-computing/why-cognitive-systems.shtml.
25. Author's interview with David McQueeney.
26. Author's interview with Ginni Rometty.
27. Author's interview with Ginni Rometty.
28. Kelly and Hamm, *Smart Machines*, 3.
29. "Accelerating the Co-Evolution," Doug Engelbart Institute, http://www.dougengelbart.org /about/co-evolution.html; Thierry Bardini, *Bootstrapping: Douglas Engelbart, Coevolution, and the Origins of Personal Computing* (Stanford, 2000).
30. Nick Bilton, *Hatching Twitter* (Portfolio, 2013), 203.
31. Usually misattributed to Thomas Edison, although there is no evidence he ever said it. Often used by Steve Case.
32. Yochai Benkler, "Coase's Penguin, or, Linux and the Nature of the Firm," *Yale Law Journal* (2002).
33. Steven Johnson, "The Internet? We Built That," *New York Times*, Sept. 21, 2012.
34. Author's interview with Larry Page. The quote form Steve Jobs comes from an interview I did with him for my previous book.
35. Kelly and Hamm, *Smart Machines*, 7.

PHOTO CREDITS

Page

6 *Lovelace:* Hulton Archive/Getty Images
Lord Byron: © The Print Collector/Corbis
Babbage: Popperfoto/Getty Images

21 *Difference Engine:* Allan J. Cronin
Analytical Engine: Science Photo Library/Getty Images
Jacquard loom: David Monniaux
Jacquard portrait: © Corbis

34 *Bush:* © Bettmann/Corbis
Turing: Wikimedia Commons/Original at the Archives Centre, King's College, Cambridge
Shannon: Alfred Eisenstaedt/The LIFE Picture Collection/Getty Images

57 *Stibitz:* Denison University, Department of Math and Computer Science
Zuse: Courtesy of Horst Zuse
Atanasoff: Special Collections Department/Iowa State University
Atanasoff-Berry Computer: Special Collections Department/Iowa State University

69 *Aiken:* Harvard University Archives, UAV 362.7295.8p, B 1, F 11, S 109
Mauchly: Apic/Contributor/Hulton Archive/Getty Images
Eckert: © Bettmann/Corbis
ENIAC in 1946: University of Pennsylvania Archives

86 *Aiken and Hopper:* By a staff photographer / © 1946 The Christian Science Monitor (www.CSMonitor.com). Reprinted with permission. Also courtesy of the Grace Murray Hopper Collection, Archives Center, National Museum of American History, Smithsonian Institution.
Jennings and Bilas with ENIAC: U.S. Army photo
Jennings: Copyright © Jean Jennings Bartik Computing Museum—Northwest Missouri State University. All rights reserved. Used with permission.
Snyder: Copyright © Jean Jennings Bartik Computing Museum—Northwest Missouri State University. All rights reserved. Used with permission.

109 *Von Neumann:* © Bettmann/Corbis
Goldstine: Courtesy of the Computer History Museum
Eckert and Cronkite with UNIVAC: U.S. Census Bureau

130 *Bardeen, Shockley, Brattain:* Lucent Technologies/Agence France-Presse/Newscom
First transistor: Reprinted with permission of Alcatel-Lucent USA Inc.
Shockley Nobel toast: Courtesy of Bo Lojek and the Computer History Museum

158 *Noyce:* © Wayne Miller/Magnum Photos
Moore: Intel Corporation
Fairchild Semiconductor: © Wayne Miller/Magnum Photos

170 *Kilby:* Fritz Goro/ The LIFE Picture Collection/ Getty Images
Kilby's microchip: Image courtesy of Texas Instruments
Rock: Louis Fabian Bachrach
Grove, Noyce, Moore: Intel Corporation

200 *Spacewar:* Courtesy of the Computer History Museum
Bushnell: © Ed Kashi/VII/Corbis

216 *Licklider:* Karen Tweedy-Holmes
Taylor: Courtesy of Bob Taylor
Larry Roberts: Courtesy of Larry Roberts

239 *Davies:* National Physical Laboratory © Crown Copyright / Science Source Images
Baran: Courtesy of RAND Corp.
Kleinrock: Courtesy of Len Kleinrock
Cerf and Kahn: © Louie Psihoyos/Corbis
262 *Kesey:* © Joe Rosenthal/San Francisco Chronicle/Corbis
Brand: © Bill Young/San Francisco Chronicle/Corbis
Whole Earth Catalog cover: Whole Earth Catalog
279 *Engelbart:* SRI International
First mouse: SRI International
Brand: SRI International
292 *Kay:* Courtesy of the Computer History Museum
Dynabook: Courtesy of Alan Kay
Felsenstein: Cindy Charles
People's Computer Company cover: DigiBarn Computer Museum
307 *Ed Roberts:* Courtesy of the Computer History Museum
Popular Electronics *cover:* DigiBarn Computer Museum
312 *Allen and Gates:* Bruce Burgess, courtesy of Lakeside School, Bill Gates, Paul Allen, and Fredrica Rice
Gates: Wikimedia Commons/Albuquerque, NM police department
Microsoft team: Courtesy of the Microsoft Archives
349 *Jobs and Wozniak:* © DB Apple/dpa/Corbis
Jobs screenshot: YouTube
Stallman: Sam Ogden
Torvalds: © Jim Sugar/Corbis
382 *Brand and Brilliant:* © Winni Wintermeyer
Von Meister: The Washington Post/Getty Images
Case: Courtesy of Steve Case
404 *Berners-Lee:* CERN
Andreessen: © Louie Psihoyos/Corbis
Hall and Rheingold: Courtesy of Justin Hall
435 *Bricklin and Williams:* Don Bulens
Wales: Terry Foote via Wikimedia Commons
Brin and Page: Associated Press
466 *Lovelace:* Hulton Archive/Getty Images
489 *Vitruvian Man:* © The Gallery Collection/Corbis

TIMELINE CREDITS (IN CHRONOLOGICAL ORDER)

Lovelace: Hulton Archive/Getty Images
Hollerith: Library of Congress via Wikimedia Commons
Bush (first image): © Bettmann/Corbis
Vacuum tube: Ted Kinsman/Science Source
Turing: Wikimedia Commons/Original at the Archives Centre, King's College, Cambridge
Shannon: Alfred Eisenstaedt/The LIFE Picture Collection/Getty Images
Aiken: Harvard University Archives, UAV 362.7295.8p, B 1, F 11, S 109
Atanasoff: Special Collections Department/Iowa State University
Bletchley Park: Draco2008 via Wikimedia Commons
Zuse: Courtesy of Horst Zuse
Mauchly: Apic/Hulton Archive/Getty Images
Atanasoff-Berry Computer: Special Collections Department/Iowa State University
Colossus: Bletchley Park Trust/SSPL via Getty Images
Harvard Mark I: Harvard University
Von Neumann: © Bettmann/Corbis
ENIAC: U.S. Army photo
Bush (second image): © Corbis
Transistor invention at Bell Labs: Lucent Technologies/Agence France-Presse/Newscom
Hopper: Defense Visual Information Center
UNIVAC: U.S. Census Bureau
Regency radio: © Mark Richards/CHM
Shockley: Emilio Segrè Visual Archives / American Institute of Physics / Science Source

Fairchild Semiconductor: © Wayne Miller/Magnum Photos
Sputnik: NASA
Kilby: Fritz Goro/The LIFE Picture Collection/Getty Images
Licklider: MIT Museum
Baran: Courtesy of RAND Corp.
Spacewar: Courtesy of the Computer History Museum
First mouse: SRI International
Kesey: © Hulton-Deutsch Collection/Corbis
Moore: Intel Corporation
Brand: © Bill Young/San Francisco Chronicle/Corbis
Taylor: Courtesy of Bob Taylor
Larry Roberts: Courtesy of Larry Roberts
Noyce, Moore, Grove: Intel Corporation
Whole Earth Catalog cover: Whole Earth Catalog
Engelbart: SRI International
ARPANET nodes: Courtesy of Raytheon BBN Technologies
4004: Intel Corporation
Tomlinson: Courtesy of Raytheon BBN Technologies
Bushnell: © Ed Kashi/VII/Corbis
Kay: Courtesy of the Computer History Museum
Community Memory: Courtesy of the Computer History Museum
Cerf and Kahn: © Louie Psihoyos/Corbis
Popular Mechanics cover: DigiBarn Computer Museum
Gates and Allen: Bruce Burgess, courtesy of Lakeside School, Bill Gates, Paul Allen, and Fredrica Rice
Apple I: Ed Uthman
Apple II: © Mark Richards/CHM
IBM PC: IBM/Science Source
Gates with Windows disc: © Deborah Feingold/Corbis
Stallman: Sam Ogden
Jobs with Macintosh: Diana Walker/Contour By Getty Images
WELL logo: Image courtesy of The WELL at www.well.com. The logo is a registered trademark of the Well Group Incorporated.
Torvalds: © Jim Sugar/Corbis
Berners-Lee: CERN
Andreessen: © Louie Psihoyos/Corbis
Case: Courtesy of Steve Case
Hall: Courtesy of Justin Hall
Kasparov: Associated Press
Brin and Page: Associated Press
Williams: Courtesy of Ev Williams
Wales: Terry Foote via Wikimedia Commons
IBM Watson: Ben Hider/Getty Images

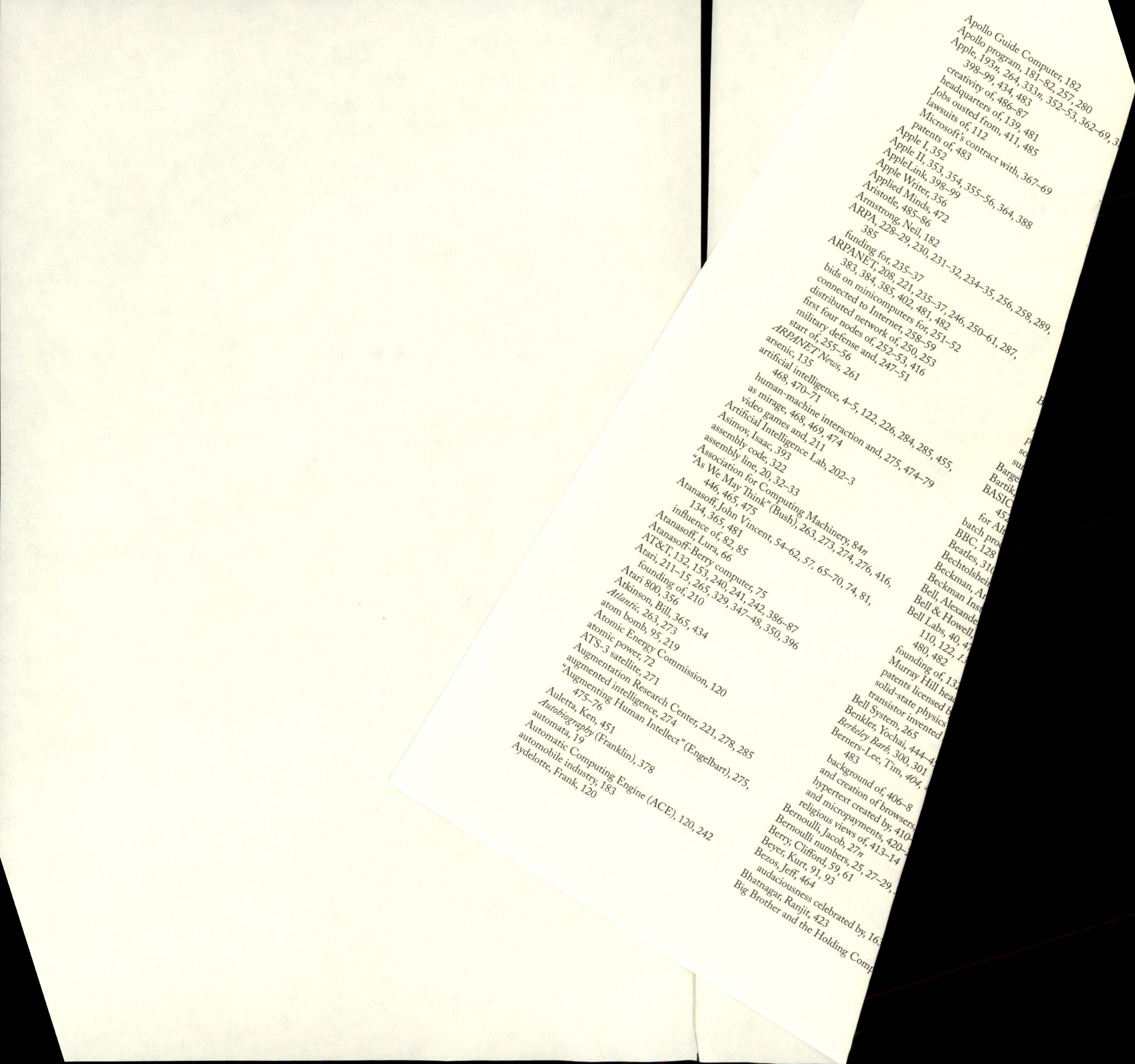

ABOUT THE AUTHOR

Walter Isaacson, the CEO of the Aspen Institute, has been the chairman of CNN and the managing editor of *Time* magazine. He is the author of *Steve Jobs; Einstein: His Life and Universe; Benjamin Franklin: An American Life;* and *Kissinger: A Biography,* and is the coauthor, with Evan Thomas, of *The Wise Men: Six Friends and the World They Made*. He and his wife live in Washington, DC.

INDEX

Fairchild Semiconductor: © Wayne Miller/Magnum Photos
Sputnik: NASA
Kilby: Fritz Goro/The LIFE Picture Collection/Getty Images
Licklider: MIT Museum
Baran: Courtesy of RAND Corp.
Spacewar: Courtesy of the Computer History Museum
First mouse: SRI International
Kesey: © Hulton-Deutsch Collection/Corbis
Moore: Intel Corporation
Brand: © Bill Young/San Francisco Chronicle/Corbis
Taylor: Courtesy of Bob Taylor
Larry Roberts: Courtesy of Larry Roberts
Noyce, Moore, Grove: Intel Corporation
Whole Earth Catalog cover: Whole Earth Catalog
Engelbart: SRI International
ARPANET nodes: Courtesy of Raytheon BBN Technologies
4004: Intel Corporation
Tomlinson: Courtesy of Raytheon BBN Technologies
Bushnell: © Ed Kashi/VII/Corbis
Kay: Courtesy of the Computer History Museum
Community Memory: Courtesy of the Computer History Museum
Cerf and Kahn: © Louie Psihoyos/Corbis
Popular Mechanics cover: DigiBarn Computer Museum
Gates and Allen: Bruce Burgess, courtesy of Lakeside School, Bill Gates, Paul Allen, and Fredrica Rice
Apple I: Ed Uthman
Apple II: © Mark Richards/CHM
IBM PC: IBM/Science Source
Gates with Windows disc: © Deborah Feingold/Corbis
Stallman: Sam Ogden
Jobs with Macintosh: Diana Walker/Contour By Getty Images
WELL logo: Image courtesy of The WELL at www.well.com. The logo is a registered trademark of the Well Group Incorporated.
Torvalds: © Jim Sugar/Corbis
Berners-Lee: CERN
Andreessen: © Louie Psihoyos/Corbis
Case: Courtesy of Steve Case
Hall: Courtesy of Justin Hall
Kasparov: Associated Press
Brin and Page: Associated Press
Williams: Courtesy of Ev Williams
Wales: Terry Foote via Wikimedia Commons
IBM Watson: Ben Hider/Getty Images